ROTATING FLUIDS IN ENGINEERING AND SCIENCE

James P. Vanyo
Department of Mechanical and Environmental Engineering
Department of Geological Sciences
University of California, Santa Barbara

DOVER PUBLICATIONS, INC.
Mineola, New York

Published in Canada by General Publishing Company, Ltd., 30 Lesmill Road, Don Mills, Toronto, Ontario.

Bibliographical Note

This Dover edition, first published in 2001, is an unabridged reprint of the work originally published by Butterworth-Heinemann, Stoneham, Mass., in 1993.

Library of Congress Cataloging-in-Publication Data

Vanyo, James P.
Rotating fluids in engineering and science / James P. Vanyo.
p. cm.
Includes bibliographical references and index.
ISBN 0-486-41704-2 (pbk.)
1. Fluid mechanics. 2. Rotating masses of fluids. I. Title.

TA357 .V36 2001
620.1'06—dc21

2001017222

Manufactured in the United States of America
Dover Publications, Inc., 31 East 2nd Street, Mineola, N.Y. 11501

Contents

Appendices

Preface

Rotating Fluids in Engineering and Science was written to help make rotating fluid phenomena more accessible to students and researchers in the field of fluid mechanics. The book is designed to be used as an upper-division text and/or beginning graduate text. It is also intended for practicing engineers and scientists engaged in research or in design and development projects requiring a knowledge of rotating fluid theory. The book leads the reader systematically through rotating flow theory and applications, and shows that the theory and applications can be mastered easily by students and practicing professionals. It is hoped that the text will help bridge the gap between undergraduate preparation in fluid mechanics, and the understanding needed to benefit from professional journal articles and more advanced monographs in this field.

The reader is assumed to be familiar with partial differential equations and vector mathematics, although important parts of these are reviewed in Appendix A. The reader is also assumed to have had a junior level undergraduate course or course sequence in fluid mechanics. Part I (Chapters 1 through 5) provides a review of the basic fluid theory needed in the remainder of the text, including definitions of terminology and notation. Part II (Chapters 6 through 13) extends fluid theory to concepts, theories, and equations specific to rotating fluids, including frequent references to vorticity and vortices. Because it is sometimes difficult to apply equations to quantitative applications, each chapter in Part II includes solved quantitative examples.

Part III (Chapters 14 through 22) presents important application areas that rely heavily on rotating fluid theory for their understanding. An attempt is made to include enough of each application area so that each chapter can be read independent of other texts, although the chapters are not intended to replace a need for other books and reference manuals. Each chapter in Part III includes a list of suggested projects and study assignments for groups and/or individuals. They vary in difficulty and type; some are analytical while others are numerical or experimental.

Constant density and viscosity flows are emphasized, although thermal and compressibility phenomena are included. Data and quantitative examples are presented in both dimensional and nondimensional forms. Most dimensional quantities are expressed in SI and cgs units, although other units commonly seen in U.S. applications are also illustrated, e.g., ft, lbf, psi, gpm, and mph. A table of conversion factors is included in Appendix D.1. Symbols follow the normal conventions. Vectors are printed in bold face characters, e.g., **V** is vector velocity, V is scalar speed, and unit vectors are designated with a caret, e.g., $\hat{\mathbf{k}}$. The author combines laboratory demonstrations and student projects with lectures. The text continues this format by including experimental and practical results to support discussions of theory. It also includes references to advanced texts and journals as an aid in formulating professional level research projects.

The University of California, Santa Barbara, follows a quarter system of instruction, ten lecture weeks followed by an examination week. As an undergraduate text for students who have completed a two quarter fluids sequence, the author devotes one week to Part I *Fluid Mechanics Review*, then five weeks to Part II *Rotating Fluid Theory*, and the remaining four weeks to four or five chapters, of most interest to that group of students, selected from Part III *Rotating Fluid Applications*. The emphasis is on helping students make the transition from an undergraduate pattern of weekly, precisely defined, homework problem sets to a more professional level, i.e., applying theory to open-ended complex problem areas, often including a need to make reasonable approximations to achieve success.

Several quizzes during the first 6 weeks, rather than homework sets, are used to promote knowledge of relevant theory and equations; these are followed by individual or small group projects selected/assigned from Part III. Again, the emphasis is to get the students into projects typical of real-world work experiences as quickly as possible. Their interest will be stimulated, and they will, on their own, restudy the Part I and II chapters as they develop their projects. Written and oral project reports replace a final examination. A semester presentation would benefit from additional time for more detail and greater depth, especially in selection and development of application projects. When using the text in a graduate course, much of Part II is covered as review, and the text is augmented with additional source material from professional journals and advanced monographs.

Rotating Fluids in Engineering and Science was developed over the last 22 years by the author at the University of California, Santa Barbara. The format and content of the text, as it developed, was in response to, and aided by discussions with, the many students who have taken my courses on rotating fluids or who assisted with laboratory projects. Two students, R. Hadley and S. Stojanovich, checked solutions for many of the example problems in Part II; J. Byram helped with the experiments of Figures 1.4 and 1.5; and P. Wilde suggested and solved Example 9.3. S. McLean assisted by reviewing material on oceanography, and L. Pauley made other valuable suggestions.

I want to thank all contributors for their help; the editorial staff and reviewers of Butterworth–Heinemann for their guidance; and I especially wish to thank Christine Townsley for the many, many hours of careful, patient, and cheerful assistance in typing several preliminary drafts, and finally in preparing the text in camera-ready form.

Professor James P. Vanyo
University of California
Santa Barbara

PART I

Fluid Mechanics Review

Chapter 1

Rotating Fluid Phenomena

Many areas of engineering and science involve the rotation of various objects; in science the object sometimes is the earth; in engineering it might be a turbine rotor or a space vehicle. In most cases the object also involves the rotation of internal or external fluids. Sometimes a rotating fluid is the principal phenomena of interest; at other times the fluid is merely an unwanted participant in the motion. In either event, success or failure of the analysis can depend critically on understanding and predicting rotating fluid phenomena.

The basic theory of fluid rotation and vorticity distinguishes between vorticity and curved (e.g., circular) translation of fluid elements. Figure 1.1 illustrates smooth uniform flow of a viscous liquid in a channel. This laminar flow, when fully developed, has a parabolic velocity distribution. Ink or dye slowly injected in the flow would move in straight lines indicating straight streamlines (fluid motion). However, small objects placed in the flow would also rotate as they move, indicating that the flow has vorticity, i.e., the infinitesimal fluid elements rotate as they translate along straight lines. Figure 1.2 illustrates a flow field called an *inviscid vortex* where all fluid elements move in circular paths. However, small objects here would not rotate, indicating a fluid that is not rotating, merely translating in circular paths. The two flows illustrate two extremes, one that has straight pathlines but fluid element rotation, and the second that has circular pathlines but fluid elements which do not rotate. Viscosity in the first flow produces the fluid element rotation called *vorticity*, which is absent in the second flow. Rotation, vorticity, and circulation are described and quantified in Chapter 8. Figure 1.3 shows a smoke ring. Its motion is often modelled as a toroidal inviscid vortex. Any cross section through the toroid is approximately an inviscid vortex as in Figure 1.2.

Figure 1.4 shows water in a cylinder. In the left photograph, both the cylinder and the water are stationary, and all the colored water, slightly less dense than the clear water, is in the top one-eighth of the cylinder. In the right photograph, the cylinder has been impulsively accelerated to a constant angular velocity, and the water is gradually being "spun up" to the angular velocity of the cylinder. Liquid spin-up is

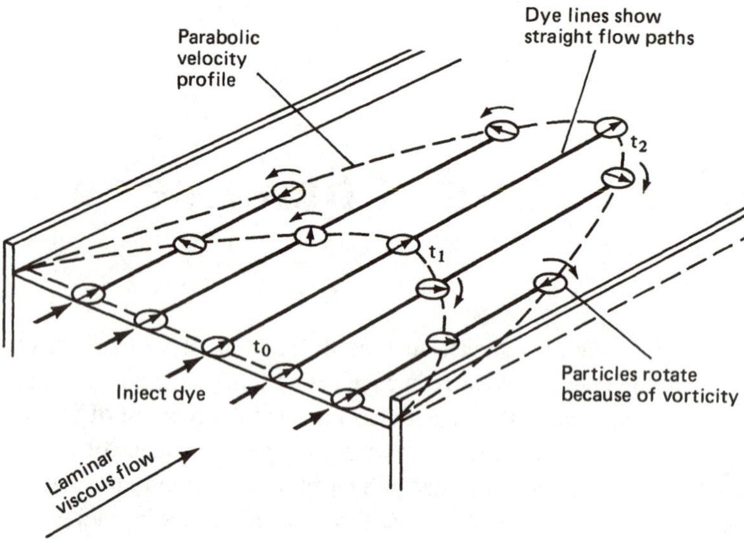

Figure 1.1: Laminar flow of a viscous fluid in a straight channel moves along straight streamlines. The fluid elements rotate, however, because of viscosity. The streamlines can be observed by injecting ink into the flow, and the rotation (vorticity) can be observed by the rotation of small corks placed in the flow.

achieved about 1% by viscous interaction at the cylinder side walls and about 99% by a viscous secondary flow at the cylinder bottom. Centrifugal force inside this very thin (almost invisible) spinning bottom boundary layer moves clear water outward and then up along the outside cylinder wall. Colored water in the interior is drawn downward until all the water will be pumped outward through the very thin, bottom boundary layer. In this experiment, a 2% buoyancy of the colored water is opposing the pumping action and causes the boundary between the clear and colored water to be tapered rather than cylindrical. More complex rotating fluid phenomena, such as internal waves and vortex stretching, are involved during a reverse spin-down process.

Rotating masses of fluid exhibit other unusual properties. Figure 1.5 shows the difference between flow patterns created by suddenly dumping a quantity of similar density colored water into nonrotating water, (top two photographs), and flow patterns created by the same act, but using rotating water as shown in the bottom two photographs. Typical turbulent eddies are generated in the nonrotating water. Such random motions are not possible in a rotating liquid; instead, permissible flows have a distinctly two-dimensional property as shown in the photographs.

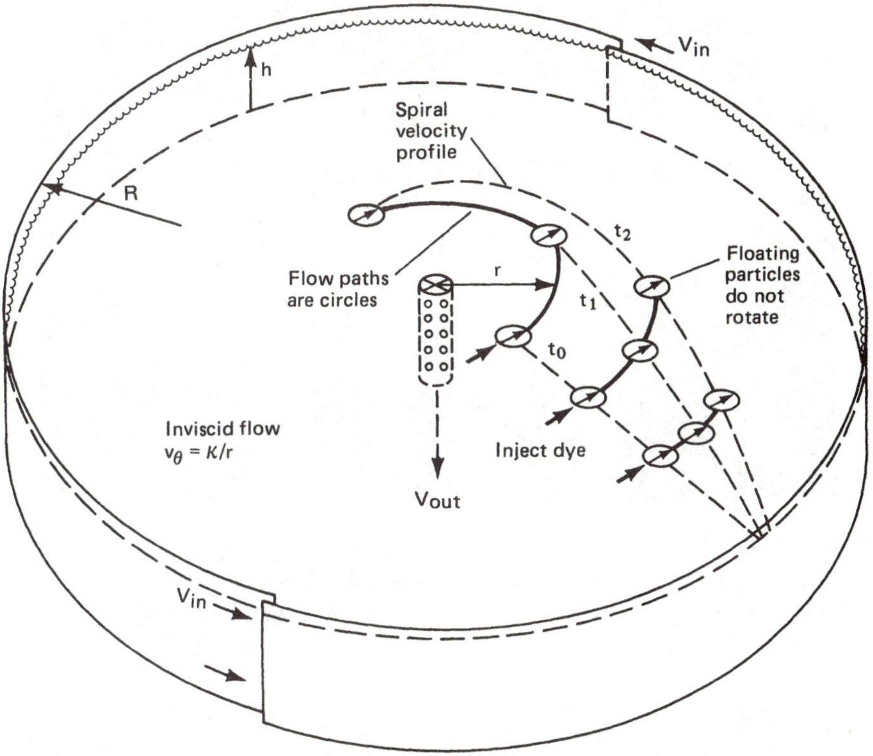

Figure 1.2: *An inviscid vortex flow has circular streamlines. In a perfect such flow, corks would move in circular paths but would not rotate, exactly the reverse of Figure 1.1. This experiment can only be approximated in the laboratory because fluid viscosity will introduce some vorticity, the fluid surface will not be flat, and a secondary flow along the bottom will pump liquid radially inward. A very small* V_{in} *and* V_{out} *is needed in the laboratory to maintain the flow.*

Rotating fluid theory helps explain important application areas in engineering and science. These include many subtle fluid-structure interactions that produce vorticity and secondary flows. For example, when a viscous fluid moves through a bend in a pipe or channel, differences in velocity and pressure between the central portion and the boundary layers near solid surfaces induce secondary flows similar to those on the bottom surface of the cylinder in Figure 1.4. These secondary flows cause energy losses in engineering applications and erosion when the channel is a river bed. Figure 1.6 shows a typical result. Bottom secondary flows and sediment move transversely from the outside of a bend toward the inside of that bend. Any slight departure from

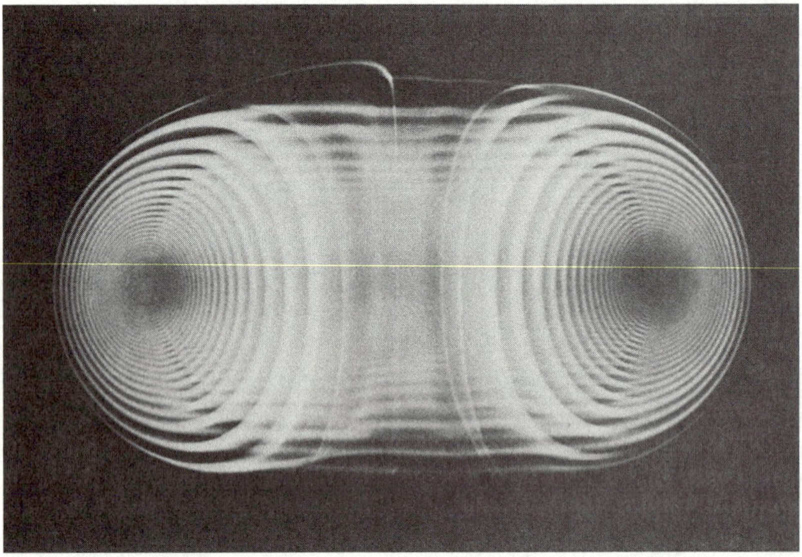

Figure 1.3: *Air flow in this cross section of a smoke ring approximates the vortex flow of Figure 1.2. Toroidal vortex patterns illustrate important fluid theories and are a part of typical turbulent flows. Reproduced with permission from Magarvey and MacLatchy (1964).*

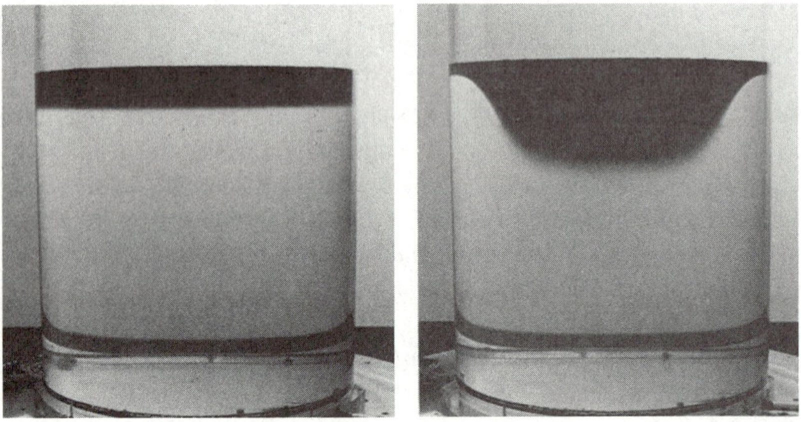

Figure 1.4: *On the left, the cylinder, the clear water, and a top layer of colored water are motionless. As the cylinder begins to rotate, at right, clear water is pumped through a thin layer on the rotating bottom surface radially outward and up along the cylinder wall displacing the top (still nonspinning) colored water inward and down. The radial flow along the bottom surface is referred to as a secondary flow. The colored water is about 2% buoyant. Photos by Vanyo and Byram.*

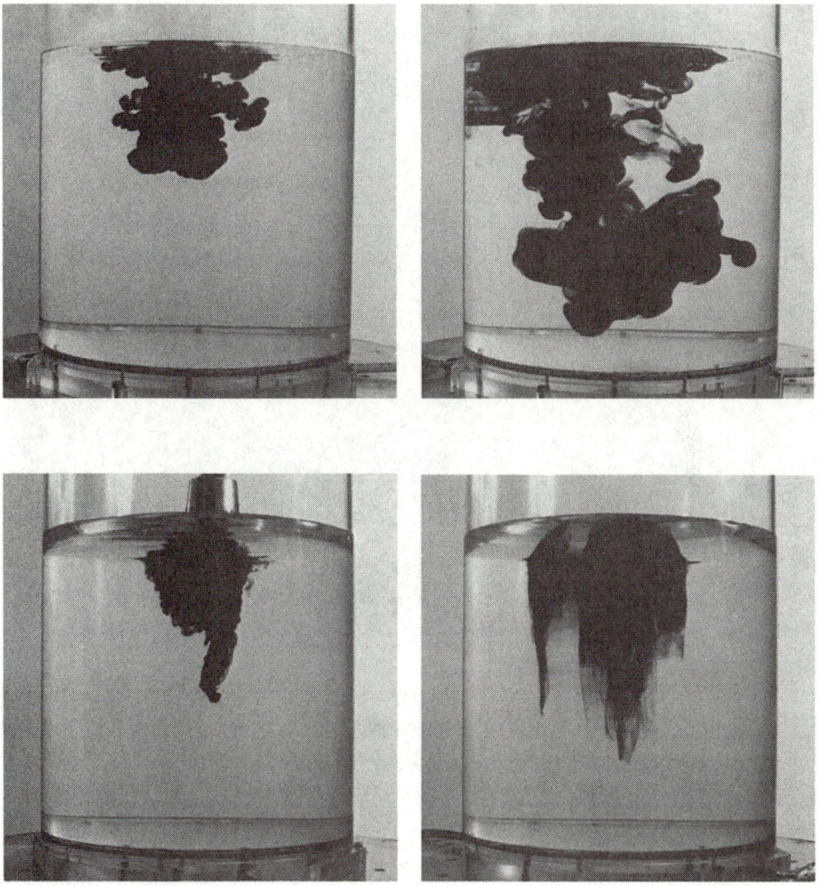

Figure 1.5: *Turbulence induced by dropping colored water into nonrotating water (top two photographs) includes toroidal patterns as in Figure 1.3. It differs markedly from turbulence induced in water rotating with constant angular velocity as shown in the bottom two photographs. Turbulence is severely constrained by the characteristic two-dimensional (2-D) nature of a rotating fluid. Photos by Vanyo and Byram.*

straight line motion becomes more and more pronounced, gradually reaching the condition shown in Figure 1.6. If a flood condition occurs, the river may overflow the meandering channel and erode a nearly straight channel to repeat a new cycle.

Vorticity, generated especially in boundary layers, is typical of all real fluid flows. Integration of vector vorticity over a finite surface leads to the concept of circulation along the curve enclosing the surface. This in turn has been shown to be related to lift

Figure 1.6: This photograph shows the typical meander pattern for rivers flowing down a slightly inclined wide valley. Secondary currents, as in Figure 1.4, along the bottom carry sediment from the outside edge of curves to the inside. The Chena River just east of Fairbanks, AK, is shown in the photograph. Photo by Vanyo.

of moving objects. Although lift of an aircraft wing is related to a difference in pressure top and bottom, lift can be expressed also as circulation about the wing represented by an equivalent vortex coincident with the wing. This vortex cannot simply end when the wing tip is reached. It is deflected rearward to form wing tip vortices that are visible at appropriate speeds and atmospheric conditions as shown in Figure 1.7. The parallel vortices interact and regroup as a series of toroids similar to the smoke ring of Figure 1.3.

An efficient wing design minimizes wasted vortex motion. Soaring birds, as in Figure 1.8, have very efficient wing tip designs for slow flight conditions. They apparently can feel and manipulate wing tip vortices using their individually controllable wing tip feathers. The bird in Figure 1.8 is a California condor in flight. Note in particular how carefully the condor has manipulated its wing tip feathers, both to reduce drag and to control its flight direction. Some long range modern aircraft, e.g., the new Boeing 747-400, mimic the condor by adding large "winglets" tilted upward at each wing tip.

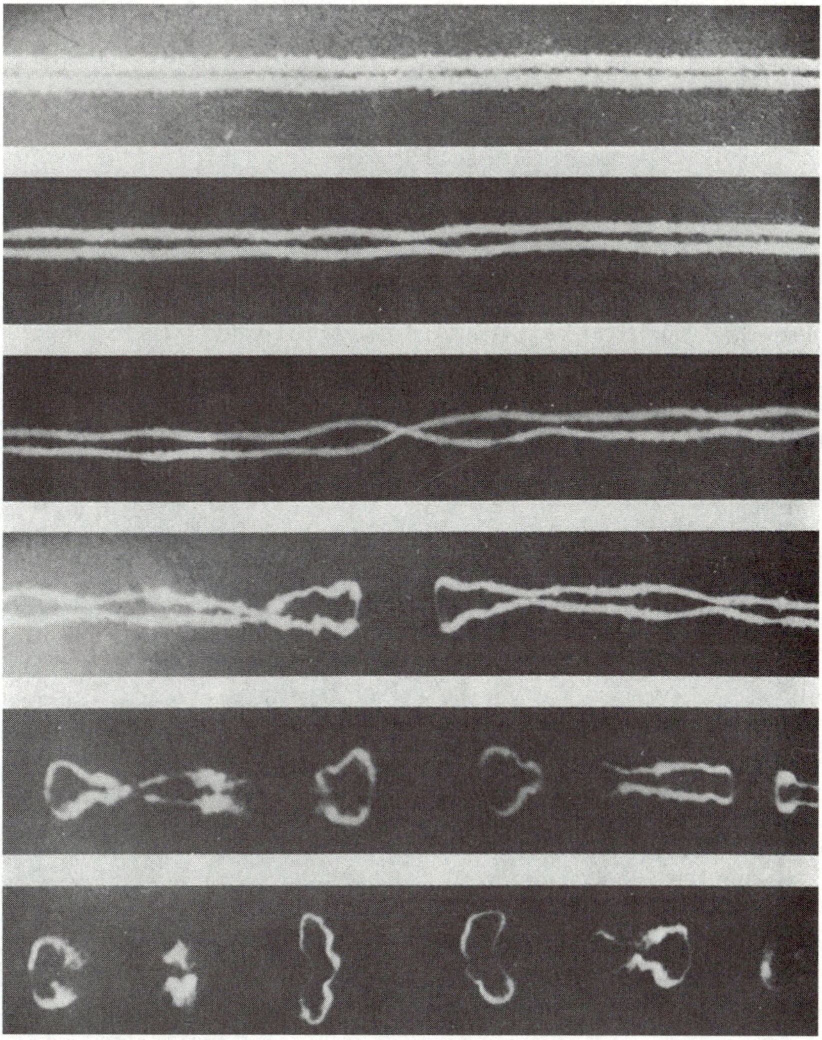

Figure 1.7: *Condensation trails sometimes make wing tip vortices visible. In part they are extensions of lift circulation patterns about the wing, but they also represent evidence of unnecessary energy losses. These were produced by a B-47 aircraft and illustrate an instability of parallel vortices; they degenerate into a series of toroidal vortices. Photos at 15 second intervals. Reproduced with permission from Crow (1970).*

Figure 1.8: *Soaring birds like this California condor increase their flight efficiency by manipulating wing tip vortices with individually controllable wing tip feathers. Photo by D. Clendenen, Condor Recovery Program, U.S. Fish and Wildlife.*

Figure 1.9: *A centrifugal pump admits fluid along the axis and spins it to a high rotational speed. Centrifugal force then induces a pressure differential and outflow at the periphery. The inlet pipe and flange (not shown) attach to the front of the casing at final assembly. Courtesy AC Compressor Corp.*

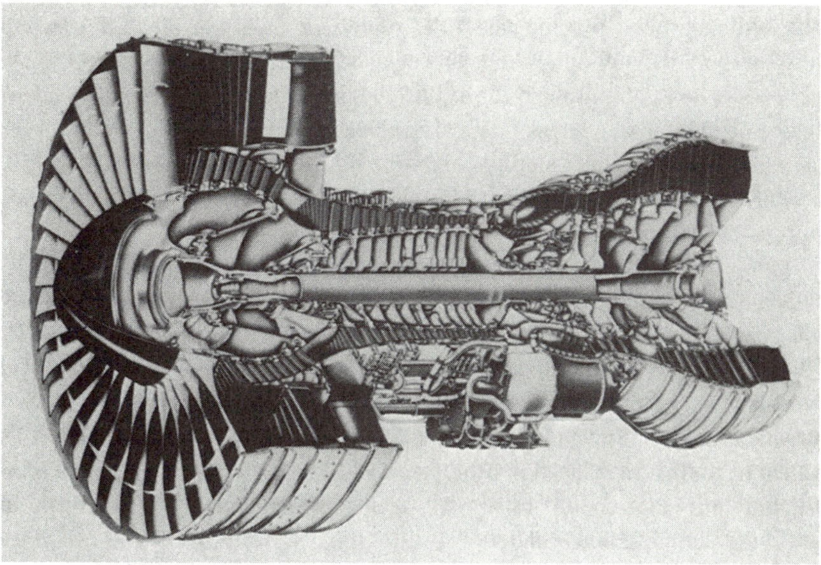

Figure 1.10: *Air in a turbofan jet engine (axial flow turbine) is compressed by the rotating blades at the left. Combustion occurs in the central portion, and part of the energy drives the turbine blades at the right. The remaining energy provides thrust by forcing gas rearward at high velocity. The turbine powers both the compressor and the fan. Courtesy Pratt & Whitney.*

Curved motion, viscosity, boundary layers, secondary flows, lift, and drag are design features common to most rotating machinery. The relevant fluid may only be air in contact with the spinning armature of an electric motor, representing an energy loss, or it may be the fluid in a rotary pump. The centrifugal pump impeller and housing shown in Figure 1.9 draws in nonrotating fluid along its axis, spins it at high speed, and uses centrifugal force to create fluid pressure and flow velocity. Within the pump the fluid moves in nearly spiral paths as it moves from the inlet to the outlet.

An axial flow turbine has alternate rows of stationary and rotating vanes and can be used either as a motor or a pump (compressor). The turbofan jet engine in Figure 1.10 is one type of axial flow turbine. It draws in and compresses air in the inlet compressor section, continuously injects and burns fuel in the central combustion section, and uses a portion of the energy of the exhaust to power the rear portion (turbine) of the jet engine. The turbine portion drives the inlet compressor section and fan. The fan helps provide propulsion. The remainder of the energy is used to eject the exhaust gases at high velocity to give additional forward propulsion. Fluid motion along curved paths and fluid vorticity have both positive and negative design implications in these applications.

Important applications of rotating fluids occur in many vehicles that contain liquids, often as fuel. In some cases, the vehicle is massive enough, relative to the quantity of fuel, that motion of the vehicle can be prescribed independent of the fluid. In other cases, e.g., an oil tanker or a space vehicle, the liquid mass may exceed the vehicle structural mass. On take-off, a space vehicle may be over 90% liquid fuel by mass, and communication satellites, upon being placed in orbit, typically are over 50% liquid by mass. Space vehicles present unusual problems because their motions are unconstrained by land, water, or air.

Figure 1.11 shows a communication satellite in its orbit configuration. It was manufactured by Ford Aerospace for the India Space Agency and was placed in orbit by a Delta rocket. During the final orbit insertion maneuver the rocket third stage (PAM) and the satellite were spun up to achieve gyroscopic attitude stability. At this point approximately 60% of the satellite's total mass was liquid fuel. Figure 1.12 shows a communication satellite manufactured by ERNO Raumfahrttechnik for Telecom (Germany). It also was placed in orbit by a Delta rocket and is shown here in launch configuration fastened on top of its PAM (payload assist module). Phenomena similar to, but more complex than, that shown in Figure 1.4 occur inside the spinning fuel tanks and have the potential for destabilizing the attitude (orientation) of the satellite. Satellites have become inoperative because of an inability to predict these rotating fluid-structure interactions correctly. These satellites were correctly analyzed and designed, and performed perfectly. Figure 1.13 is a NASA photograph of astronauts manually reorienting a satellite relative to their space shuttle. Continued space exploration will present many unusual situations where an ability to predict rotational motions of vehicles containing large quantities of liquids will be essential for success.

Rotating geophysical fluids are usually related to and dependent upon the earth's rotation. For example, the earth has a liquid core whose radius is slightly more than half the earth's outside radius; its mass is approximately one-third the earth's total mass. It is assumed to be molten iron with small amounts of other elements. This provides a situation very similar to that of liquid fuel in a spinning communication satellite. Part of the complexity of both applications is that neither the earth nor a satellite spins about a fixed axis. The spin axis of each wobbles (precesses) and, in so doing, continuously changes direction. Under this condition the internal liquid is continuously being perturbed.

Figure 1.14 shows a laboratory experiment used to analyze liquid motions in a container that spins and precesses at various rates. The transparent tank shown in the photograph has the same nonspherical shape as the earth's mantle-core boundary. The angle between spin and precession shown in the photograph (23.5°) matches the earth's forced precession angle. While many variables can be matched during experiments, not all can. Dimensionless ratios and scaling techniques can often resolve such experimental difficulties. Even if the earth's axis did not wobble, the earth's spin rate would restrict thermal convection, turbulence, and internal waves in the liquid core to prescribed patterns and magnitudes.

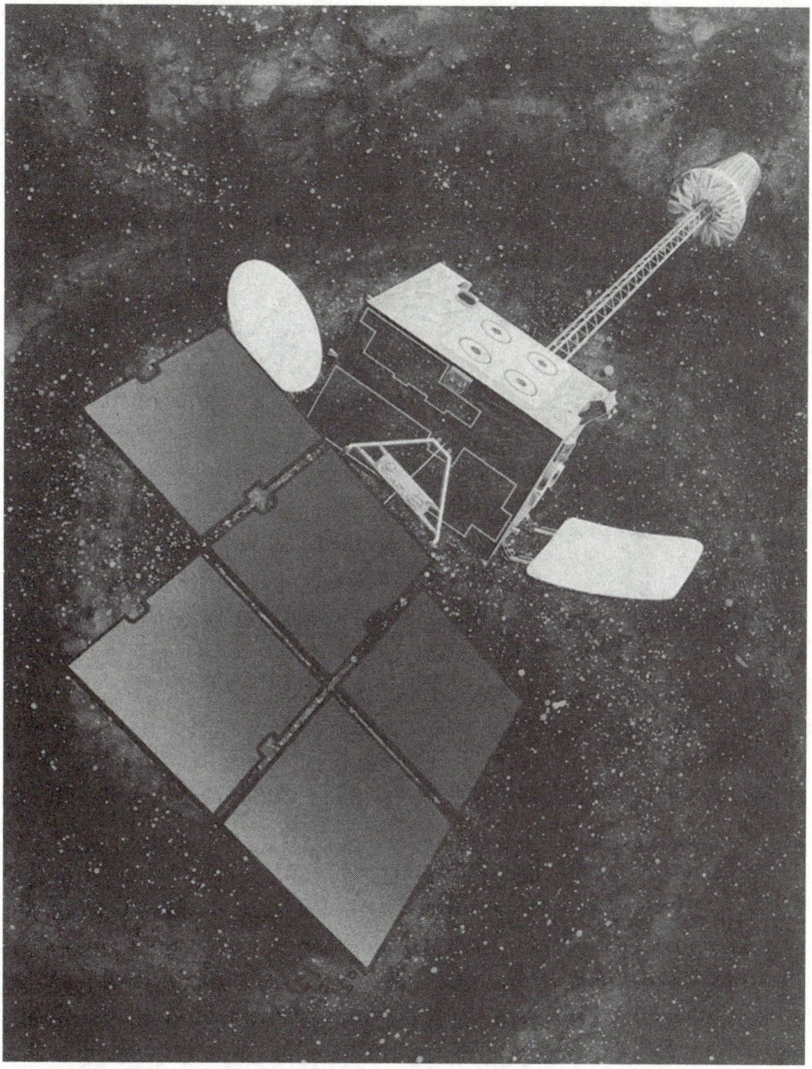

Figure 1.11: *The successful orbit insertion and operation of this communication satellite was the result of careful analysis, experimentation, and design on many levels. Rotating flow analyses were important because the satellite contained large amounts of liquid fuels. Courtesy Department of Space, Government of India.*

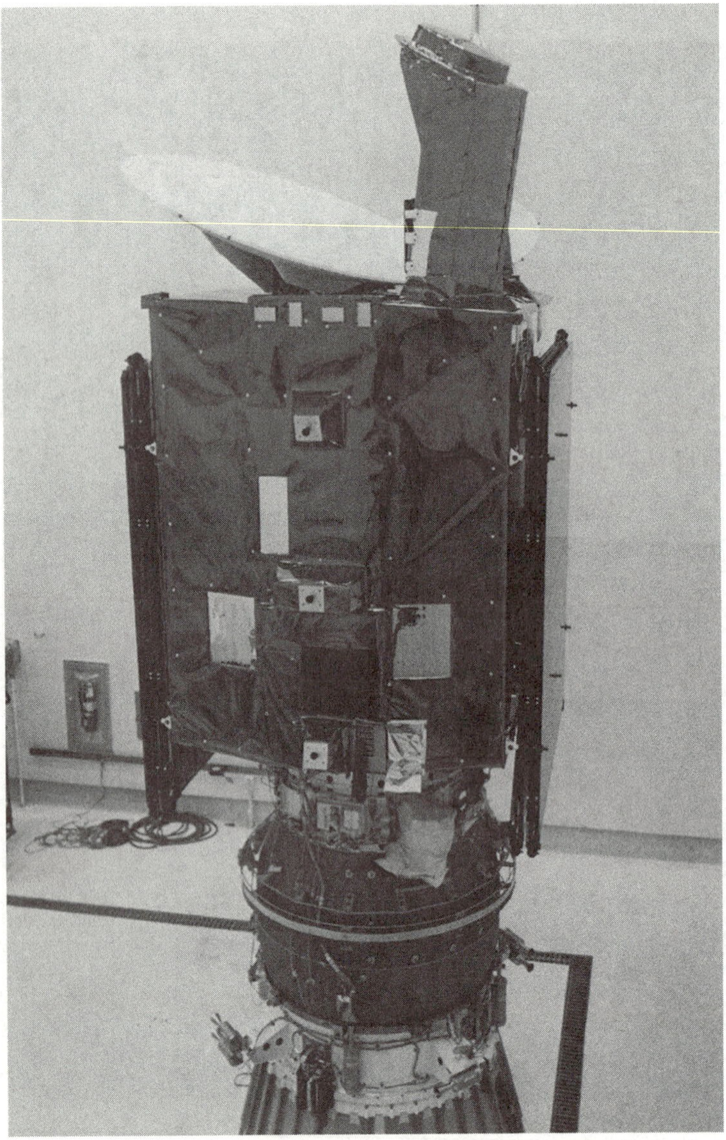

***Figure 1.12:** This communication satellite is shown in its launch configuration attached to its third-stage rocket. This satellite and the satellite in Figure 1.11 were in this configuration when spinning. Courtesy ERNO Raumfahrttechnik, Bremen, Germany, and McDonnell Douglas.*

Figure 1.13: *In this photo a satellite is maneuvered into position by astronauts on a space shuttle. Enclosed liquids greatly complicate maneuvers in space that might otherwise seem rather simple. Photo courtesy NASA.*

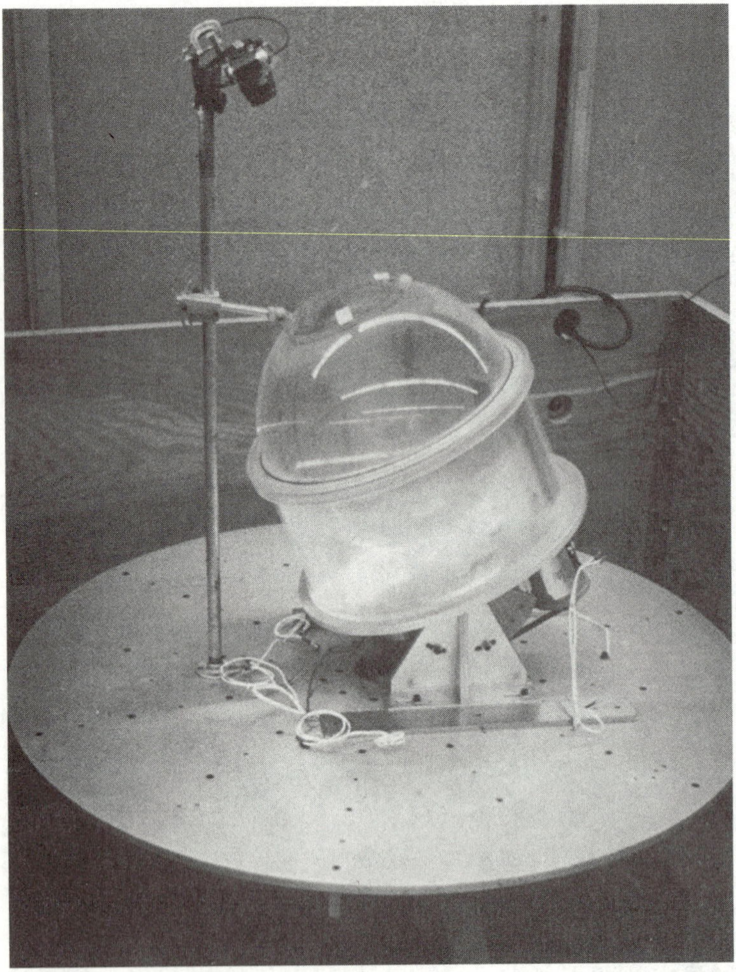

Figure 1.14: *The earth is filled out to half its radius with molten iron and minerals. This transparent tank, filled with liquid, can be used to model both the earth's interior and liquid fuels in communication satellites that spin and wobble. Photo by Vanyo.*

Rotating fluid theory historically developed during attempts at understanding and predicting fluid flow phenomena on the earth's surface, especially large-scale atmospheric and oceanic flows. Figure 1.15 shows the earth viewed from space. The continent of Africa fills the upper-left quadrant of the photograph. Medium (meso-) scale cloud patterns are obvious. Major components of large-scale flows do not vary day or night, summer or winter, but are not easily made visible. These long-term

Figure 1.15: *Some features of earth atmospheric flow patterns are made visible by clouds in this photograph from Apollo 17. Solar heat provides the necessary energy, but Coriolis phenomena due to the earth's rotation control the flow patterns. Photo courtesy NASA.*

flow components are caused by the average flow of heat from the earth's hot equatorial region to the cold polar regions. The earth's rotation causes these major north-south atmospheric flows to deflect east or west, as viewed from the earth's surface, to produce the east-west trade winds. These trade winds were known and used by early mariners.

Surface winds blow across the broad expanses of the oceans causing surface ocean currents 100 or more meters deep. These currents measure up to thousands of kilometers wide and typically move at speeds of a few kilometers per day. A few, such as the narrow, deep Gulf Stream and the Kuroshiro (Japanese) Current, move at speeds up to 120 km/day. Shear stresses caused by the wind and Coriolis phenomena caused by the earth's rotation induce rotational patterns in each of the five major oceans, clockwise in the Northern Hemisphere and counterclockwise in the Southern Hemisphere. These rotating patterns are called *gyres*. The motion can be seen in infrared satellite photographs that distinguish water temperatures and by measurements from research ships. Figure 1.16 shows the warm North Equatorial Current moving westward across the Atlantic Ocean, then bending north to become the Gulf Stream, completing its circle (gyre) as a flow east to England and then south toward northern

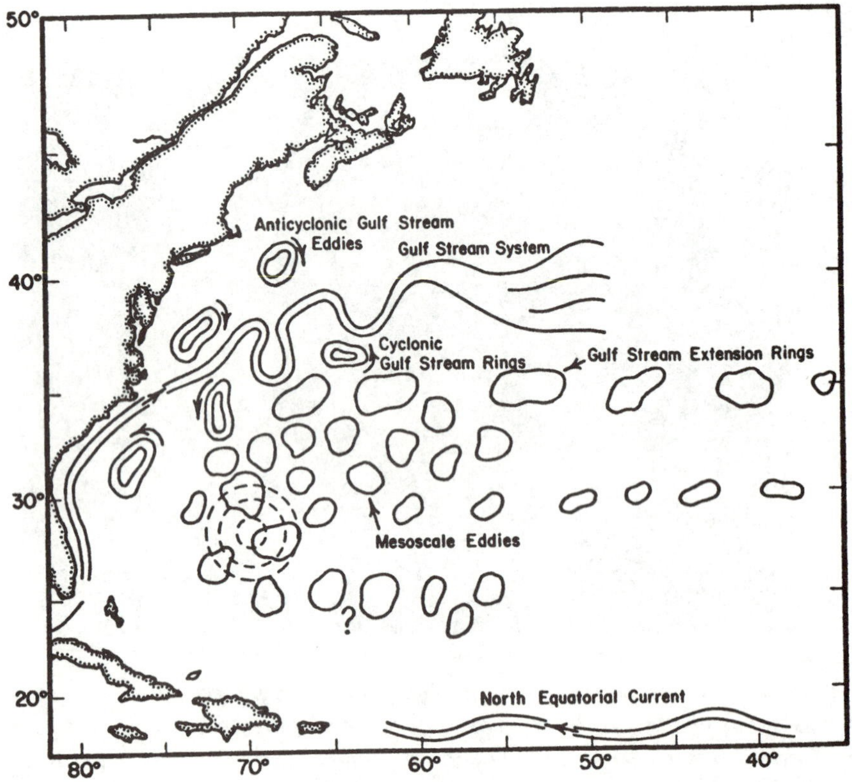

Figure 1.16: *Oceanic surface currents are driven by the wind but controlled by Coriolis phenomena and continental boundaries. The net effect is to cause oceans in the Northern Hemisphere to rotate clockwise and those in the Southern Hemisphere to rotate counterclockwise. The Gulf Stream is the intensified western edge of the North Atlantic gyre shown here with typical vortex rings. Reproduced with permission from Munk (1987).*

Africa. Figure 1.16 also shows vortex motions produced along the boundaries of the Gulf Stream as it moves north and east.

Some atmospheric motions are not caused directly by thermal convection or earth rotation but rather by vorticity produced in regions of wind shear. All combinations of effects are possible. Small whirlwinds behind the edges of buildings are produced randomly by wind gusts. Large-scale winds associated with movement of cold or warm fronts produce flow patterns that combine earth rotation and shear vorticity effects. Some of the more spectacular are the large-scale (up to 1000 km) dust storms such as those of central Asia and northern Africa. The major flow is produced by energy extracted from the earth's overall atmospheric motion; however, its destructive ability is intensified by wind shear vorticity.

***Figure 1.17:** Air over warm oceans rises, aided by the latent heat of evaporation of rain, and draws in air over a 1000 km region. Earth vorticity (rotation) of the atmosphere is concentrated, here producing hurricane Gilbert as viewed from a NOAA satellite. Courtesy Coastal Studies Institute, Louisiana State University.*

Tropical cyclones (called *hurricanes* in the Atlantic and *typhoons* in the Pacific) and tornadoes have easily earned the classification "intense atmospheric vortices." However, except that they both consist of air rotating at very high speed and are extremely destructive, they have little in common. Note in Figure 1.15 the immature typhoon (hurricane) forming over the ocean between India and Africa. Figure 1.17 shows hurricane Gilbert (September 1988) viewed from a NOAA weather satellite. The hurricane covers the entire Caribbean from Florida in the upper right to the Yucatan Peninsula in the lower left. The major diameter of this hurricane is greater than 1000 km, and its eye is 40 km in diameter. Gilbert had 145 mph winds and the lowest central pressure ever recorded for a hurricane. A tornado is rarely over a few hundred meters in diameter and would be too small to be visible in this hurricane photograph.

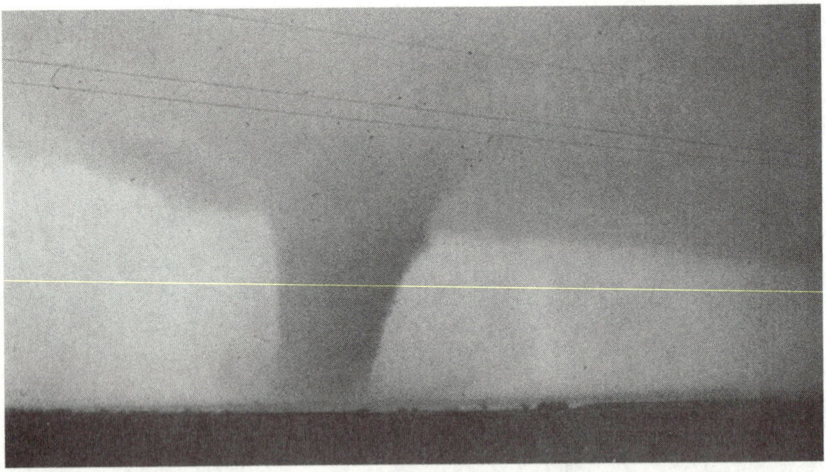

***Figure 1.18:** Tornadoes are typically 10 to 100 m in diameter with rotating winds sometimes estimated at near the speed of sound. This tornado formed in 1979 near Seymour, TX. Courtesy National Severe Storms Laboratory.*

A hurricane is caused by large-scale thermal convection over a region of warm (>26.5°C) ocean water. Convection is augmented by release of latent heat during condensation of rain. The hurricane's angular momentum (rotation) is extracted directly from the earth's angular momentum (rotation) during convergence of air toward its low-pressure center. Because its rotation is extracted directly from the earth's rotation, its direction of rotation agrees with the earth's, counterclockwise viewed from the Northern Hemisphere and clockwise viewed from the Southern Hemisphere. Tangential wind speeds near the center often reach 200 to 300 km/hr. When a tropical cyclone (hurricane) passes over land or cool water, it loses its energy source and soon dissipates.

Tornadoes receive their energy and their angular momentum from energy and vorticity (rotation) produced in and stored by other phenomena, e.g., hurricanes, squall lines, thunderstorms, or even volcanoes or fire storms. Tornadoes usually descend from overhead clouds as in the Figure 1.18 photograph of a massive tornado that occurred near Seymour, TX, on April 10, 1979. When the vortex funnel reaches the earth's surface, it usually accumulates water, dust, or debris. Over land it is called a *tornado*. When over water it is called a *waterspout* as in Figure 1.19. The waterspout shown here was observed by the author at Santa Barbara on October 1, 1976. It started as a low (300 m high), thick (75 m diameter) column in the ocean about 500 m from the University of California, Santa Barbara campus, and then lengthened as it moved east about 10 km to where it was photographed. It soon dissipated. Tornadoes (waterspouts) can rotate in either direction and tend to be brief in duration. Some evidence suggests maximum tangential winds near supersonic velocities, but measured velocities are less than 400 km/hr.

Figure 1.19: *Waterspouts are tornadoes that occur over water. This photograph by D. Meaney shows a rare waterspout off the coast of Santa Barbara, CA, in 1976. © Dennis Meaney 1976.*

The figures are explained in more detail in Parts II and III. Mathematical models are included in most cases along with experimental results. References to more advanced and more specialized texts and to the professional literature are included for those wishing to explore subjects more thoroughly.

Chapter 2

Mass and Momentum Conservation

2.1 Eulerian Mechanics

A principal task of fluid mechanics is the analysis and understanding of distortion. The ability of a fluid element to distort without limit raises the question of identity, *How can we identify and monitor the motion of a specific quantity of fluid?* Nearly a century after Newton derived **F** = m**A**, Euler resolved the question by adopting a formulation for the kinematics of fluids that is different than Newton's approach. It was yet another century before Navier and Stokes resolved the question of internal fluid forces, including viscosity, and completed the derivation of the fluid equivalent of Newton's **F** = m**A**.

Newton's concept of identifying specific particles or objects and following their motion through space was formalized by Lagrange and is often referred to as the Lagrangian formulation of mechanics. Euler developed a different approach to fluid mechanics that identifies fixed locations in space and analyzes the motion of whatever fluid is in that location at that time. Where the fluid comes from or goes to is a secondary matter.

Figures 2.1a and 2.1b compare the Eulerian and Lagrangian formulations. Figure 2.1a shows the Lagrangian formulation for particle mechanics. Position **R** of the particle is the principal variable. Velocity is obtained as the first derivative of **R** and acceleration as the second derivative. The coordinates x and y define **R** of the particle so dx/dt = u and dy/dt = v. Figure 2.1b illustrates the Eulerian formulation. Here x and y are the locations of Eulerian grid points. The Eulerian grid is fixed so dx/dt = 0 and dy/dt = 0. Velocity or momentum become the principal variables.

2.2 Mass Conservation

The principle of conservation of mass in particle dynamics is mathematically expressed

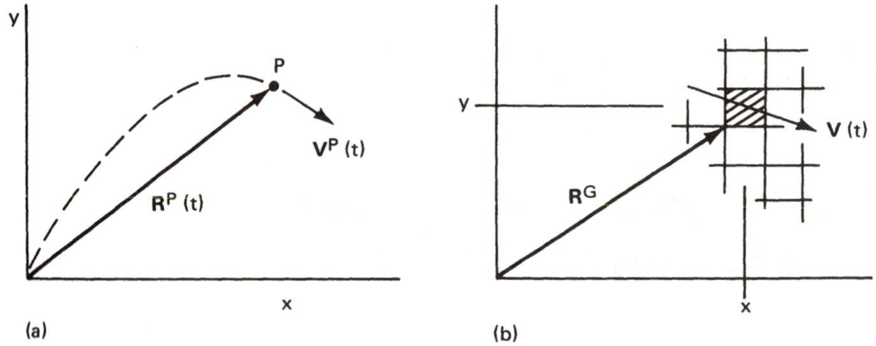

Figure 2.1: *Lagrangian vs. Eulerian formulations of mechanics. (a) In Lagrangian mechanics attention is placed on the trajectory* $\mathbf{R}^P(t)$ *of a specific mass element (P). Here (x, y) are coordinates of P, and* $d\mathbf{R}/dt = \mathbf{V}$. *(b) In Eulerian mechanics attention is placed on motion of fluid* $\mathbf{V}(t)$ *past fixed grid locations* $\mathbf{R}^G$. *Here the (x, y) grid locations (G) are fixed, and* $d\mathbf{R}^G/dt = 0$.

as m = constant, or m = m(t). Newton's law of motion, $\mathbf{F} = d(m\mathbf{V})/dt$, becomes

$$\mathbf{F} = m\mathbf{A}$$

for m = constant, or

$$\mathbf{F} = m\mathbf{A} + \dot{m}\mathbf{V}$$

for m = m(t). Conservation of mass in the analysis of fluids is defined using the Eulerian formulation by the simple statement, *The net change of mass within a given volume is equal to the difference between the quantity of mass entering the volume and that leaving.*

Figure 2.2a shows an application of the conservation of mass theorem for a finite control volume ($\forall$). The mass conservation statement here is expressed quantitatively as

$$\frac{\partial}{\partial t} \int_{\forall} \rho \, d\forall + \int_S \rho(\mathbf{V} \cdot d\mathbf{S}) = 0 \tag{2.1}$$

This equation states that the time rate of change of total mass within the control volume $\partial(\int_{\forall} \rho \, d\forall)/\partial t$ is equal to the net flux of mass through the control volume surface $\int_S \rho(\mathbf{V} \cdot d\mathbf{S})$. The term $(\mathbf{V} \cdot d\mathbf{S})$ quantifies the volumetric flow rate integrated (or summed) over the relevant portions of the surface, and ρ (mass per unit volume) is the fluid property transferred across the surface.

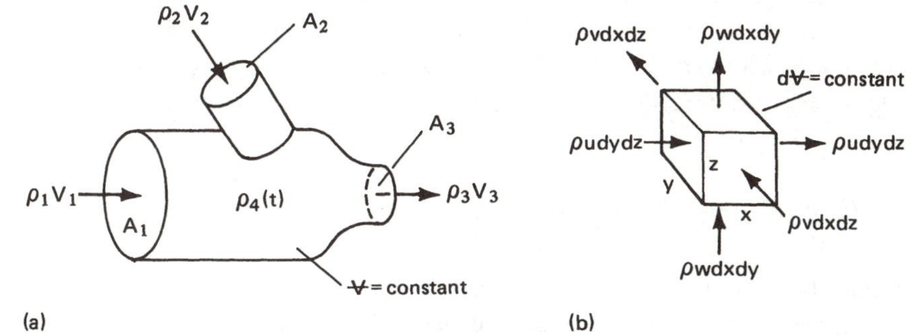

Figure 2.2: *Conservation of mass in Eulerian mechanics. (a) Mass flow through a fixed volume enclosed by the walls of a pipe and control surfaces A_1, A_2, and A_3. (b) Mass flow through an infinitesimal volume $d\forall = dx\,dy\,dz$.*

Figure 2.2b applies the same concept to an infinitesimal volume $d\forall$. The mass flow rate through the left wall using (2.1) is [ρ(u × area)]. The mass flow rate difference between the left wall and the right wall per unit time is $\Delta(\rho u)/\Delta x$ or in differential form $\partial(\rho u)/\partial x$. Summation of the net flows in the x, y, and z directions must equal the total change of mass (density) within $d\forall$ per unit time

$$\frac{\partial \rho}{\partial t} + \frac{\partial(\rho u)}{\partial x} + \frac{\partial(\rho v)}{\partial y} + \frac{\partial(\rho w)}{\partial z} = 0 \tag{2.2}$$

If density is constant over time and space, this reduces to

$$\frac{\partial u}{\partial z} + \frac{\partial v}{\partial y} + \frac{\partial w}{\partial z} = 0 \tag{2.3}$$

Each of (2.2) and (2.3) can be written as a shorter vector equation by use of the del operator and vector velocity as described in Appendix A. Equation (2.2) can now be written as

$$\frac{\partial \rho}{\partial t} + \nabla \cdot (\rho \mathbf{V}) = 0 \tag{2.4}$$

and (2.3) for ρ constant becomes

$$\nabla \cdot \mathbf{V} = 0 \tag{2.5}$$

Equation (2.5) is referred to as the continuity equation for an incompressible fluid. Note that the common use of the word *incompressible* to mean $\rho \neq \rho(x, y, z, t)$ is ambiguous. *Incompressible* refers to a substance whose density does not change with changes in pressure; a fluid can be incompressible but still be a function of (x, y, z, t), for example, an incompressible fluid whose density changes with variations in temperature or with variations in salinity. A variation with respect to time requires that the term $\partial\rho/\partial t$ in (2.4) be retained, and a variation in space requires the term $\nabla \cdot (\rho \mathbf{V})$.

2.3 Force and Momentum

In most applications, the conservation of mass equation is used to solve for flow rates and velocities entering and leaving the volume. After complete flow information is obtained, flow dynamics can then be analyzed. Use again is made of the Eulerian formulation of mechanics. A specific control volume is chosen (finite or infinitesimal). Net force acting on fluid in that volume is equated to acceleration of that fluid--not acceleration of the control volume, which is inertially fixed here, but acceleration of the fluid in the control volume.

All control volumes in this book will be assumed to have fixed size and shape. The professional literature includes the somewhat more difficult case of flexible or deforming volumes, e.g., a large, thin-walled liquid fuel tank in a launch vehicle during acceleration. The use of accelerating control volumes will be introduced in Part II.

The Lagrangian path of a specific quantity of fluid is rarely needed, but fluid acceleration (or rate of change of momentum) is needed for use in an equation that is the equivalent to $\mathbf{F} = m\mathbf{A}$. In the finite control volume of Figure 2.2a, force due to changes of momentum within the volume and momentum entering or leaving the volume are computed using

$$\mathbf{F} = \frac{\partial}{\partial t} \int_{\text{¥}} \rho \mathbf{V} \, d\text{¥} + \int_{S} \rho \mathbf{V} \, (\mathbf{V} \cdot d\mathbf{S}) \tag{2.6}$$

The first term on the right-hand side quantifies the time rate of change of total momentum within the volume. The second term quantifies the net flux of momentum $(\rho\mathbf{V})$ through the control volume surface. The term $(\mathbf{V} \cdot d\mathbf{S})$ again quantifies the volumetric flow rate of the fluid property, here $(\rho\mathbf{V})$, integrated (or summed) over the relevant portions of the surface.

The equation for net force on the fluid in the differential control volume of Figure 2.2b also can be analyzed using Euler's formulation of mechanics. To distinguish the calculation of time derivatives in Eulerian mechanics from those in Lagrangian mechanics, the symbol D/Dt is used rather than d/dt. The quantity D/Dt is called the substantive derivative and is computed in two parts: (1) the convective part, and (2) the local or instantaneous part.

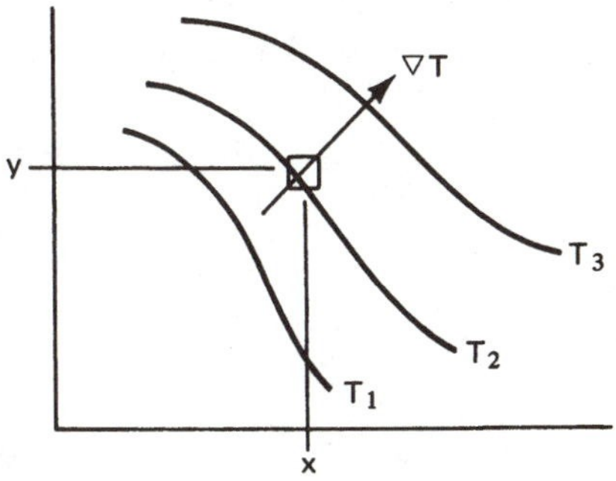

Figure 2.3: *The gradient of T computes the spatial rate of change of T and its direction at each point **R** (shown here in 2-D with $T_3 > T_2 > T_1$).*

To understand convective changes, it is useful to review the concept of a gradient. Figure 2.3 shows lines of constant scalar temperature (T) over some region T(x, y). The gradient of T is defined as the direction and magnitude of the steepest increase in T at each point (x, y). The gradient is a vector and is computed in three-dimensional (3-D) rectangular coordinates as

$$\nabla T = \frac{\partial T}{\partial x}\hat{\mathbf{i}} + \frac{\partial T}{\partial y}\hat{\mathbf{j}} + \frac{\partial T}{\partial z}\hat{\mathbf{k}} \tag{2.7}$$

The gradient in other coordinate systems, e.g., polar or spherical, has additional terms, but means the same thing. Equation (2.7) describes the variations of T over 3-D space; if T varied only in the x direction, the gradient would reduce to (dT/dx).

A dot product $\mathbf{V} \cdot (\nabla T)$ gives the component of $\mathbf{V}$ in the direction of ∇T times the magnitude of ∇T. In a 1-D case this is simply u(dT/dx), with units of degrees per unit time. In Cartesian coordinates it is permissible to rewrite $\mathbf{V} \cdot (\nabla T)$ as $\mathbf{V} \cdot \nabla T$ or as $(\mathbf{V} \cdot \nabla)T$ where

$$\mathbf{V} \cdot \nabla = u\frac{\partial}{\partial x} + v\frac{\partial}{\partial y} + w\frac{\partial}{\partial z} \tag{2.8}$$

The convective change in T is then

$$(\mathbf{V} \cdot \nabla)\, T = u\frac{\partial T}{\partial x} + v\frac{\partial T}{\partial y} + w\frac{\partial T}{\partial z} \tag{2.9}$$

Note that $\mathbf{V} \cdot \nabla$ is entirely different than $\nabla \cdot \mathbf{V}$ of the continuity equation (2.4).

If temperature distribution is not constant over time, a local (or instantaneous) operator $\partial/\partial t$ is also necessary to include transient phenomena. Combining local and convective parts yields the Eulerian substantive derivative

$$\frac{D}{Dt} = \left[\frac{\partial}{\partial t} + \mathbf{V} \cdot \nabla\right] = \frac{\partial}{\partial t} + u\frac{\partial}{\partial x} + v\frac{\partial}{\partial y} + w\frac{\partial}{\partial z} \tag{2.10}$$

The substantive derivative can be applied to any fluid property to compute the time rate of change of that property. It includes both the convective change (due to **V** and ∇) and the instantaneous or local change (due to $\partial/\partial t$).

The x component of velocity (u) is a fluid property, and Eq. (2.10) can be used to calculate the x component of acceleration

$$\frac{Du}{Dt} = \frac{\partial u}{\partial t} + (\mathbf{V} \cdot \nabla)u = \frac{\partial u}{\partial t} + u\frac{\partial u}{\partial x} + v\frac{\partial u}{\partial y} + w\frac{\partial u}{\partial z} = A_x \tag{2.11}$$

In a similar way Dv/Dt and Dw/Dt provide A_y and A_z. The three scalar equations can be combined into one vector equation

$$\frac{D\mathbf{V}}{Dt} = \frac{\partial \mathbf{V}}{\partial t} + (\mathbf{V} \cdot \nabla)\mathbf{V} = \mathbf{A} \tag{2.12}$$

as the final expression for calculating acceleration in the Eulerian formulation of mechanics.

Figure 2.4 shows DV/Dt for a 3-D Eulerian grid point (G) whose location is $\mathbf{R}^G$. If this grid is fixed relative to inertial space then

$$\mathbf{F} = \rho\mathbf{A} = \rho\frac{{}^{n}D\mathbf{V}}{Dt} = \rho\left[\frac{\partial \mathbf{V}}{\partial t} + (\mathbf{V} \cdot \nabla)\mathbf{V}\right] \tag{2.13}$$

is the Eulerian formulation of Newton's second law. The n in ${}^{n}D\mathbf{V}/Dt$ is used to make explicit that the derivative must be taken relative to an inertial (Newtonian) frame. In these equations, **F** is measured in force per unit volume because momentum is expressed as momentum per unit volume ($\rho\mathbf{V}$). When **F** is restricted to pressure forces ($-\nabla p$) and gravity forces (ρg) the resulting equation is called *Euler's equation* and is written as

$$\rho\left[\frac{\partial \mathbf{V}}{\partial t} + (\mathbf{V} \cdot \nabla)\mathbf{V}\right] = \rho \mathbf{g} - \nabla p \tag{2.14}$$

The interpretation of $-\nabla p$ is similar to that of ∇T in Figure 2.3. The direction and magnitude of changes in pressure, with respect to an Eulerian grid are given by the gradient of pressure

$$\nabla p = \frac{\partial p}{\partial x}\hat{\mathbf{i}} + \frac{\partial p}{\partial y}\hat{\mathbf{j}} + \frac{\partial p}{\partial z}\hat{\mathbf{k}} \tag{2.15}$$

An increase in pressure over distance results in a force field in the reverse direction (i.e., from high pressure to low pressure) and therefore pressure force per unit volume is

$$\mathbf{F}_p = -\nabla p \tag{2.16}$$

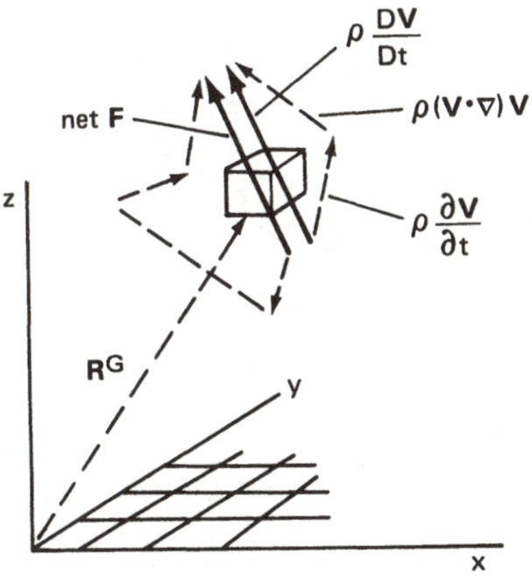

Figure 2.4: Force, mass, and acceleration of fluid at location $\boldsymbol{R}^G$ *in Eulerian mechanics. Vector* **F** *equals vector* $\rho\, DV/Dt$ *at each grid point* $\boldsymbol{R}^G$, *provided the grid is not accelerating.*

Figure 2.4 illustrates a possible pair of local and convective components for DV/Dt and four components (unnamed) for **F**. Gravity and pressure forces are discussed above. Forces due to shear stresses on the surface of each fluid element are discussed next. Additional forces (electric, magnetic, surface tension, etc.) can exist, but are not necessary for most applications of this text.

2.4 Navier-Stokes Equations

In the study of fluids, the symbol μ is called the coefficient of dynamic viscosity. It is used in the definition of a Newtonian fluid where

$$\tau = \mu \frac{du}{dy} \tag{2.17}$$

This linear equation states that shear stress (τ) at (x, y, z, t) in a fluid is equal to the product of μ (a fluid property) and the gradient of velocity at that point. If shear stress within the fluid involves phenomena inconsistent with this equation, the fluid is said to be non-Newtonian. Most engineering and geophysical fluids, including air and water under normal conditions, are Newtonian.

During the 1800s several fluid flow problems were solved on the basis of (2.17), observation of real flows, and logic, even though more formal mathematical models were then available. Two of these solutions are remembered by the names of the persons who solved them. Couette, by experiment and analysis of lubricated surfaces, solved the problem of flow between two closely spaced surfaces as shown in Figure 2.5. The name *simple Couette flow* is applied to the equivalent flat plate solution. Hagen, a hydraulic engineer, and Poiseuille, a physician studying blood flow, again by a combination of experiment and analysis, solved the problem of viscous fluid flow driven by a pressure gradient through a circular pipe (or blood vessel) as given in Figure 2.6.

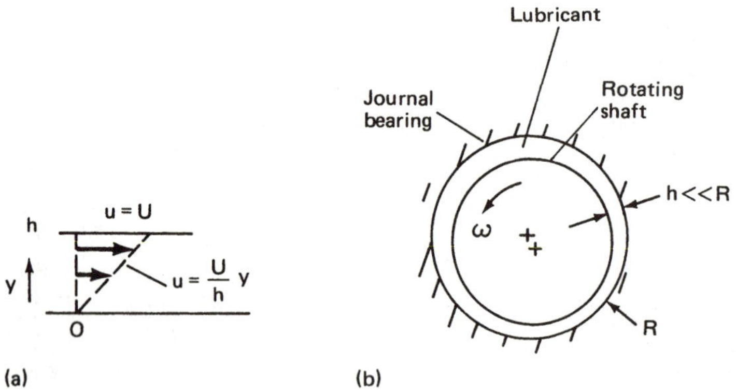

Figure 2.5: *Couette flow with $\tau = \mu\, du/dy$. (a) Parallel plates show simple Couette flow. (b) When $h << R$, flow of oil in a journal bearing approximates simple Couette flow.*

Navier initiated an extension of Euler's 3-D equations of fluid motion to include the general phenomena of fluid viscosity. The derivation was completed by Stokes and resulted in the *Navier-Stokes* equations of fluid flow. These equations are the basis for all modern flow analyses. The complete derivation is beyond the scope of this text, but a summary can provide understanding that will be useful in later chapters.

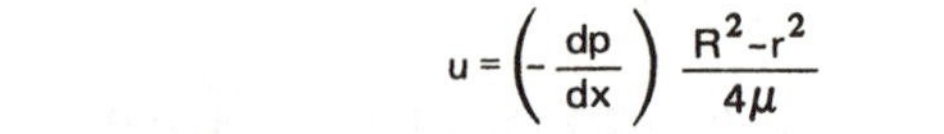

$$u = \left(-\frac{dp}{dx}\right)\frac{R^2 - r^2}{4\mu}$$

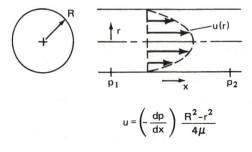

Figure 2.6: *Hagen-Poiseuille flow in a circular pipe. In the steady state, an externally applied dp/dx is balanced by viscous shear stress. The velocity profile $u(r)$ forms a paraboloid.*

Figures 2.7a and 2.7b illustrate stress in a Newtonian fluid. Figure 2.7a shows the 2-D representation for shear stress in Couette flow, and Figure 2.7b shows a 3-D fluid element used to define total stress (pressure and shear) at some point (x, y, z, t). Force per unit area (stress) on each face is resolved into three components, e.g., σ_{xx}, τ_{xy}, and τ_{xz} on the positive x face. The difference or gradient of these stresses between the positive and negative faces of the element is the net force (per unit area) on the element. When the similar forces for the y and z faces also are resolved into x, y, z components, there are nine quantities, which are arranged in matrix form as

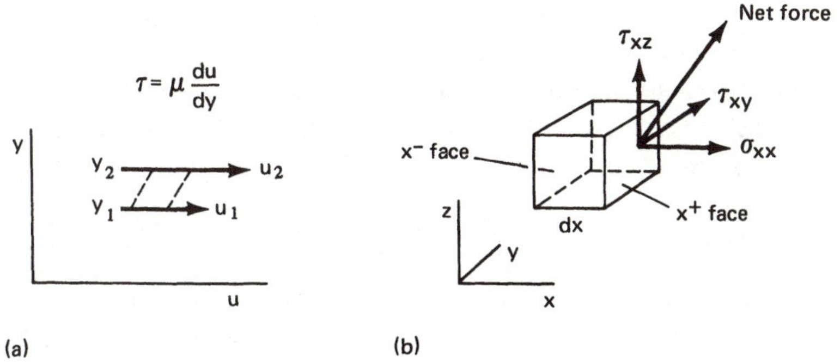

Figure 2.7: *Stress in a Newtonian flow. (a) Shear stress τ is due to a velocity gradient times a coefficient of viscosity μ. (b) The 3-D stress tensor includes pressure stresses (σ) and shear stresses (τ) at each point (x, y, z). Only stresses on the x^+ face are shown here.*

$$\text{Stress Tensor} = \begin{bmatrix} \sigma_{xx} & \tau_{xy} & \tau_{xz} \\ \tau_{yx} & \sigma_{yy} & \tau_{yz} \\ \tau_{zx} & \tau_{zy} & \sigma_{zz} \end{bmatrix} \tag{2.18}$$

Stress at point (x, y, z) and at time (t) is represented by one tensor (2.18) with nine elements. The nine elements define a single symmetric tensor (e.g., $\tau_{yx} = \tau_{xy}$), which describes the 3-D state of stress in the fluid at that point in space and time. The gradient of the diagonal elements ($\sigma_{xx}, \sigma_{yy}, \sigma_{zz}$) leads to normal (pressure) stresses, and the gradient of the off-diagonal terms leads to shear stresses.

The gradient of the stress tensor equals force per unit area as follows

$$\text{Total } \frac{\mathbf{F}}{\mathrm{A}} = \frac{\partial \sigma_{xx}}{\partial x}\hat{\mathbf{i}} + \frac{\partial \tau_{xy}}{\partial y}\hat{\mathbf{j}} + \frac{\partial \tau_{xz}}{\partial z}\hat{\mathbf{k}} \qquad \leftarrow \text{All forces on x face}$$

$$+ \frac{\partial \tau_{yx}}{\partial x}\hat{\mathbf{i}} + \frac{\partial \sigma_{yy}}{\partial y}\hat{\mathbf{j}} + \frac{\partial \tau_{yz}}{\partial z}\hat{\mathbf{k}} \qquad \leftarrow \text{All forces on y face}$$

$$+ \frac{\partial \tau_{zx}}{\partial x}\hat{\mathbf{i}} + \frac{\partial \tau_{zy}}{\partial y}\hat{\mathbf{j}} + \frac{\partial \sigma_{zz}}{\partial z}\hat{\mathbf{k}} \qquad \leftarrow \text{All forces on z face}$$

↑ All forces in x direction — ↑ All forces in y direction — ↑ All forces in z direction (2.19)

The next step is to relate this fluid stress to the fluid motion causing the stress. A tensor can only be equated to other tensors, so a symmetric nine-element tensor is needed to define the state of fluid motion causing the symmetric nine-element stress tensor (written as a matrix here). Further, this tensor should simplify to (2.17) in the case of simple Couette flow. Consider a small change **dV** in the velocity field over some small distance (dx, dy, dz). This is written as

$$d\mathbf{V} = \frac{\partial \mathbf{V}}{\partial x}dx + \frac{\partial \mathbf{V}}{\partial y}dy + \frac{\partial \mathbf{V}}{\partial z}dz$$

But **V** in turn has three components (u, v, w), and the above vector equation can be expanded into three scalar equations that can then be repackaged into a single matrix equation as

$$\begin{bmatrix} du \\ dv \\ dw \end{bmatrix} \begin{matrix} = \\ = \\ = \end{matrix} \begin{bmatrix} \frac{\partial u}{\partial x} & \frac{\partial u}{\partial y} & \frac{\partial u}{\partial z} \\ \frac{\partial v}{\partial x} & \frac{\partial v}{\partial y} & \frac{\partial v}{\partial z} \\ \frac{\partial w}{\partial x} & \frac{\partial w}{\partial y} & \frac{\partial w}{\partial z} \end{bmatrix} \begin{bmatrix} dx \\ dy \\ dz \end{bmatrix} \tag{2.20}$$

The matrix of (2.20), composed of the nine partial derivatives of velocities relative to displacements, is the velocity gradient tensor. This tensor includes both the effects of fluid rotation, which do not include distortion or stress, and the effects of fluid distortion rate, which do cause stress. Further, the distortion rate portion includes compression (of volume) and rate of strain (of shape).

The velocity gradient tensor (2.20) is separated into its distortion part (symmetric) and its rotation part (skew-symmetric) as follows. Clearly, for any two quantities A and B (scalar, vector, or tensor), it is true that

$$A = \frac{1}{2}(A + B) + \frac{1}{2}(A - B)$$

Applied to the velocity gradient tensor of (2.20), with A the velocity gradient tensor and B its transpose (exchange rows and columns), this gives for (A + B)/2

$$\frac{1}{2}\begin{bmatrix} 2\frac{\partial u}{\partial x} & (\frac{\partial u}{\partial y} + \frac{\partial v}{\partial x}) & (\frac{\partial u}{\partial z} + \frac{\partial w}{\partial x}) \\ (\frac{\partial u}{\partial y} + \frac{\partial v}{\partial x}) & 2\frac{\partial v}{\partial y} & (\frac{\partial v}{\partial z} + \frac{\partial w}{\partial y}) \\ (\frac{\partial u}{\partial z} + \frac{\partial w}{\partial x}) & (\frac{\partial v}{\partial z} + \frac{\partial w}{\partial y}) & 2\frac{\partial w}{\partial z} \end{bmatrix} = [\dot{\varepsilon}] \tag{2.21}$$

and for (A – B)/2

$$\frac{1}{2}\begin{bmatrix} 0 & (\frac{\partial u}{\partial y}-\frac{\partial v}{\partial x}) & (\frac{\partial u}{\partial z}-\frac{\partial w}{\partial x}) \\ (\frac{\partial v}{\partial x}-\frac{\partial u}{\partial y}) & 0 & (\frac{\partial v}{\partial z}-\frac{\partial w}{\partial y}) \\ (\frac{\partial w}{\partial x}-\frac{\partial u}{\partial z}) & (\frac{\partial w}{\partial y}-\frac{\partial v}{\partial z}) & 0 \end{bmatrix} = [\tilde{\omega}] \tag{2.22}$$

Equation (2.22) is the rotation tensor that defines the rotation of fluid elements. Chapter 8 relates this tensor to vorticity. Equation (2.21) is the symmetric rate of strain tensor that defines distortion rate. The symmetric stress tensor (2.18) is then related to (2.21), element by element, to complete the Navier-Stokes equations. Stokes successfully did this as reported, for example, in Schlichting (1979). Details of the exact relationship continue to be debated, but experiments verify Stokes analysis as valid.

Navier and Stokes used Euler's $D(\rho \mathbf{V})/Dt$ as their expression for rate of change of momentum, and their derivation divided $\mathbf{F}$ into body forces ($\mathbf{F}_b$), pressure forces ($\mathbf{F}_p$), and shear forces ($\mathbf{F}_s$)

$$\rho\, D\mathbf{V}/Dt = \mathbf{F}_b + \mathbf{F}_p + \mathbf{F}_s$$

The complete equation is valid for $\rho(x, y, z, t)$ and $\mu(x, y, z, t)$. For almost all applications of this text, the assumptions of constant density and constant viscosity are used. The Navier-Stokes vector equation for $\rho = c$, $\mu = c$, and with gravity the only body force, is, after dividing through by ρ

$$\frac{D\mathbf{V}}{Dt} = \mathbf{g} - \frac{1}{\rho}\nabla p + \frac{\mu}{\rho}\nabla^2\mathbf{V} \tag{2.23}$$

Except for the viscous term, (2.23) is the same as Euler's equation for inviscid flow (2.14). The viscous term is significant in that it precisely quantifies internal shear due to μ, including shear along solid boundaries. It is also significant in that it adds a second order term (∇^2) to the fluid equation. In Cartesian coordinates the Laplacian ∇^2 is computed as

$$\nabla^2 = \nabla \cdot \nabla = \frac{\partial^2}{\partial x^2} + \frac{\partial^2}{\partial y^2} + \frac{\partial^2}{\partial z^2} \tag{2.24}$$

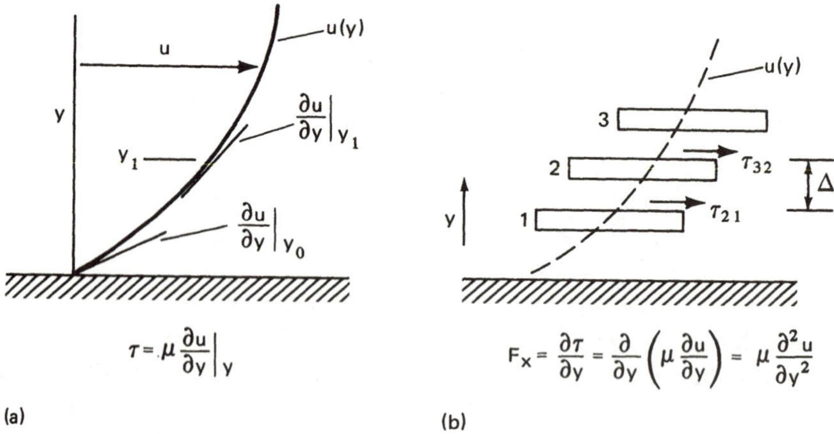

Figure 2.8: *Shear, and net force due to shear, on a Newtonian fluid element. (a) Shear stress is proportional to gradient of velocity. (b) Net force is equal to the gradient of shear stress.*

and in (2.23) is applied successively to each of the components of **V**. Again the reader is cautioned that $(\mathbf{V} \cdot \nabla)\mathbf{V}$ and $\nabla^2 \mathbf{V}$ are valid in Cartesian coordinates; they are not easily applied in other coordinate systems. Refer to Appendix A.

Figures 2.8a and 2.8b provide a physical interpretation of these terms using the one term $\mu\partial^2 u/\partial y^2$ as an example. Figure 2.8a reviews the 1-D computation of shear stress in a Newtonian fluid ($\tau = \mu\ \partial u/\partial y$), and Figure 2.8b illustrates changes of τ normal to streamlines. Substitute τ from Figure 2.8a into the expression for $\partial\tau/\partial y$ in Figure 2.8b to get

$$\partial\tau / \partial y = \partial\left(\mu \frac{\partial u}{\partial y}\right)/\partial y \ = \ \mu\, \partial^2 u/\partial y^2$$

This gradient of shear represents net force per unit volume on element 2 due to viscous interaction with elements 1 and 3. Three of the nine possible terms of $\nabla^2\mathbf{V}$, namely $\partial^2 u/\partial x^2$, $\partial^2 v/\partial y^2$, and $\partial^2 w/\partial z^2$, are not as easy to interpret. In many flows these terms are zero or negligible as will be seen in later sections.

Equation (2.23) and the continuity equation (2.5) comprise a set of four scalar equations in four unknowns for a constant density, constant viscosity flow. In rectangular coordinates and with the z axis vertical, the four equations are

$$\frac{\partial u}{\partial t}+u\frac{\partial u}{\partial x}+v\frac{\partial u}{\partial y}+w\frac{\partial u}{\partial z}=-\frac{1}{\rho}\frac{\partial p}{\partial x}+\nu\left(\frac{\partial^2 u}{\partial x^2}+\frac{\partial^2 u}{\partial y^2}+\frac{\partial^2 u}{\partial z^2}\right) \tag{2.25a}$$

$$\frac{\partial v}{\partial t}+u\frac{\partial v}{\partial x}+v\frac{\partial v}{\partial y}+w\frac{\partial v}{\partial z}=-\frac{1}{\rho}\frac{\partial p}{\partial y}+\nu\left(\frac{\partial^2 v}{\partial x^2}+\frac{\partial^2 v}{\partial y^2}+\frac{\partial^2 v}{\partial z^2}\right) \tag{2.25b}$$

$$\frac{\partial w}{\partial t}+u\frac{\partial w}{\partial x}+v\frac{\partial w}{\partial y}+w\frac{\partial w}{\partial z}=g-\frac{1}{\rho}\frac{\partial p}{\partial z}+\nu\left(\frac{\partial^2 w}{\partial x^2}+\frac{\partial^2 w}{\partial y^2}+\frac{\partial^2 w}{\partial z^2}\right) \tag{2.25c}$$

$$\frac{\partial u}{\partial x}+\frac{\partial v}{\partial y}+\frac{\partial w}{\partial z}=0 \quad \text{continuity} \tag{2.25d}$$

The ratio μ/ρ appears so often in the study of viscous flow phenomena that it has been given a special name (kinematic viscosity) and symbol (ν)

$$\nu=\frac{\mu}{\rho} \tag{2.26}$$

It is called the *kinematic viscosity* because its units (L^2/T) consist only of kinematical variables. Physically this ratio represents the rate at which a fluid can diffuse and/or suppress its own internal motions. Air, for example, has a dynamic viscosity (μ) roughly 1/50 that of water, but because the ratio of their densities is even greater, the kinematic viscosity (ν) of air is about 15 times that of water. Therefore, a disturbance caused by motion of an object in air will suppress itself sooner and over a shorter distance than an equivalent motion in water.

Equations (2.25a,b,c,d) are a set of four, nonlinear (because of $u\partial u/\partial x$ type terms), second order (because of $\partial^2 u/\partial x^2$ type terms), partial (four independent variables x, y, z, t) differential equations. Because there are the same number of unknowns (u, v, w, p) as there are equations, solutions are theoretically possible. But in reality the set of equations are so difficult that they have never been solved for general problems where each term is nonnegligible. Fortunately, there are many practical problems where most or many terms are zero, or small enough to be neglected. In these cases, solutions are practical; some of the solutions are exact, others are approximate.

2.5 Dimensionless Equations and Ratios

Variables in the preceeding equations can be expressed in System International (SI), engineering (US), or any other consistent set of physical units. However, it is possible to define variables, for the same equations, that have no physical units. Use of these dimensionless variables has proved extremely useful in the study of fluid mechanics. For example, consider the four equations (2.25a,b,c,d). If we assume that the gravity term can be neglected (e.g., a constant density fluid with no free surface), then these four equations have four dependent variables (u, v, w, p), four independent variables (x, y, z, t), and four parameters (ρ, μ, L, U). The four parameters are density, viscosity, a length characteristic of the problem, and a velocity characteristic of the problem. Reynolds converted these equations to dimensionless form, and in the process reduced the four parameters to one parameter. This one parameter is a dimensionless ratio of the four original parameters and is called the *Reynolds number*

$$R_e = \frac{\rho LU}{\mu} = \frac{LU}{\nu} \tag{2.27}$$

Solution of a set of four, nonlinear, second order, partial differential equations is notoriously difficult, and the Navier-Stokes equations, in four parameters, would have to be solved over and over for each and every change in any of the four parameters. The same problem, after converting to dimensionless variables, requires only the one parameter (R_e). The actual value of ρ, μ, L, or U are of no concern, only their ratio, the Reynolds number. The entire range of possibilities for drag (F_D) on any sphere in any Newtonian fluid is shown graphically by the one curve in Figure 2.9. The graph is used by computing R_e for ρ, μ, L, U of a problem of interest. For flow past a sphere, L is the diameter (D) of the sphere, and U is the speed of the fluid relative to the sphere. Enter R_e as the abscissa, and read C_D as the coefficient of drag. Note that C_D also is dimensionless for the same reason that R_e is dimensionless. Actual drag force on the sphere, F_D (in physical units), is obtained using the following equation

$$C_D = \frac{F_D}{\frac{1}{2}\rho U^2 A} \tag{2.28}$$

where A is the cross sectional area of the sphere. Figure 2.9 also includes the drag coefficient for flow past a circular cylinder for later reference. Use of dimensionless variables and parameters is so great an advantage that the transition from dimensional to dimensionless quantities is well justified. As a result, most texts, handbooks, reference tables, etc., make extensive use of them.

The Navier-Stokes equations can be made dimensionless as follows. Equations (2.23) and (2.5) with DV/Dt expanded and g removed are

$$\frac{\partial \mathbf{V}}{\partial t} + (\mathbf{V} \cdot \nabla)\,\mathbf{V} = -\frac{1}{\rho}\nabla p + \nu\,\nabla^2 \mathbf{V} \tag{2.29a}$$

$$\nabla \cdot \mathbf{V} = 0 \tag{2.29b}$$

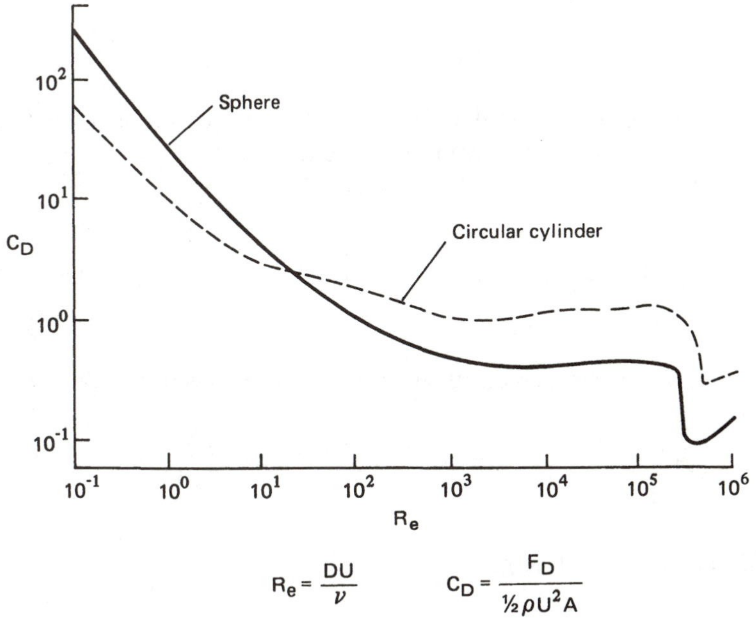

Figure 2.9: *Drag coefficient (C_D) as a function of Reynolds number (R_e) for flow past a sphere and for flow past a circular cylinder of unit length, each of diameter (D).*

Velocity is made dimensionless by dividing **V** by some speed characteristic of the given problem, represented here by U. Dimensionless velocity becomes

$$\mathbf{V}^* = \mathbf{V}/U$$

The del operator (∇) is composed of terms like ($\partial/\partial x$) and has units of inverse length. Del is made dimensionless using some size characteristic of the given problem, represented here by L. Dimensionless ∇ is

$$\nabla^* = \nabla / \frac{1}{L} = L\nabla$$

Time is a variable in the equation because of the operator $\partial/\partial t$. It is made dimensionless using L/U as a characteristic time, which is the time required for the fluid to move a distance L at speed U

$$t^* = t / \frac{L}{U}$$

Note that use of some other arbitrary characteristic time is not justified, because time has already been specified indirectly through L and U. A characteristic pressure is defined as

$$p^* = p/\rho U^2$$

The motivation for this is developed in Chapter 3. Density and viscosity will help to form a dimensionless ratio and need not be made dimensionless.

Substitute for **V**, ∇, t, and p in (2.29a) to get

$$\frac{\partial(U\mathbf{V}^*)}{\partial(\frac{L}{U}t^*)} + \left(U\mathbf{V}^* \cdot \frac{1}{L}\nabla^*\right)U\mathbf{V}^* = -\frac{1}{\rho}\frac{1}{L}\nabla^*(\rho U^2 p^*) + \frac{\mu}{\rho}\left(\frac{1}{L}\nabla^*\right)^2 U\mathbf{V}^*$$

The quantities U, L, and ρ are constants, and the previous equation simplifies to

$$\frac{U^2}{L}\frac{\partial \mathbf{V}^*}{\partial t^*} + \frac{U^2}{L}(\mathbf{V}^* \cdot \nabla^*)\mathbf{V}^* = -\frac{U^2}{L}\nabla^* p^* + \frac{\mu U}{\rho L^2}(\nabla^*)^2\mathbf{V}^*$$

Next, divide through by U^2/L to get

$$\frac{\partial \mathbf{V}^*}{\partial t^*} + (\mathbf{V}^* \cdot \nabla^*)\mathbf{V}^* = -\nabla^* p^* + \frac{\mu}{\rho L U}(\nabla^*)^2\mathbf{V}^*$$

By now it is easy to see that the dimensionless quantities form exactly the same equation as before, except for the parameters. All the parameters have now combined to become the inverse of the dimensionless ratio earlier called the Reynolds number

$$R_e = \rho L U/\mu$$

It is conventional to omit the asterisks since they are now unnecessary and simply to write

$$\frac{D\mathbf{V}}{Dt} = \frac{\partial \mathbf{V}}{\partial t} + (\mathbf{V} \cdot \nabla)\mathbf{V} = -\nabla p + \frac{1}{R_e}\nabla^2\mathbf{V} \qquad (2.30)$$

as the dimensionless form of (2.29a). It is easy to show by substituting for ∇^* and $\mathbf{V}^*$ in (2.29b) that the continuity equation holds whether in dimensional or dimensionless variables

$$\nabla \cdot \mathbf{V} = \nabla^* \cdot \mathbf{V}^* = 0$$

Solutions for u, v, w, p, t, etc. in dimensionless variables will be in multiples of U, ρU^2, L/U, etc. To return to conventional physical variables merely multiply by the characteristic quantity chosen for that problem. Note that dimensionless quantities like R_e are independent of any system of units. If $R_e = 786$ for a given problem in SI units, it will also equal 786 in US units, as may be verified.

The Reynolds number can also be derived as the ratio of a typical inertia force ($\rho u \, \partial u/\partial x$) to a typical viscous force ($\mu \, \partial^2 u/\partial y^2$), again by substituting in characteristic lengths and velocities. The ratio is as before

$$R_e \sim \frac{\rho u \, \partial u/\partial x}{\mu \, \partial^2 u/\partial y^2} \sim \frac{\rho U(U/L)}{\mu \, U/L^2} = \frac{\rho UL}{\mu}$$

The Froude number (F_r) is a ratio of typical inertial forces to gravity forces (ρg)

$$\frac{\rho u \, \partial u/\partial x}{\rho g} \sim \frac{U(U/L)}{g} = \frac{U^2}{gL}$$

Some researchers use U^2/gL as the Froude number, but more commonly they take its square root as F_r

$$F_r = U/(gL)^{1/2} \tag{2.31}$$

If g and μ both exist in a given problem, e.g., as in (2.23), F_r and R_e both appear in the dimensionless form of the equation. In more complex problems, additional dimensionless numbers may be required. Although the problem then may not be reduced to a one-parameter problem, the number of dimensionless parameters is still less than the number of dimensional parameters, plus the advantage of units (ratios) scaled to that problem.

The Mach number

$$M_a = U/c \tag{2.32}$$

measures of the compressibility of the fluid, where c is the speed of sound in that fluid. In this text, unless noted, $M_a << 1$, and both liquids and gases can be considered incompressible relative to pressure changes of magnitude $\rho U^2/2$.

The most important and most used of the dimensionless ratios is R_e. Consideration of the range of R_e from zero to infinity separates fluid flows into manageable regions for analysis. Consider

$R_e = 0$

$R_e << 1$

$R_e \sim 1$

$R_e >> 1$

$R_e \to \infty$

Whether R_e can equal zero is analyzed using its definition, $R_e = \rho LU/\mu$. R_e is equal to zero if $\rho = 0$ or $L = 0$, but then there is no problem. $R_e = 0$ also if $U = 0$ or if $\mu \rightarrow \infty$. If $U = 0$ the fluid analysis is a hydrostatic problem, and if $\mu \rightarrow \infty$, the material is a rigid body. The Reynolds number can approach infinity if ρ, L, or U approach infinity, none of which are realistic, or if $\mu = 0$. The case $\mu = 0$ represents an ideal or perfect fluid and reduces to Euler's equation and its solutions for inviscid flow (see Chapter 3). The case $R_e \sim 1$ represents many real flows. These are impossible to solve analytically, although certain approximations and/or simplified flows are accessible to analytical methods. Success in this region has mostly been achieved using experimental techniques and more recently using computers.

This leaves the important regions $R_e << 1$ and $R_e >> 1$. Consider first the region $R_e << 1$. This case is analyzed most easily by rearranging (2.30) to read

$$R_e \left[\frac{\partial \mathbf{V}}{\partial t} + (\mathbf{V} \cdot \nabla)\mathbf{V} \right] = - R_e \nabla p + \nabla^2 \mathbf{V} \tag{2.33}$$

The notation (<< 1) or (>> 1) is used here to assess the relative magnitude of certain terms in an equation relative to other terms. In dimensional variables, typical velocities are in the range of the characteristic velocity U, so dimensionless velocity, being their ratio, typically has a value near one. A similar argument applies to most dimensionless variables. Therefore, if some quantity is << 1, then its value is so small, relative to other terms, that it may usually be neglected. In (2.33), if $R_e << 1$, then the left-hand side, being multiplied by a very small number (R_e), can be neglected relative to $\nabla^2 \mathbf{V}$ the right-hand side.

The quantity (∇p) is also multiplied by R_e, and if p consisted only of flow induced pressures, then ($R_e \nabla p$) could also be neglected--but then there is no flow. However, p can also include externally applied pressures, which can be of any magnitude. In that case, even if $R_e << 1$, ∇p could be >>1, and the product $R_e \nabla p$ may not be neglected. The final equation for $R_e << 1$ is

$$\nabla p = \frac{1}{R_e} \nabla^2 \mathbf{V} \qquad \text{(Dimensionless)} \tag{2.34a}$$

$$\nabla p = \nu \nabla^2 \mathbf{V} \qquad \text{(Dimensional)} \tag{2.34b}$$

Small Reynolds numbers are typically achieved with

1. L small: fine sand settling through water, dust in air, blood flow through small diameter capillaries, or lubrication between close fitting parts
2. U small: a typical object moving at a very, very slow speed, say 1 cm/hr
3. μ large: thick grease

Of course in all these examples, it is the ratio, $R_e = \rho LU/\mu$, that is important. For example, even though μ may be large in some applications, it is possible that the product ρLU is even larger so that $R_e >> 1$.

Near the end of the 1800s it was becoming apparent to a small group of practical experimenters and scientists that powered flight would soon become a reality. The Navier-Stokes equations had been in use for 50 years, but solutions valid for nearly inviscid flow ($R_e >> 1$) were not yet feasible. The value of R_e for flow of air about a typical aircraft wing at realistic flow velocities is always very large, in part because of the very small value of μ for air. Because air and water are nearly inviscid ($\mu << 1$), and typical engineering and geophysical sizes and velocities large, the region $R_e >> 1$ is of great practical importance. It includes the region of boundary layer flows and of turbulence and is covered in detail in Chapter 4.

Chapter 3

Potential (Inviscid) Flow

3.1 Bernoulli Equation

A fluid flow can be represented visually by means of pathlines, streamlines, streaklines, timelines, and velocity profiles. In a flow whose pattern changes with time, the five are distinctly different. Their differences are easily apparent when viewed in a laboratory experiment, but they are difficult to illustrate in a drawing. Later examples will help to illustrate these flow visualizations.

1. *Pathlines* describe the paths of specific fluid elements shown as points as they move through the region of interest. Motion of a small neutrally buoyant particle, carried along by the fluid, defines a pathline.
2. *Streamlines* connect velocity vectors of the fluid, at a specific time t, as continuous lines. If a fluid region is filled with many small neutrally buoyant particles, a brief time exposure will show motion of each as a short line. Streamlines are drawn tangent to these and define flow patterns. A *stream tube* is a streamline with finite volume.
3. *Streaklines* connect all fluid elements that have passed a given fixed point. A streak or line formed in flowing water by slowly injecting dye at a fixed location is a streakline. In a steady-state flow, this line is also a pathline and a streamline, but not in a transient or oscillatory flow.
4. *Timelines* connect all fluid elements along a given locus of points, usually a line that crosses streamlines, at time t. In one application, a laser pulse causes the line of fluid illuminated by the laser to turn purple. As this line of purple fluid moves in the flow field it deforms, illustrating deformations of that flow field caused by velocity variations.
5. *Velocity profiles* show the distribution of flow velocities over some region of flow at time t. In steady-state flows, velocity profiles are identical to timelines. The velocity profile of the *mean* flow is often used to describe turbulent flow patterns. Streamlines and velocity profiles are especially useful for illustrating flow details in drawings.

Potential flow is characterized as a flow whose energy per unit mass is conserved along a streamline. In this class of flows, velocity at any point can be derived as a gradient of some scalar function. Potential flows occur when

1. Only conservative forces exist. Conservative forces (e.g., gravity) can always be derived as the gradient of a scalar potential function; nonconservative forces cannot be.
2. Density is constant, and therefore $\nabla \cdot \mathbf{V} = 0$.
3. The fluid is inviscid ($\mu = 0$).

As will be shown in Part II, when conditions 1, 2, and 3 are met, rotation of fluid elements (2.22) cannot change. If $\omega = 0$ initially in the fluid, it remains zero for any motion induced by conservative forces, i.e., the flow is then irrotational. This condition is more often stated as $\zeta = \nabla \times \mathbf{V} = 0$, where ζ, the vorticity, is related to fluid element rotation by $\zeta = 2\omega$. Vorticity is discussed in detail in Chapter 8 and later chapters. Some of these conditions can be violated in special applications as will be noted.

Assuming also steady state, Euler's equation (2.14) now becomes

$$(\mathbf{V} \cdot \nabla)\mathbf{V} = \mathbf{g} - \frac{1}{\rho}\nabla p \tag{3.1}$$

When flow is one-dimensional (1-D), and the flow direction is parallel to the x axis, (3.1) reduces to a scalar equation

$$u\frac{\partial u}{\partial x} = -g\frac{\partial h}{\partial x} - \frac{1}{\rho}\frac{\partial p}{\partial x} \tag{3.2}$$

Because the x axis is not assumed to be horizontal, there will generally be some change in vertical height associated with changes in x, and the term $(-g\ \partial h/\partial x)$ is necessary. The equation can now be integrated as follows

$$u\frac{du}{dx} + \frac{1}{\rho}\frac{dp}{dx} + g\frac{dh}{dx} = 0$$

$$\frac{d}{dx}\left(\frac{u^2}{2}\right) + \frac{d}{dx}\left(\frac{p}{\rho}\right) + \frac{d}{dx}(gh) = 0$$

$$\frac{d}{dx}\left(\frac{u^2}{2} + \frac{p}{\rho} + gh\right) = 0$$

$$\frac{u^2}{2} + \frac{p}{\rho} + gh = \text{constant} \tag{3.3}$$

Equation (3.3) is known as the Bernoulli equation. It states that total energy per unit mass along a streamline is conserved when $\rho = c$, $\mu = 0$, and $\partial/\partial t = 0$. Here $u^2/2$ is kinetic energy per unit mass, p/ρ is pressure energy per unit mass (compare to energy stored in a spring), and gh is potential energy per unit mass due to change in elevation. The Bernoulli equation (3.3) was derived here for flow along a straight streamline in the x direction. The derivation can easily be extended to an arbitrary streamline with (u) following any curve in 3-D space. The critical limitation is that it be applied along a specific streamline--going from a point on one streamline to a point on an adjacent streamline is not a valid application of the Bernoulli equation unless both streamlines have the same total energy per unit mass. Examples are given in later sections.

3.2 Stream Functions and Velocity Potentials

An important subclass of potential flows is restricted to 2-D flows (x, y or r, θ) or a 3-D flow that can be transformed to an equivalent 2-D flow such as flow past an axisymmetric body. The solution makes use of a stream function (ψ) and a velocity potential (ϕ).

As shown in Figure 3.1a, lines of ψ = constant coincide with streamlines. Because streamlines are always tangent to velocity vectors, no flow occurs across streamlines or across lines of ψ = constant. Therefore, the mass flow rate ($\dot{m}$) between any two streamlines (stream functions) remains constant throughout the flow field. Further if $\rho = c$, then volume flow rate (Q) also remains constant between pairs of streamlines and can be quantified by integrating ($\mathbf{V} \cdot d\mathbf{S}$) along transverse lines, say line B or line C, either of which traverses the region between ψ_2 and ψ_3. Note that lines here, such as B, represent surfaces of unit height in the 2-D flow.

In Figure 3.1a, line E is parallel to the x axis, only the v component of flow traverses it, and ($\mathbf{V} \cdot d\mathbf{S}$) between ψ_0 and ψ_1 becomes vdx. If line F, parallel to the y axis, is used, ($\mathbf{V} \cdot d\mathbf{S}$) becomes (udy). Volumetric flow rate between ψ_0 and ψ_1 can then be obtained by integrating ($\int$udy) along line F starting at point 0. The same value can be obtained by integrating $\int v(-dx)$ along line E starting at point 0. This relationship between Q, ψ, and **V** is expressed formally in Cartesian coordinates as

$$u = \frac{\partial \psi}{\partial y}, \quad v = -\frac{\partial \psi}{\partial x} \tag{3.4a}$$

In polar coordinates (r, θ) the relationship is

$$v_r = \frac{1}{r}\frac{\partial \psi}{\partial \theta}, \quad v_\theta = -\frac{\partial \psi}{\partial r} \tag{3.4b}$$

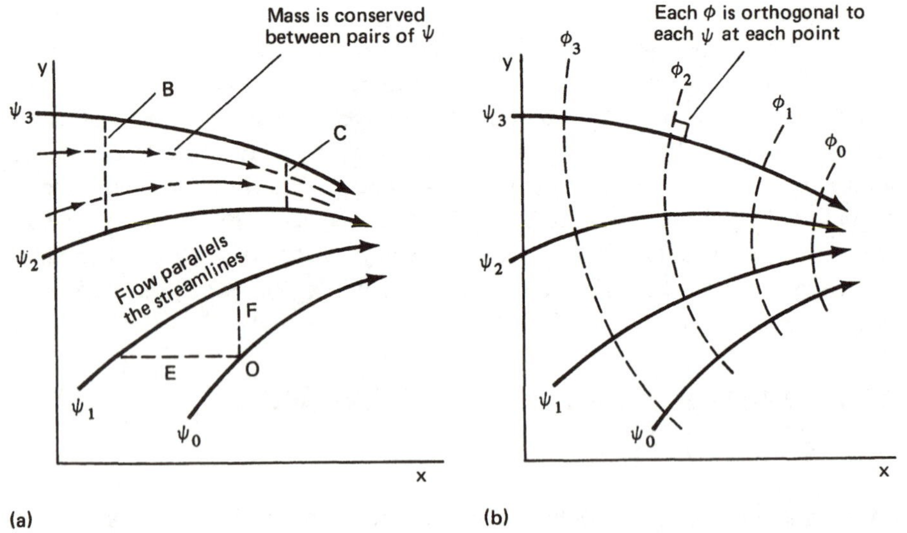

Figure 3.1: *Stream functions (ψ) and velocity potentials (ϕ) in a 2-D, $\rho = c$, $\mu = 0$ flow field. (a) Lines of constant ψ coincide with streamlines and quantify mass flow rate per unit depth between pairs of ψ. (b) ψ and ϕ form an orthonormal net. If ψ and ϕ are interchanged, they describe a flow orthogonal to the one shown.*

Most modern texts use the sign conventions of (3.4a,b); earlier texts sometimes use $u = -\partial\psi/\partial y$ and $v = \partial\psi/\partial x$. Either convention is acceptable; the change merely involves changing also the convention for the slope of ψ. In either case the magnitude of ψ_0 can be set arbitrarily; it is the gradient of ψ that yields mass flow rate (or volumetric flow rate if $\rho = c$) leading to

$$\psi_{i+1} - \psi_i = Q \text{ (between i and i + 1)} \tag{3.5}$$

Use of (3.4a) leads to **V** expressed as a type of gradient of the scalar function ψ as follows

$$\mathbf{V} = \frac{\partial\psi}{\partial y}\,\hat{\mathbf{i}} - \frac{\partial\psi}{\partial x}\,\hat{\mathbf{j}} \tag{3.6}$$

Next assume that **V** can be expressed as the negative gradient of another scalar function ϕ, the velocity potential, as

$$\mathbf{V} = -\nabla\phi \tag{3.7}$$

Again it is possible to reverse the sign in (3.7) by reversing the gradient of ϕ. Some authors do so, but (3.7), paralleling the concept of $\mathbf{F}_p = -\nabla p$, is preferred. Equation (3.7) is valid in 3-D space, but is of particular interest here in 2-D flows as

$$u = -\frac{\partial \phi}{\partial x}, \quad v = -\frac{\partial \phi}{\partial y} \tag{3.8a}$$

In polar coordinates the relationship is

$$v_r = -\frac{\partial \phi}{\partial r}, \quad v_\theta = -\frac{1}{r}\frac{\partial \phi}{\partial \theta} \tag{3.8b}$$

As with ψ, the magnitude of ϕ_o is arbitrary; only its gradient is relevant. Note that if any of $\mathbf{V}$, ψ, or ϕ are given, the other two can be obtained by differentiation or integration. Figure 3.1b illustrates these relationships; ψ and ϕ define an orthonormal net. The slope of ψ is (v/u), i.e., a streamline, and the slope of ϕ is $(-u/v)$, i.e., perpendicular to the flow.

In the case of plane (2-D), steady-state ($\partial/\partial t = 0$), incompressible ($\rho = c$), and inviscid ($\mu = 0$) flows, ψ and ϕ as a pair satisfy four unique conditions. The first two conditions use the divergence and curl of $\mathbf{V}$ as follows

$$\nabla \cdot \mathbf{V} = \frac{\partial u}{\partial x} + \frac{\partial v}{\partial y} = \frac{\partial}{\partial x}\left(\frac{\partial \psi}{\partial y}\right) + \frac{\partial}{\partial y}\left(-\frac{\partial \psi}{\partial x}\right) = 0 \tag{3.9}$$

$$\nabla \times \mathbf{V} = \frac{\partial v}{\partial x} - \frac{\partial u}{\partial y} = \frac{\partial}{\partial x}\left(-\frac{\partial \phi}{\partial y}\right) - \frac{\partial}{\partial y}\left(-\frac{\partial \phi}{\partial x}\right) = 0 \tag{3.10}$$

upon exchanging the order of differention. In addition, ψ and ϕ satisfy Laplace's equation as follows

$$\nabla \cdot \mathbf{V} = \frac{\partial u}{\partial x} + \frac{\partial v}{\partial y} = \frac{\partial}{\partial x}\left(-\frac{\partial \phi}{\partial x}\right) + \frac{\partial}{\partial y}\left(-\frac{\partial \phi}{\partial y}\right) = -\left(\frac{\partial^2 \phi}{\partial x^2} + \frac{\partial^2 \phi}{\partial y^2}\right) = 0 \tag{3.11}$$

$$\nabla \times \mathbf{V} = \frac{\partial v}{\partial x} - \frac{\partial u}{\partial y} = \frac{\partial}{\partial x}\left(-\frac{\partial \psi}{\partial x}\right) - \frac{\partial}{\partial y}\left(\frac{\partial \psi}{\partial y}\right) = -\left(\frac{\partial^2 \psi}{\partial x^2} + \frac{\partial^2 \psi}{\partial y^2}\right) = 0 \tag{3.12}$$

That they satisfy $\nabla \cdot \mathbf{V} = 0$ and $\nabla \times \mathbf{V} = 0$ is consistent with the restriction of $\rho = c$ and $\mu = 0$. That they satisfy Laplace's equation indicates that any solution of Laplace's equation in 2-D defines a possible flow field. Most of the infinity of solutions known for Laplace's equation do not define practical flow fields of interest, but many do. As seen in (3.11) and (3.12), a solution of Laplace's equation can be used either as ϕ or as ψ. Therefore, ϕ and ψ can be interchanged. In Figure 3.1b, ψ can be used as ϕ, and ϕ as ψ; each solution defines an additional orthogonal solution. A last matter of interest here is the principle of superposition based on the linearity of Laplace's equation. Superposition indicates that if ϕ_1 and ϕ_2 (or ψ_1 and ψ_2) are each solutions of Laplace's equation, then so also is $\phi = \phi_1 + \phi_2$ (or $\psi = \psi_1 + \psi_2$), i.e., solutions can be superposed to give a new solution.

Solutions often are derived in the complex plane. In a 2-D flow the complex number

$$z = x + iy \qquad (= re^{i\theta} \text{ in polar coordinates})$$

locates the fluid element. The letter w is typically used to represent the complex potential

$$w = \phi + i\psi$$

and, without proof, the complex velocity becomes

$$\upsilon = u - iv = -dw/dz$$

Speed, the magnitude of υ, is obtained as $|dw/dz|$. The letters z and w, used here to represent complex numbers, should not be confused with their use elsewhere to represent vertical direction and velocity, respectively. For a complete discussion of the solution of potential flow problems in the complex plane the reader is referred to specialized texts, e.g., Milne-Thomson (1960). Later sections will illustrate the simplicity of flow descriptions and solutions using ϕ and ψ but will omit proofs and details. Stream functions and velocity potentials for useful flows are listed in Appendix B.

3.3 Flow Past a Circular Cylinder

This section uses the velocity potential (ϕ) in polar coordinates (v_r, v_θ) to illustrate flow around a circular cylinder by combining ϕ for a sink, ϕ for a source, and ϕ for a uniform flow. Figure 3.2a illustrates a uniform flow, and Figures 3.2b and 3.2c illustrate a source and a sink, respectively. A similar solution uses the stream function ψ rather than ϕ. The velocity potential for a uniform flow in the +x direction is

$$\phi_{uf} = -Ux \tag{3.13a}$$

verified by computing

$$V = -\nabla\phi = -\left(\frac{\partial}{\partial x}\hat{i} + \frac{\partial}{\partial y}\hat{j} + \frac{\partial}{\partial z}\hat{k}\right)(-Ux) = U\hat{i} \tag{3.13b}$$

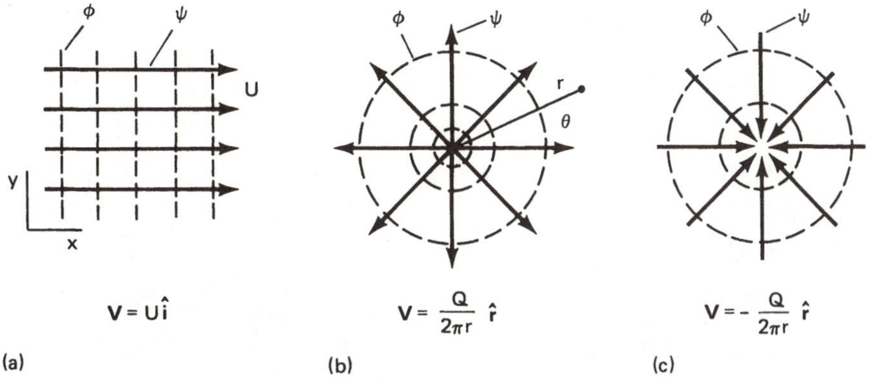

Figure 3.2: *Three elementary 2-D potential flows. (a) A flow moving to the right with uniform velocity. (b) A line source with fluid moving radially outward. c) A line sink with fluid moving radially inward.*

Velocity potentials for a sink and a source are

$$\phi_{si} = -\frac{Q}{2\pi}\,\ell n\ r, \quad \phi_{so} = \frac{Q}{2\pi}\,\ell n\ r \tag{3.14a}$$

where Q is the source strength. The gradients of ϕ_{si} and ϕ_{so} are

$$\mathbf{V} = -\frac{Q}{2\pi r}\,\hat{\mathbf{r}}\ \text{(sink)}, \quad \mathbf{V} = \frac{Q}{2\pi r}\,\hat{\mathbf{r}}\ \text{(source)} \tag{3.14b}$$

If the line source and sink are simply combined, and they have equal Q, they cancel each other. If, however, a solution is obtained, as the distance between the source and sink approaches zero, the limiting flow is called a *doublet*. Its velocity potential is

$$\phi_{do} = -k\,\frac{\cos\theta}{r} \tag{3.15}$$

with k the doublet strength. Figures 3.3a and 3.3b show development of a doublet from a source and sink.

As a final step the doublet is placed within, and opposing, a uniform stream as shown in Figure 3.4. Flow about a circular cylinder is obtained as follows. The uniform-flow velocity potential is first transformed to polar coordinates ($\phi_{uf} = -Ux = -Ur\cos\theta$) and then added to the doublet velocity potential

$$\phi_{cc} = \phi_{uf} + \phi_{do} = -Ur\cos\theta - k\,\frac{\cos\theta}{r}$$

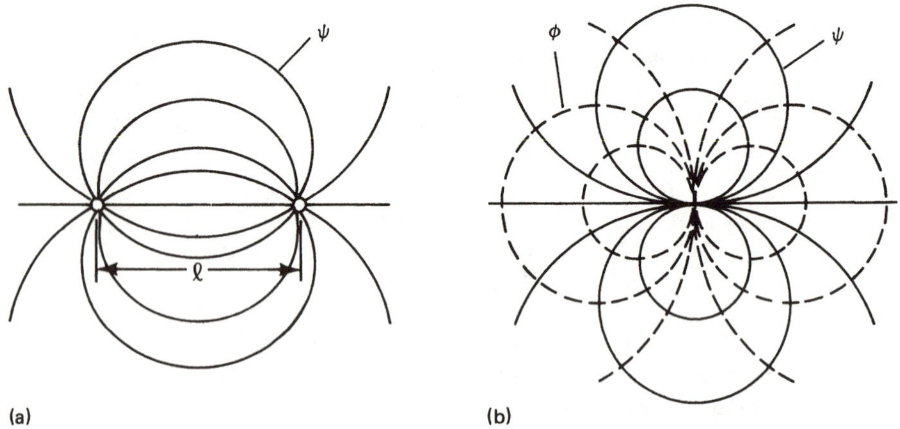

Figure 3.3: *Derivation of the flow field called a doublet. (a) A line source and a sink separated by a distance* ℓ *form the flow field shown. (b) If the source and sink coincide, they cancel, but in the limit of* $\ell \rightarrow 0$ *the flow is as shown.*

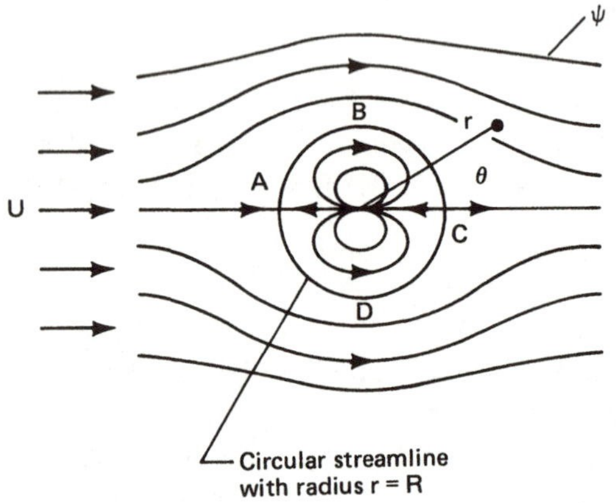

Figure 3.4: *If a doublet is superimposed on a uniform flow, one of the streamlines is a perfect circle. Flow exterior to the circle then provides the solution for incompressible inviscid flow past a circular cylinder.*

The velocity field is obtained as the negative gradient of ϕ_{cc}. First obtain

$$v_r = -\frac{\partial \phi_{cc}}{\partial r} = \left(U - \frac{k}{r^2} \right) \cos\theta$$

and observe that on the circle $(r = \sqrt{k/U})$, the radial velocity is zero for all θ. This is the exact condition imposed by placing a cylinder of radius $(R = \sqrt{k/U})$ in an inviscid free stream. Therefore, the velocity potential ϕ_{cc} can be used to represent flow about a cylinder with radius R. The flow field satisfies all requirements for a 2-D potential flow, and it satisfies the boundary conditions imposed by the cylinder. With k evaluated now as $(k = R^2U)$ the final solution is

$$\phi_{cc} = -U\cos\theta \left(r + \frac{R^2}{r} \right) \tag{3.16a}$$

$$v_r = U\cos\theta \left(1 - \frac{R^2}{r^2} \right) \tag{3.16b}$$

$$v_\theta = -U\sin\theta \left(1 + \frac{R^2}{r^2} \right) \tag{3.16c}$$

As noted previously, $v_r = 0$ at all points on the cylinder including the front, top, and rear. At the front and rear ($\theta = 180°$ and $0°$), $\sin\theta = 0$, and therefore, $v_\theta = 0$ also. At the top (or bottom), $\sin\theta = \pm 1$, and v_θ is evaluated as $\pm 2U$. In Figure 3.4, at the cylinder top $\mathbf{V} = v_\theta = -2U\hat{\boldsymbol{\theta}}$, or $\mathbf{V} = 2U\hat{\mathbf{i}}$ in (x, y) coordinates. At the bottom, $+2U\hat{\boldsymbol{\theta}}$ again gives $\mathbf{V} = 2U\hat{\mathbf{i}}$ in (x, y) coordinates.

At the cylinder surface, $v_r = 0$, and the pressure distribution at $r = R$ can be computed using v_θ and the Bernoulli theorem (3.3). The Bernoulli theorem is only valid along a given streamline, but because every streamline here starts in the same uniform flow (neglecting gravity), all streamlines in this flow field have the same energy, and $(p + \rho v^2/2)$ at any point can be equated to $(p + \rho v^2/2)$ at any other point. Designating static pressure in the uniform flow as p_o and following the streamline at the cylinder surface

$$p + \rho v_\theta^2 / 2 = p_o + pU^2 / 2$$

$$p - p_o = \frac{\rho}{2}\left(U^2 - v_\theta^2\right) = \frac{\rho}{2}\left(U^2 - 4U^2 \sin^2\theta\right)$$

$$p = p_o + \frac{\rho U^2}{2}\left(1 - 4\sin^2\theta\right) \tag{3.17}$$

$$\text{At } \theta = 0° \text{ and } 180°, \quad p = p_o + \rho U^2/2$$

$$\text{At } \theta = 90° \text{ and } 270°, \quad p = p_o - 3\rho U^2/2$$

Pressures at the front and rear of the cylinder represent the pressure obtained by converting all the flow energy to pressure. These are called *stagnation points*. The lowest pressures (at the top and bottom) correspond to the highest flow velocities as shown in Figure 3.4.

This flow is steady state, i.e., $\partial \mathbf{V}/\partial t = 0$ at all points, but $(\mathbf{V} \cdot \nabla)\mathbf{V} \neq 0$, and convective acceleration and deceleration do occur. Fluid approaching the front of the cylinder decelerates from a velocity of U to a velocity of zero in response to (consistent with) the increase in static pressure from p to p + $\rho U^2/2$. The fluid then accelerates from zero velocity to 2U as it moves around the cylinder to the top and bottom. This is consistent with a net pressure decrease of

$$\left(p + \rho U^2/2\right) - \left(p - 3\rho U^2/2\right) = 2\rho U^2$$

Fluid again decelerates to zero at the rear (pressure increases), but once there, accelerates back to a velocity U in the uniform flow behind the cylinder (pressure decreases). All of these convective accelerations/decelerations $(\mathbf{V} \cdot \nabla)\,\mathbf{V}$ are associated with pressure gradients (∇p). The acceleration term $\partial \mathbf{V}/\partial t$ also would be nonzero if the free-stream velocity U were some function of time, U(t), or if the cylinder were being accelerated.

In this flow, pressure integrated front to rear (points A to C in Figure 3.4) over the top equals pressure integrated over the bottom so no force transverse to the flow direction exists. Also, pressure integrated top to bottom (B to D) over the front equals that over the rear, indicating drag is zero. d'Alembert noted this, and it is called *d'Alembert's paradox*. It is a paradox because it is easily observed that drag on a cylinder in a real flow field is not zero. The flaw is in the assumption that the fluid is inviscid (or equivalently here, irrotational, i.e., $\zeta = 0$).

In a real flow, viscosity causes drag to occur in two ways. One, the horizontal component of shear stress at the surface provides a net drag force. A second and usually much larger drag force is called *form drag* and is due to a difference in pressure front to rear. Because of shear, energy is dissipated in a thin layer of fluid adjacent to the cylinder. Energy is lost between the front stagnation point and the top and bottom, so velocity at $\theta = 90°$ and $270°$ is less than 2U. More energy is lost in the flow between the top and bottom points and the rear stagnation point. Therefore, p at the rear will be much less than p at the front, leading to a net force (form drag) aligned with the net flow direction. Chapter 4 analyzes this flow in further detail. Addition of a line vortex to flow about a cylinder to create lift is deferred to later chapters. Inviscid fluids and irrotational flows both are mathematical models that are never exactly correct for real flows but often are very good approximations to some parts of real flows. An inviscid flow often provides a first step toward achieving a real fluid solution.

3.4 Pressure Components

This section briefly analyzes the concept of pressure, its more common causes, and compares thermodynamic pressure, hydrostatic pressure, stagnation pressure, dynamic pressure, static pressure in a moving fluid, and pressure as it appears in the stress tensor. In each of these types (causes) of pressure field, distinctions must be made between *absolute* and *gage* pressure and between *balanced* and *unbalanced* pressure differentials. Absolute pressure is measured relative to a vacuum and can never be negative. Gage pressures are measured relative to some arbitrary reference pressure, and might be above (positive) or below (negative) the reference pressure. A pressure differential may be balanced by some other force field in which case it does not induce fluid motion. For example, pressure variation with depth in a calm motionless lake is dp/dz. This pressure differential does not induce motion because it is balanced by the weight of an element of fluid ($\rho g\, dz$). An unbalanced pressure differential is a net force and causes fluid acceleration. Balanced pressure gradients can usually be removed from the Navier-Stokes and/or Euler's equations. In many flow analyses the pressure differential caused by gravity and the associated liquid weight are equal and opposite; they balance and can be removed.

Thermodynamic pressure in a gas is defined (derived) as the force per unit area due to gas molecules striking and rebounding from that unit area. This derivation was one of the first successes of statistical mechanics and the atomic (not nuclear) model of physical phenomena. This pressure must be proportional to the number and mass of the molecules in a given volume (density, ρ) and the velocity of the molecules (temperature, T). The product (ρT) is then multiplied by a constant (R) to obtain the perfect gas law ($p = \rho RT$). In liquids and solids thermodynamic pressure is due to intermolecular (and even smaller-scale) forces. Detail derivations of thermodynamic pressure are given in advanced specialized texts on that topic.

Hydrostatic pressure in a fluid is the force per unit area that a nonmoving (static) fluid exerts on itself or on the surface of a submerged object or the walls of a container. The concept assumes a static or motionless perfect fluid, i.e., a substance incapable of sustaining shear or tension stress. Hydrostatic pressure in most applications is generated in one of three ways: (1) a fluid in a gravitational field or during acceleration; (2) a contained compressible fluid; and (3) a fluid, even one assumed incompressible, in a constrained volume subject to an externally applied force, e.g., a hydraulic brake system. Hydrostatic pressure in a fluid is independent of direction (isotropic).

The concepts of static pressure in a moving fluid, stagnation pressure, and dynamic pressure are most easily applied to an incompressible and inviscid potential flow where fluid energy per unit volume is conserved. The Bernoulli equation then applies as $\rho v^2/2 + p + \rho gh = c$. In this analysis it will be convenient to remove all pressure differentials that do not induce motion. In many applications, e.g., a filled container of liquid or flight of an airplane, that part of p due to ρgh balances with ρgh and can be removed. The Bernoulli equation (now omitting ρgh) can be applied along the central streamline in the circular cylinder flow shown in Figure 3.4. Pressure measured in the unperturbed uniform flow (far removed from the cylinder) by a gage moving with the flow at velocity (U) is the static pressure (P). At the stagnation point A in Figure 3.4, flow velocity is zero and kinetic energy converts to pressure. This pressure equivalent of the kinetic energy is called the *dynamic pressure*. The sum of static and dynamic pressure is the *stagnation pressure* (p_o) and is obtained from the Bernoulli equation evaluated between the uniform flow (U, P) and the flow about the cylinder (v, p)

$$\rho\frac{v^2}{2} + p = \rho\frac{U^2}{2} + P$$

At a stagnation point v = 0, and

$$p_o = P + \rho\frac{U^2}{2} \tag{3.18}$$

The dynamic pressure is $\rho U^2/2$, the static pressure is P, and the stagnation pressure is p_o.

The Bernoulli equation can be used for nearly inviscid fluids (air and water) to estimate velocity by measuring pressure differences. Two devices that use this principle are a venturi meter and a pitot tube as illustrated in Figures 3.5a and 3.5b. However, if the fluid is highly viscous, enough energy may be dissipated in the flow that the correlation between pressure and velocity is lost, and the Bernoulli equation then may not be even approximately valid.

Only the gradient of pressure appears in the Navier-Stokes equations, not pressure itself. As a result, constant pressures, $p \neq f(x, y, z)$, do not affect fluid flow. They appear in solutions only as constants of integration. When gravity provides a pressure gradient, it often can be dropped, as noted earlier. This leaves only those portions of $(-\nabla p)$ due to dynamic pressure, any externally applied pressure gradients, and stress

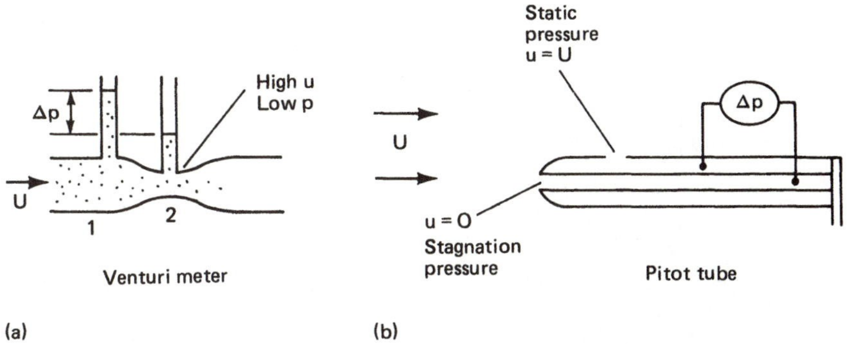

Figure 3.5: *The velocity measuring devices shown here measure Δp as a way to estimate U by assuming the Bernoulli equation applies. Because $\mu \neq 0$ and because of other small errors, a correction factor is always needed. (a) A venturi meter uses a known contraction (A_2/A_1), and Δp between points 1 and 2, to estimate U. (b) A pitot tube measures Δp between stagnation and static pressure to estimate U.*

tensor components. Stress tensor pressures complicate the utility of simple pressure concepts. *Stress tensor pressure* is defined in the Navier-Stokes equations as

$$-p = \left(\sigma_{xx} + \sigma_{yy} + \sigma_{zz}\right) / 3 \tag{3.19}$$

using the diagonal terms of (2.18). With this definition, stress tensor pressure has the same value as thermodynamic pressure in the limit of zero velocity (all $\tau_{ij} = 0$, and $\sigma_{xx} = \sigma_{yy} = \sigma_{zz}$). For solutions to typical problems, the $-\nabla p$ term represents ordinary pressure gradients, internal to the flow or externally applied. Other methods for generating pressures and pressure gradients, for example, centrifugal forces in a rotating fluid, are introduced as needed in later sections. Further details relating stress to rate of strain in a viscous fluid can be found in advanced fluid texts; see, e.g., Schlichting (1979).

Chapter 4

Boundary Layers and Turbulence

4.1 Introduction and Two Solutions by Stokes

The potential (inviscid) flow solutions of Chapter 3 place no restrictions on flows tangent to a solid boundary. However, all real fluids are observed to adhere to boundary surfaces (the no-slip condition) so that their tangential velocities are zero at the boundary surfaces (i.e., walls). In many practical engineering and geophysical flows, the transition from zero tangential velocity at the wall to the free-stream velocity occurs in a very thin layer called the *boundary layer*. This boundary layer is typically so thin (~ millimeters) that prior to the early 1990s it was difficult to measure. In 1904, Prandtl derived his boundary layer equations to explain this transitional region. His equations were criticized initially because they omitted many terms of the full set of Navier-Stokes equations. It was some years later before his equations were shown to predict surface drag and velocity flow profiles with great accuracy.

Stokes, prior to Prandtl, had obtained two exact solutions to the Navier-Stokes equations for viscous incompressible flow, both of which indicate that viscous effects are confined to a thin region near a solid surface in a low-viscosity fluid. Stokes solutions are called *exact solutions*, not because the solutions may or may not be exact numerical or analytical statements, but rather because the differential equations upon which the solutions are based are exact. That is, all the terms omitted from the full equations are omitted because they are exactly zero. When terms are omitted because they are assumed negligible relative to other terms, as Prandtl did, the remaining terms form only an approximate set of equations. Any solutions must then carry some suspicion, *How different would the solution be if all the nonzero terms had been retained?*

Stokes's first problem assumes an infinite flat plate with a viscous fluid extending to infinity on one side and seeks the solution for the induced laminar flow in the fluid

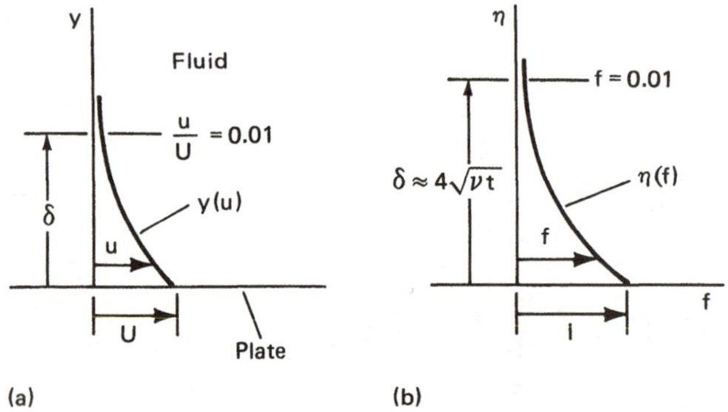

Figure 4.1: *Stokes's first problem of a flat surface suddenly accelerated to a velocity U. Viscosity diffuses momentum of the surface into the fluid. (a) The problem in dimensional variables u(y). Note that the drawing actually shows y(u), as is customary, so the drawing will look like a flow in the x-direction. (b) The problem after converting to dimensionless variables f(η), shown here as η(f).*

when the plate is suddenly accelerated to a constant velocity (U) in its plane as in Figure 4.1a. The continuity and the Navier-Stokes equations (2.25a,b,c,d) are to be simplified and solved. Only u exists if x is chosen as the flow direction, and v and w are then exactly zero. Removal of v and w from the continuity equation (2.25d) leaves only $\partial u/\partial x = 0$. In (2.25b) each term involving v and w is zero, and therefore, $\partial p/\partial y = 0$. In (2.25c) only ρg and $\partial p/\partial z$ remain, indicating hydrostatic equilibrium in the z direction. Lastly, in (2.25a) all terms with v or w are dropped leaving

$$\frac{\partial u}{\partial t} + u\frac{\partial u}{\partial x} = -\frac{1}{\rho}\frac{\partial p}{\partial x} + \nu\left(\frac{\partial^2 u}{\partial x^2} + \frac{\partial^2 u}{\partial y^2} + \frac{\partial^2 u}{\partial z^2}\right)$$

The continuity equation result of $\partial u/\partial x = 0$ causes the u $\partial u/\partial x$ term and the $\partial^2 u/\partial x^2$ term also to be zero. It is conventional in this problem to designate y as the distance from the wall. Since now there are no changes occurring in the z direction (perpendicular to the page here), $\partial^2 u/\partial z^2$ is zero also. In the case of a circular cylinder, a pressure gradient in the flow direction would exist; however none exists for flow past a flat plate. Also, no externally applied pressure gradient is specified, so $\partial p/\partial x = 0$.

All that remains of the full set of Navier-Stokes equations and the continuity equation is the final exact differential equation

$$\frac{\partial u}{\partial t} = \nu\frac{\partial^2 u}{\partial y^2} \tag{4.1}$$

commonly known as the diffusion equation. In Stokes's problem, momentum of the plate is diffusing into the fluid. Momentum change of the fluid at some point, $\partial(\rho u)/\partial t$, is determined by diffusion of momentum in the y direction, $\mu\, \partial^2 u/\partial y^2$. The boundary conditions for the suddenly accelerated plate problem are, at $t \leq 0$, $u = 0$ for all y, and at $t > 0$, $u = U$ at the wall, and $u = 0$ at infinity. Solution is augmented here by transforming variables so that the partial differential equation is changed to an ordinary differential equation. Such transformations, when they exist, are discovered partly by logic, partly by perseverance, and partly by luck. The successful transformation here is

$$\eta = \frac{y}{2\sqrt{\nu t}}, \quad \frac{u}{U} = f(\eta)$$

which, after substitution for y and u into (4.1), leads to

$$\frac{d^2 f}{d\eta^2} + 2\eta \frac{df}{d\eta} = 0$$

and a solution

$$f = \frac{u}{U_o} = \operatorname{erfc} \eta = 1 - \frac{2}{\sqrt{\pi}} \int_0^\eta e^{-\eta^2}\, d\eta \tag{4.2}$$

The complementary error function (erfc η) is tabulated in standard mathematics reference books. The abscissa is $f = u/U$ and the ordinate is η. The solution is graphed in Figure 4.1b as $\eta(f)$ rather than $f(\eta)$ to correspond with the physical appearance of the flow. Both η and f are dimensionless, in that they are ratios of two quantities that have the same physical dimensions. It is easy to verify that $\sqrt{\nu t}$ has the same dimensions as y, and although $\sqrt{\nu t}$ is not a real distance, the quantity provides a useful *characteristic length* for making (y) dimensionless. In this problem, the mass of fluid that has been affected by motion of the plate lies mostly within a distance of

$$\delta \sim \sqrt{\nu t} \tag{4.3}$$

from the plate, and it "grows with time" as $\sqrt{t}$. The distance $y = \delta$ is defined as the distance from the wall to a point where the flow is essentially undisturbed by the wall, taken typically to be $u = 0.01\, U$ as shown in Figures 4.1a and 4.1b.

Stokes's second problem concerns laminar flow when the same plate is oscillated in its plane with velocity $u = U \sin \omega t$, where ω is angular frequency. Reduction of the full set of equations again leads to the diffusion equation (4.1) but this time with different boundary conditions. At $y = 0$, $u(t) = U \cos \omega t$, and at $y = \infty$, $u = 0$. If the oscillatory motion of the wall is assumed to have been active for a long enough time, a limiting solution is achieved

$$u(y, t) = Ue^{-ky} \cos (\omega t - ky) \tag{4.4a}$$

with $k = \sqrt{\omega/2\nu}$. If y is made dimensionless using k^{-1}, then $\eta = y/k^{-1} = y/\sqrt{2\nu/\omega}$ and

$$f = \frac{u}{U} = e^{-\eta} \cos (\omega t - \eta) \tag{4.4b}$$

In Figure 4.2, $\eta(f)$ is shown rather than $f(\eta)$ so that again the graph corresponds to the physical appearance of the flow.

The cosine term in (4.4b) oscillates with time, while the exponential term forces u to approach zero as y approaches infinity. It is easy to verify that for typical fluids and frequencies

$$\delta \sim \sqrt{\nu/\omega} \tag{4.5}$$

is usually very small. The distance δ is called the *depth of penetration* of the viscous wave. As the momentum of the plate diffuses into the fluid, fluid very near the wall is moved back and forth with the plate. At further distances from the wall, the alternating flow directions diffuse until at distance δ they nearly cancel each other. Fluid beyond δ is essentially isolated from the motion of the wall.

In Stokes's first problem $\delta \sim \sqrt{\nu t}$, and δ increases with time. In his second problem $\delta \sim \sqrt{\nu/\omega}$, and δ has a fixed height. A solution similar to $\sqrt{\nu/\omega}$ appears in the analysis of boundary layers in rotating fluid problems. The size of δ, for Stokes's two solutions, is typically very small for most engineering and geophysical flows. This knowledge was available to Prandtl.

4.2 Prandtl Boundary Layer Equations

Prandtl derived an approximate set of differential equations based on the Navier-Stokes equations when $R_e >> 1$. Here $R_e = \rho x U/\mu$, with x distance from a surface leading edge as in Figure 4.3. The Prandtl model uses experimental evidence that, only a small distance from a solid surface, the flow field can often be described very precisely by an inviscid flow solution, but at the surface the flow velocity is zero. The flow velocity transitions from the zero, solid-surface velocity to the free-stream velocity in a very thin layer (δ) near the surface as predicted by Stokes's two solutions, both of which predict $\delta \sim \sqrt{\nu}$.

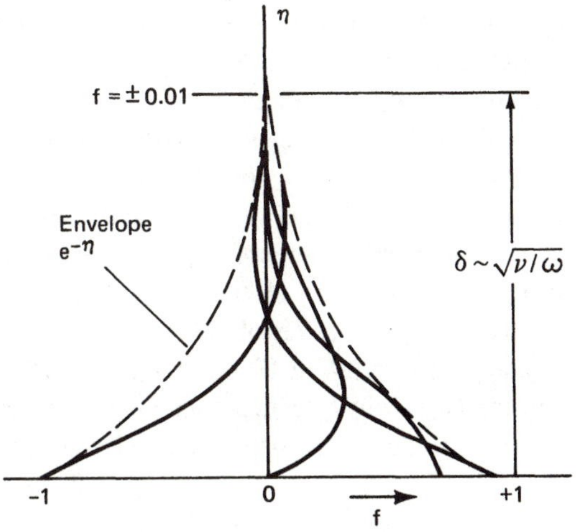

Figure 4.2: Stokes's second problem of a flat surface oscillated in its plane with frequency ω. The solution is shown in dimensionless variables f(η). The fluid above δ, where $-0.01 \le f \le 0.01$, is essentially isolated from motion of the surface.

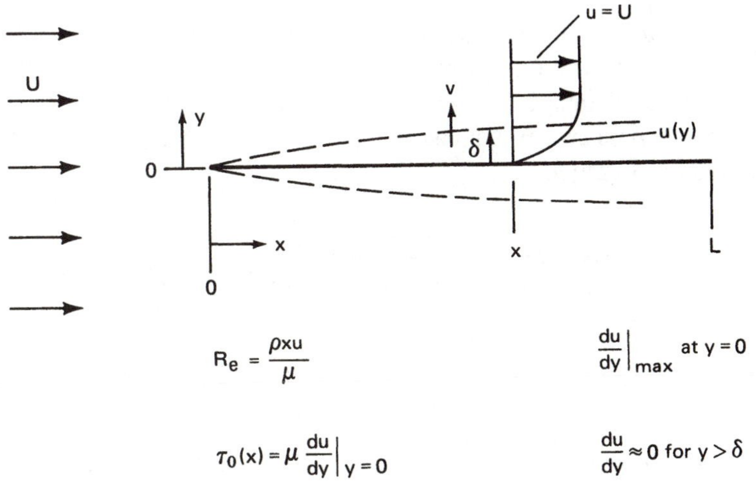

Figure 4.3: Flow past a flat surface (plate) at $R_e >> 1$ showing δ(x) and boundary layer development.

Figure 4.3 illustrates the phenomena of flow past a flat plate aligned with the flow for $R_e >> 1$. The velocity gradient ($du/dy \sim U/\delta$) near the plate is very large, and although μ may be small, $\tau = \mu\, du/dy$ can lead to a significant drag force when it is integrated over the plate area. This agrees with the observed fact that there is drag on a flat plate aligned with a flow field. Further, the solution to the Prandtl equation for this problem is in nearly perfect agreement, not only with the total drag force, but also with the velocity profile u(y) within the boundary layer.

The following summarizes derivation of the two-dimensional (2-D) dimensionless boundary layer equations. As shown in Figure 4.3, the x axis is parallel to the plate, and the y axis is perpendicular to the plate. With U and x as characteristic velocity and length, respectively, assign each a value of one in order to assess the relative magnitude of other quantities relative to U and x. Because it is known that the thickness of the boundary layer will be much smaller than x (say 0.01 x or even 0.001 x), if $x \sim 1$, then $\delta << 1$. Next, use the continuity equation $\partial u/\partial x = -\partial v/\partial y$ in the sense of

$$\left(\frac{\partial u}{\partial x} \sim \frac{\Delta u}{\Delta x} \sim \frac{U}{x} \sim \frac{1}{1}\right)$$

to get

$$\frac{\partial u}{\partial x} = -\frac{\partial v}{\partial y}$$

quantified as

$$\frac{1}{1} \sim \frac{v}{\delta}$$

Therefore, v at the top of the boundary layer must also be of order δ for v/δ to be of order one. Prandtl's analysis seeks only a crude evaluation of the relative magnitude of terms to see if some terms might be dropped relative to other terms.

Now place these approximations for x, y, u, and v in the x component Navier-Stokes equation. Equation

$$\frac{\partial u}{\partial t} + u\frac{\partial u}{\partial x} + v\frac{\partial u}{\partial y} = -\frac{\partial p}{\partial x} + \frac{1}{R_e}\left(\frac{\partial^2 u}{\partial x^2} + \frac{\partial^2 u}{\partial y^2}\right)$$

becomes

$$\frac{1}{1} + 1\frac{1}{1} + \delta\frac{1}{\delta} = \text{see text} \quad \left(\frac{1}{1^2} + \frac{1}{\delta^2}\right)$$

Because there is known to be drag in the x direction, this equation cannot just disappear; at least one of the two viscous terms must be retained. Nothing is known about $\partial p/\partial x$, while $\partial^2 u/\partial x^2 \sim 1$. At this point $\delta^2 u/\partial y^2$ dominates the problem. Its value is $(1/\delta^2) >> 1$, while the acceleration terms are of order one. A solution can only exist if at least one of the viscous terms is of the same general size as the acceleration terms. The other viscous term can be neglected if it is much smaller, but not if it is much larger. This can be accomplished if $R_e >> 1$, i.e., $R_e^{-1} << 1$, say $R_e^{-1} \sim \delta^2$. Then

$$\frac{1}{R_e}\frac{\partial^2 u}{\partial y^2} \sim \delta^2(\frac{1}{\delta^2}) \sim 1$$

but this means also that

$$\frac{1}{R_e}\frac{\partial^2 u}{\partial x^2} \sim \delta^2\left(\frac{1}{1^2}\right) << 1$$

and should be neglected.

The same value of R_e must be used in the y component equation. Equation

$$\frac{\partial v}{\partial t} + u\frac{\partial v}{\partial x} + v\frac{\partial v}{\partial y} = -\frac{\partial p}{\partial y} + \frac{1}{R_e}\left(\frac{\partial^2 v}{\partial x^2} + \frac{\partial^2 v}{\partial y^2}\right)$$

becomes

$$\frac{\delta}{1} + 1\frac{\delta}{1} + \delta\frac{\delta}{\delta} = \text{see text} \quad \delta^2\left(\frac{\delta}{1^2} + \frac{\delta}{\delta^2}\right)$$

When this is done, each term except $\partial p/\partial y$ is seen to be of size δ or smaller, which says that $\partial p/\partial y$ is equal to a sum of very small quantities. Therefore, $\partial p/\partial y$ also is of order δ, and all of the y component equation can be neglected. Pressure is then only a function of x, and the final (approximate) differential equation is

$$\frac{\partial u}{\partial t} + u\frac{\partial u}{\partial x} + v\frac{\partial u}{\partial y} = -\frac{dp}{dx} + \frac{1}{R_e}\frac{\partial^2 u}{\partial y^2} \tag{4.6a}$$

in dimensionless variables. As a dimensional equation it becomes

$$\frac{\partial u}{\partial t} + u\frac{\partial u}{\partial x} + v\frac{\partial u}{\partial y} = -\frac{1}{\rho}\frac{dp}{dx} + \nu\frac{\partial^2 u}{\partial y^2} \tag{4.6b}$$

Although simpler than the Navier-Stokes equations, the Prandtl boundary layer equation and the 2-D continuity equation are still a pair of coupled, nonlinear, second-order, partial differential equations. The dependent variables are (u, v) with independent variables (x, y, t). In dimensionless form the only parameter is R_e, and it must be much greater than one (say 100 or more). The value of dp/dx is obtained from its value in the external flow field. The same pressure gradient (dp/dx) is assumed valid inside the boundary layer because ($\partial p/\partial y$) was assumed negligible. Note that (4.6a,b) are not exact descriptions of the problem, but rather only approximate descriptions in that many negligible (but not zero) terms were discarded. For this reason Prandtl's result was originally considered to be impractical.

4.3 Flat Plate and Cylinder Solutions

The first major success in solving (4.6b) was due to Blasius, a student of Prandtl, who solved the case of steady flow past a flat plate. For this flow, $\partial u/\partial t = 0$ and $dp/dx = 0$. His solution is as follows. The continuity equation is automatically satisfied if a stream function ψ is used as a variable in place of u and v as in (3.9). The final result is one equation in one variable (ψ), given here in physical variables

$$\left(\frac{\partial\psi}{\partial y}\right)\frac{\partial^2\psi}{\partial x\partial y}-\left(\frac{\partial\psi}{\partial x}\right)\frac{\partial^2\psi}{\partial y^2}=\nu\frac{\partial^3\psi}{\partial y^3} \tag{4.7a}$$

Blasius then devised a *similarity* transform where dimensionless distance from the wall (η) is given by $\eta = y/(\nu x/U)^{1/2}$ with $f = \psi/(\nu x U)^{1/2}$ as a dimensionless stream function. Use of η and f transforms (4.7a) from a partial differential equation into an ordinary differential equation

$$ff''+2f'''=0 \tag{4.7b}$$

called the Blasius equation, where

$$f' = df/d\eta = u/U \tag{4.7c}$$

Figure 4.4 shows a numerically obtained u/U as a function of dimensionless distance from the wall (η). The Blasius transformation is called a *similarity transform* because it transforms the curve u(y) for each x into exactly the same, i.e., similar, dimensionless curve valid for any value of x.

Figures 4.5 and 4.6 show applications to two cases where $dp/dx \neq 0$. In each case when $R_e >> 1$, the flow field is analyzed as two parts, the flow inside the boundary layer, described by the Prandtl equation (4.6b), and the flow outside the boundary layer, described by Euler's equations (3.1). Use of Euler's equations outside the boundary layer does not indicate that the fluid there has no viscosity--it is the same fluid as that inside the boundary layer. The difference is that inside the boundary layer the value of $\partial u/\partial y$ is large enough that ($\tau = \mu\partial u/\partial y$) is not negligible, while outside the layer, the value of $\partial u/\partial y$ is so small that $\mu\partial u/\partial y$ is negligible.

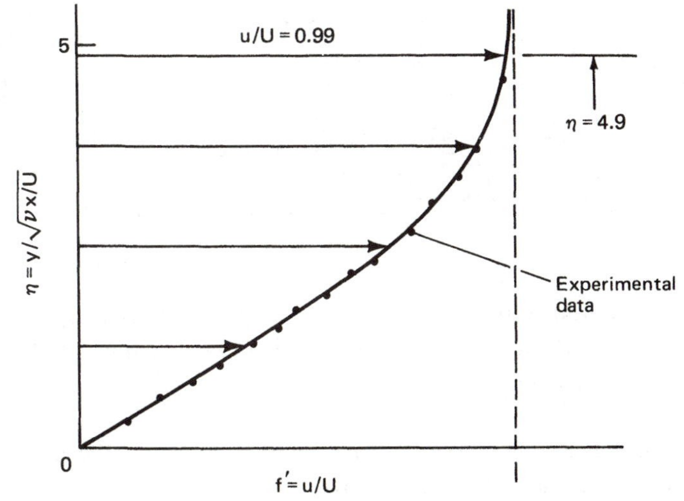

Figure 4.4: *The Blasius solution of the Prandtl boundary layer equations for viscous flow past a flat plate ($R_e >> 1$), shown in dimensionless variables.*

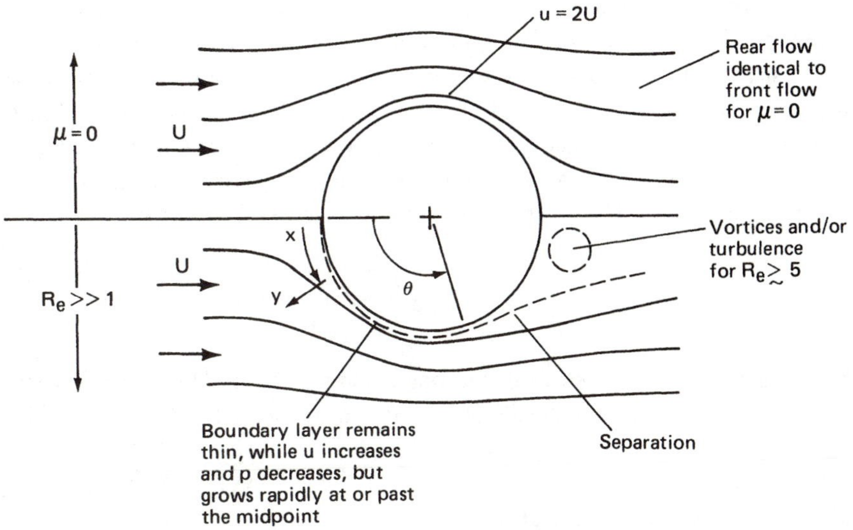

Figure 4.5: *Flow about a circular cylinder with potential (inviscid) flow over the cylinder top for comparison with flow of a slightly viscous fluid at $R_e >> 1$ below the cylinder bottom. The difference is especially apparent at the rear of the cylinder.*

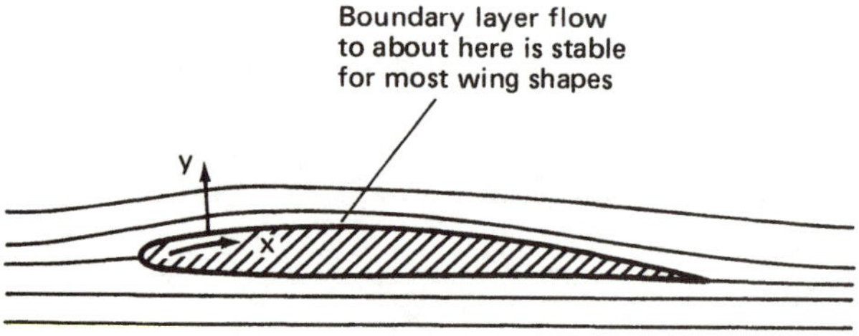

Figure 4.6: *Flow past a wing shape at $R_e >> 1$. A laminar boundary layer is too thin to be illustrated. At higher velocities, higher-lift wing shapes, and at higher angles of attack, separation and turbulence occur.*

Solution of either the cylinder or the wing shape requires first the Euler inviscid solution for flow about the body yielding dp/dx at the edge of the boundary layer. Sometimes (4.6b) for steady-state flows is given in the form

$$u\frac{\partial u}{\partial x} + v\frac{\partial u}{\partial y} = U\frac{dU}{dx} + \nu\frac{\partial^2 u}{\partial y^2} \tag{4.8}$$

where (4.6b) has been evaluated in the free stream just outside the boundary layer by assuming $y = \delta$, $u = U$, $v = 0$, $\partial^2 u/\partial y^2 = 0$, and therefore, $-dp/dx = \rho\, U\, dU/dx$.

An analytical solution, or at least a very good approximation, to boundary layer phenomena is possible whenever U(x) or p(x) is available from an inviscid flow solution. Blasius used the known solution for inviscid flow around the surface of a circular cylinder, (3.16b,c), to solve for boundary layer development starting from the front stagnation point. The potential flow solution can be expressed as

$$U(x) = 2U_\infty \sin\frac{x}{R} = 2U_\infty \sin\theta$$

where x and θ are measured from the stagnation point, and U_∞ is the free-stream velocity. The sine function can be represented as a power series. Then, after defining $\eta = y(2U_\infty/R\nu)^{1/2}$, use of the power series for sin θ, and a stream function, a solution is obtained. An interesting result is the prediction that the boundary layer will cause the main flow to separate from the contour of the cylinder at $\phi \approx 109°$.

Figure 4.5 illustrates separation of the flow field past a circular cylinder. If $\mu \neq 0$ and $R_e >> 1$, a boundary layer must form behind the front stagnation point. Energy is dissipated for flow within this layer but not in the nearly stress-free flow outside the boundary layer. A potential flow solution for flow outside the boundary layer yields

$U(x) = 2U_\infty$ at the cylinder midpoint. Flow inside the layer has $u < 2U_\infty$ at the cylinder midpoint and lacks the energy per unit mass to return to the same pressure at the rear that it had at the front stagnation point. Flow just outside the layer with velocity $2U_\infty$ at the midpoint has more energy and reaches a higher pressure past the midpoint than pressure inside the layer. A flow reversal occurs, and the external flow separates from the contour of the object. The region between the object and the external flow can be filled with various patterns of vortices or turbulence, depending on the magnitude of R_e. Pressure integrated over the rear is now less than that over the front. This pressure difference is called *form drag* ($\sim \rho U^2$) and is typically much larger than skin friction ($\sim \mu\ U/\delta$) for blunt bodies. Flow details about the wing shape in Figure 4.6 are covered in more detail in Chapter 16. Success in predicting separation requires a combination of analyses, experiments, and numerical procedures.

4.4 Momentum Integral Equation

A momentum integral solution to the boundary layer equations was developed by von Kármán, another student of Prandtl. It is a variation of a general class of methods used in science and engineering that succeeds in expressing a desired unknown as a function of some integral. In an integral method, the exact curve enclosing the integral, here u(y), is replaced with some curve that only approximates the real curve in shape and area, but is a function that is easy to integrate. Polynomials, sine functions, and exponentials are commonly used. Results accurate to several percent are common. Reference data and measuring devices are not always that accurate, so integral approximate solutions can be very useful.

The momentum integral method is based on an equation that states, *Loss of flow momentum in steady-state boundary layer flow is equal to the sum of wall shear stress and applied pressure gradient.* This is the same statement as Newton's law which states, *The sum of applied forces equals rate of change of momentum.* The equation is usually written as

$$\frac{\tau_w}{\rho} = \frac{d}{dx}\left(U^2\delta_2\right) + \delta_1\, U\frac{dU}{dx} \tag{4.9}$$

where τ_w provides shear stress at the wall, $[\rho d(U^2\delta_2)/dx]$ is momentum loss gradient, and $(\rho\delta_1 UdU/dx)$ is pressure gradient. The term $(\rho UdU/dx)$ replaces $(-dp/dx)$ of Prandtl's boundary layer equation as noted in (4.8).

The variable δ_1 is called the *displacement boundary layer thickness*, and the variable δ_2 is called the *momentum boundary layer thickness*. Each represents a more exact flow quantity than does δ, which only represents distance to where the velocity field equals an arbitrary 99% of the free-stream velocity outside the boundary layer. δ_1 quantifies exactly the amount that the external flow is displaced outward as the boundary layer grows in thickness, and δ_2 quantifies exactly a thickness of free-stream flow equivalent to the loss of momentum in the boundary layer. Both δ_1 and δ_2 are functions of x only.

Figures 4.7a and 4.7b illustrate the computations for δ_1 and δ_2. The shaded area in Figure 4.7a, from y = 0 to ∞, is the mass flow rate loss within the boundary layer, given by $\rho \int_0^\infty (U - u)\, dy$. A height δ_1 in the free stream is selected to give the same mass flow defect, therefore

$$\rho \delta_1 U = \rho \int_0^\infty (U - u)\, dy$$

$$\delta_1(x) = \int_0^\infty \left(1 - \frac{u}{U}\right) dy \qquad (4.10)$$

This is the distance the external flow is deflected away from the wall at that point x. Any external pressure gradient $(-dp/dx)$ acts on this height δ_1 to give pressure force per unit flow width of $(-\delta_1\, dp/dx)$, or its equivalent here of $(\rho\, \delta_1\, U\, dU/dx)$.

Figure 4.7b illustrates the computation for δ_2, similar to that of δ_1, except δ_2 measures the momentum flux loss in the boundary layer. Momentum flux loss is given by $\rho \int_0^\infty u(U - u)\, dy$, and δ_2 is selected to give the same momentum flux defect in fluid moving at velocity U. Therefore

$$\rho\, \delta_2\, U^2 = \rho \int_0^\infty u(U - u)\, dy$$

$$\delta_2(x) = \int_0^\infty \frac{u}{U}\left(1 - \frac{u}{U}\right) dy \qquad (4.11)$$

The quantity $d(U^2\, \delta_2)/dx$ then represents the rate that momentum is lost within the boundary layer per unit width.

Both δ_1 and δ_2, as defined by (4.10) and (4.11), involve integrations of u(y) from y = 0 to ∞. This cannot be done because u(y) is not known, however any curve that has the approximate shape and characteristics of u(y) will do almost as well--often within a few percent. Figure 4.8 shows several such curves, a straight line (not too good) and a sine wave and a cubic polynomial (both very good).

An energy boundary layer thickness (δ_3) that represents energy lost in the boundary layer is defined as

$$\delta_3 = \int_0^\infty \frac{u}{U}\left[1 - \left(\frac{u}{U}\right)^2\right] dy \qquad (4.12)$$

It has had less utility in flow analyses than the displacement and momentum boundary layer thicknesses. Most modern fluid texts provide solved examples, see, e.g., Fox and McDonald (1992) or Schlichting (1979).

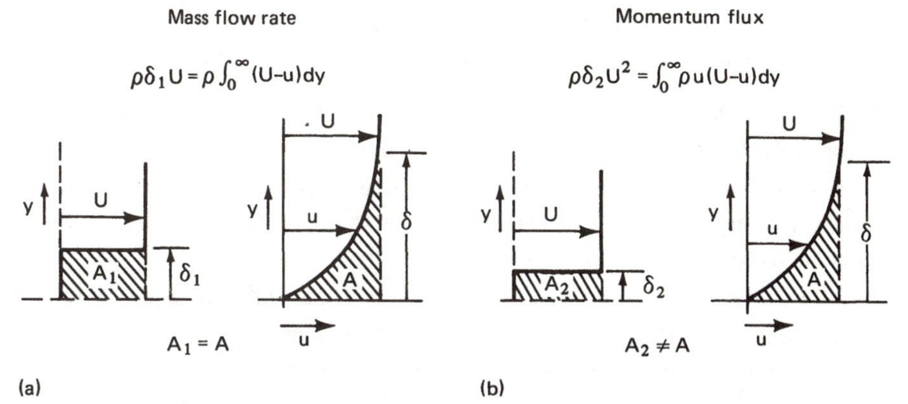

Figure 4.7: *(a) The displacement boundary layer thickness (δ_1) estimates mass flow loss due to the boundary layer (area A) by using a layer of fluid in the free stream with the same mass flow loss (area A_1). (b) The momentum boundary layer thickness (δ_2) estimates the momentum flux deficit in the boundary layer flow by using a layer of fluid in the free stream with the same momentum flux ($\rho\delta_2 U^2$).*

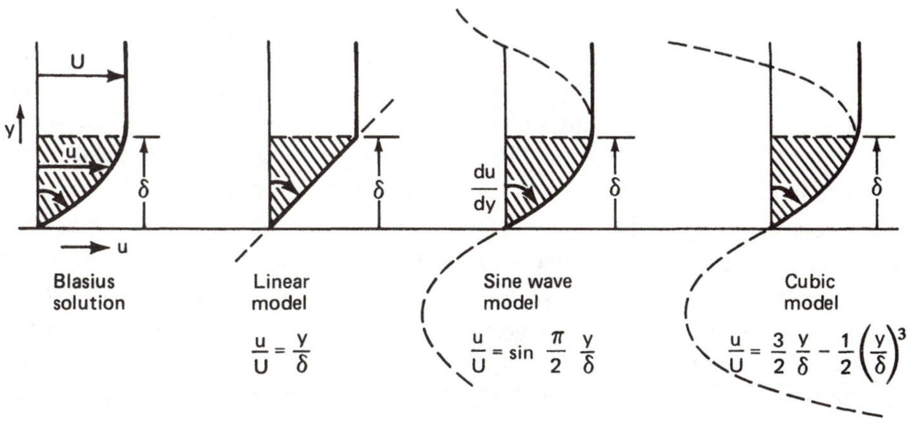

Figure 4.8: *Boundary layer models used in the momentum integral equation to approximate a real boundary layer. Only the portions $0 \leq y \leq \delta$ are used. For best results the model should closely represent a realistic u(y), e.g., a sine function or cubic polynomial model.*

4.5 Turbulence and Stability

Five mathematical models developed for analysis of fluid flows have now been presented and analyzed. These are

1. *Mass continuity.* Motion of a fluid is constrained to those motions consistent with conservation of mass (continuity equation).
2. *Velocity potentials and stream functions.* These, especially when combined with the continuity equation, yield detailed solutions and express the solutions as scalar functions (ϕ and/or ψ) rather than vector functions (**V**).
3. *Euler's equations.* Motion of an inviscid fluid subjected to body forces and surface pressure forces is determined by Euler's equivalent of Newton's law of motion and the mass conservation equation.
4. *The Navier-Stokes equations.* These equations provide a model for real fluids having the property of viscosity. A velocity potential no longer exists because viscosity causes energy losses and also causes the fluid elements to rotate--to have vorticity. Variable density and variable viscosity are included in the most general form of the Navier-Stokes equations.
5. *The boundary layer equations.* These equations analyze motion near a solid surface in $R_e >> 1$ flows. Viscous forces are significant only in a very thin layer of fluid near the surface. As with models 1, 2, 3, and 4, only laminar flows have been analyzed so far.

Nonlaminar flows are referred to as turbulent flows. They are not smoothly varying flows, and their streamlines, pathlines, and streaklines are not locally parallel. The turbulent (also called disorderly, random, or chaotic) motion can continue to the smallest scale, presumably to the size of molecular spacing where the turbulent energy is dissipated by viscosity. Viscosity is involved both in the creation of turbulence and in its dissipation. Whether a flow is laminar or turbulent is not relevant for model 1. The constant-density inviscid fluid models (2 and 3) have no mechanism to produce instability and turbulence. Models 4 and 5 have a mechanism (viscosity). In small Reynolds number flows the dissipative effects of viscosity dominate the instabilities caused by viscosity. Turbulence appears only in those flows where the kinetic energy of the mean motion, when perturbed, cannot be restored to a laminar condition by available restoring forces. Viscosity is the restoring force of interest here; other restoring forces will be included in later chapters.

Reynolds discovered that flow is laminar in circular pipes when $R_e = \rho LU/\mu \lesssim 2300$, with L the diameter of the pipe and U the mean flow velocity. In normal pipe flows, disturbances in the entering flow or irregularities of the pipe surface always exist, so $R_e > 2300$ typically indicates turbulent flow. Turbulence persists when energy transfers from the main flow field are adequate to supply the energy dissipated by viscosity in the turbulent flow. Other values of R_e are appropriate for other flows, for example, for flow over a flat plate the boundary layer becomes turbulent when $R_e > 5 \times 10^5$ approximately. Here L in the Reynolds number is distance from the leading edge of the plate, and U is the free-stream velocity. Figures 4.9 and 4.10 illustrate laminar and turbulent flow phenomena in circular pipes and near flat plates.

The mass conservation equation and the Navier-Stokes equation continue to be valid for turbulent flows, but the flow field can be so complicated that exact solutions are impossible--even on the fastest modern computers. Various approximate models have been devised to seek solutions useful for at least portions of some flows. The most common approach is to separate the mean flow (averaged over time) from the fluctuations that are characteristic of turbulence. Each variable is expressed as the sum of an average ($\overline{u}$) and a fluctuation (u')

$$u = \overline{u} + u', \quad v = \overline{v} + v', \quad w = \overline{w} + w', \quad p = \overline{p} + p'$$

Most modern texts use this convention of u', v', w', and p' for small fluctuations, although primes are also used elsewhere to represent derivatives. After insertion into the Navier-Stokes equations and the mass conservation equation, and use of mathematical rules for averaged quantities, the Navier-Stokes equations, in mean flow variables, simplify to

$$\overline{u}\frac{\partial \overline{u}}{\partial x} + \overline{v}\frac{\partial \overline{u}}{\partial y} + \overline{w}\frac{\partial \overline{u}}{\partial z} + \underbrace{\frac{\partial \overline{u'u'}}{\partial x} + \frac{\partial \overline{u'v'}}{\partial y} + \frac{\partial \overline{u'w'}}{\partial z}}_{\text{Reynolds stresses}} = -\frac{1}{\rho}\frac{\partial \overline{p}}{\partial x} + \nu\left(\frac{\partial^2 \overline{u}}{\partial x^2} + \frac{\partial^2 \overline{u}}{\partial y^2} + \frac{\partial^2 \overline{u}}{\partial z^2}\right) \tag{4.13}$$

with equivalent y and z and continuity equations.

If all terms containing fluctuations, e.g., ($\overline{u'v'}$), are collected from the x, y, and z equations and arranged in matrix form, they become the Reynolds stress tensor. They describe an additional transverse momentum exchange mechanism available to turbulent flows in addition to viscosity. These terms are typically moved to the right-hand side of the equation and combined with the Navier-Stokes viscous stress tensor. In this model of turbulence there are more unknowns than equations, and additional relationships are needed (the problem of "closure"). Navier and Stokes used the definition of a Newtonian fluid (stress proportional to velocity gradient, e.g., $\tau = \mu \, \partial u/\partial y$) in relating their stress tensor (τ_{xy}, τ_{xz}, τ_{yz}, etc.) to the flow variables ($\partial u/\partial y$, $\partial u/\partial z$, $\partial v/\partial x$, etc.) so that they had four equations in four unknowns (u, v, w, p). Similar methods have been tried for turbulence, that is, relate the fluctuations ($\overline{u'u'}$, $\overline{u'v'}$, $\overline{u'w'}$, etc.) to the mean flow variables ($\partial\overline{u}/\partial x$, $\partial\overline{u}/\partial y$, $\partial\overline{u}/\partial z$, etc.), but never in a way as useful as had been done for viscous stresses.

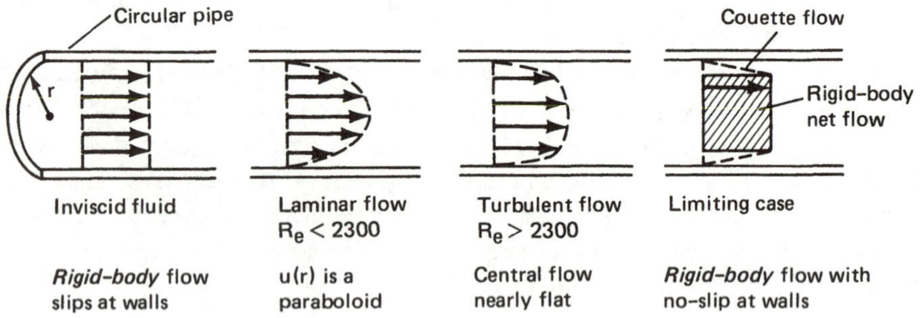

Figure 4.9: *Laminar and turbulent pipe flow. Drag inside the pipe equals the pressure gradient (dp/dx) needed to maintain a steady-state flow.*

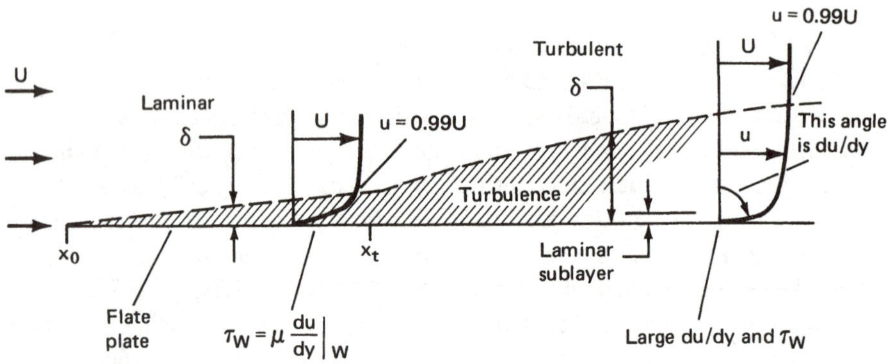

Figure 4.10: *Laminar and turbulent flow past a flat plate. When $x = x_t$ in the equation $R_e = \rho x_t U/\mu \approx 5 \times 10^5$, flow near the surface becomes turbulent.*

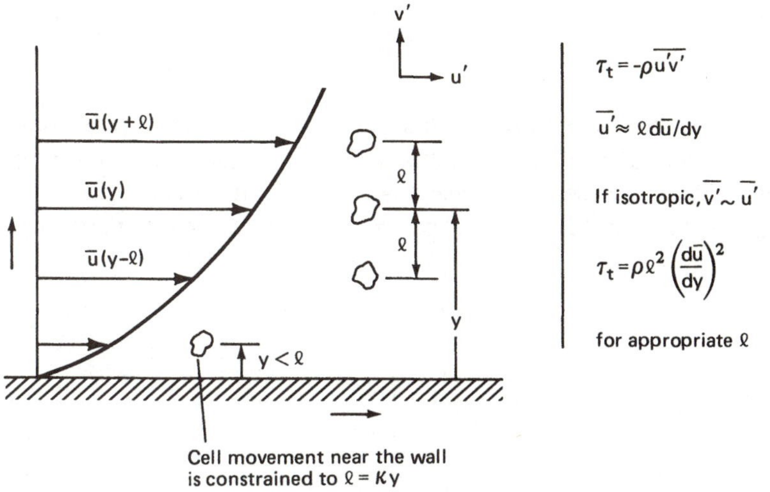

***Figure 4.11:** Outline of the Prandtl mixing length hypothesis for turbulence. Several models and theories exist for estimating the size of (ℓ).*

Prandtl related the fluctuating terms to the mean flow variables using a *mixing length* theory. Figure 4.11 illustrates the physical mechanism envisioned by Prandtl to support his mixing length hypothesis. The model assumes cells of fluid of varying sizes constantly move in random directions relative to the mean flow. A cell of fluid moving from $(y - \ell)$ to (y) will carry with it the mean flow velocity at $(y - \ell)$. At (y), this flow has a value $(u' = -\ell\, d\bar{u}/dy)$. If turbulence is assumed independent of direction, then $v' \sim u'$ and

$$-\rho\overline{u'v'} \approx \rho\ell^2 \left|\frac{d\bar{u}}{dy}\right| \frac{d\bar{u}}{dy} \tag{4.14}$$

where an absolute value of one $d\bar{u}/dy$ is taken so the other $d\bar{u}/dy$ can preserve the sign of the momentum transfer, either plus or minus. To complete his model, Prandtl assumed $\ell = \kappa y$, noting that very near a wall (y small) only small eddies were possible, with larger and larger eddies being possible with increasing distance from the wall. His model is useful because many experiments indicate nearly constant values for κ. Therefore, data obtained from one set of experiments can be applied to other flow situations with some accuracy.

After substitution of (4.14) and $\ell = \kappa y$ into (4.13) for the flow of Figure 4.11 and integrating, the defining equation becomes

$$\mu\frac{d\bar{u}}{dy} + \rho\kappa^2 y^2 \left|\frac{d\bar{u}}{dy}\right| \frac{d\bar{u}}{dy} = \tau_o + y\frac{\partial\bar{p}}{\partial x} \tag{4.15}$$

One limiting case exists very near the wall resulting in

$$\mu \, d\bar{u}/dy = \tau_o \tag{4.16}$$

This is the laminar sublayer that must always exist near a very smooth wall. It need not exist for a rough wall, where the average surface irregularities protrude beyond the laminar sublayer thickness. A second case of interest looks at a region slightly more distant from the wall. The first and last terms of (4.15) are neglected, leaving

$$\rho \kappa^2 y^2 \left| \frac{d\bar{u}}{dy} \right| \frac{d\bar{u}}{dy} = \tau_o$$

Assuming positive $d\bar{u}/dy$, this leads to

$$\frac{1}{u^*} \frac{d\bar{u}}{dy} = \frac{1}{\kappa y}$$

where $u^* = (\tau_o/\rho)^{1/2}$ is called the *friction velocity*. After integration and some manipulation

$$\frac{\bar{u}}{u^*} = G + \frac{1}{\kappa} \ell n \left(\frac{\rho u^* y}{\mu} \right) \tag{4.17a}$$

more often written as

$$u^+ = G + \frac{1}{\kappa} \ell n \, y^+ \tag{4.17b}$$

Here u^+ is the mean flow made dimensionless with the friction velocity, and y^+ has the form of a Reynolds number based on the friction velocity and distance from the surface. Experiments indicate $\kappa \approx 0.40$ to 0.42 and $G = 5.5$ to 5.8.

Finally, for y^+ in the range 70 to 700, (4.17b) becomes the well-known 1/7 power law approximation for turbulent flow over a smooth flat plate

$$u^+ = 8.74 \, (y^+)^{1/7} \tag{4.18}$$

The Prandtl model is partially successful, as are other models, but none of them are useful for all turbulent flows. As a consequence, mathematical predictions of turbulent phenomena are always modified and adjusted to experimental data.

Another model for turbulence, the *route to chaos model*, uses a recursion formula that, when iterated in a computer, makes the transition from an orderly output to a chaotic or random output. It is similar in some ways to a transition from laminar to turbulent flow. Note that neither the mixing length model nor the route to chaos model attempt exact descriptions of the turbulent mechanism. None of these models are precise enough that solutions to the equations should be interpreted as physically realizable flows in the absence of other verification.

A general mathematical approach for predicting instability and the consequent development of turbulence is called the *method of small disturbances*. It begins by making an assumption that the flow field can be represented by a mean flow with a disturbance superimposed on it. The method differs from Reynolds approach in that an assumption is made, u = U + u', v = V + v', w = W + w', and p = P + p', where U, V, W, and P represent laminar flow values prior to a disturbance, and u',v',w', and p' are perturbations applied to the flow. Averages are not taken. Instead, perturbations of some wavelength and frequency are assumed, and the analysis investigates whether the perturbation will grow or decay. For a 2-D flow past a flat plate, with u = U + u', v = v', w = 0, and p = P + p', the analysis continues as follows. A stream function representing an oscillatory disturbance is assumed as

$$\psi(x, y, t) = \phi(y)\, e^{i(\alpha x - \beta t)} \tag{4.19}$$

with ϕ and β both complex. Here ϕ is the magnitude of ψ (not a velocity potential), and α is the wave number and β is the frequency that characterize the perturbation.

After insertion into the x and y component Navier-Stokes equations and considerable algebra, the Orr-Sommerfeld equation is derived

$$\underbrace{(U - C)\left(\phi'' - \alpha^2\phi\right) - U''\phi}_{\text{Inertial terms}} = \underbrace{\frac{i}{\alpha R_e}\left(\phi'''' - 2\alpha^2\phi'' + \alpha^4\phi\right)}_{\text{Viscous terms}} \tag{4.20}$$

where primes denote (d/dy), and $C = \beta/\alpha$, having a real part (C_r) and an imaginary part (C_i). The imaginary part determines whether an initial disturbance is amplified ($C_i > 0$) or damped ($C_i < 0$). The value of $C_i = 0$ defines the neutral condition dividing regions of stability and instability. R_e is computed as Ub/ν or $U\delta/\nu$, depending on the application, channel width (b) or boundary layer thickness (δ). This approach has been successful for many flows, including flow over a flat plate. Stability criteria unique to specific topics are presented in later chapters. Further details and limitations are available in specialized texts. See, e.g., Schlichting (1979).

Chapter 5

Wave Theory

5.1 Introduction and Definitions

This section reviews beginning wave theory starting with the theory of mechanical oscillations, extending it next to various waves in a fluid continuum, and then briefly reviewing shock waves. Chapter 13 extends wave theory to a unique type of wave phenomena that appears in rotating fluids.

A mechanical oscillator consisting of a mass and a spring or a mass and gravitational force, as in Figures 5.1a and 5.1b, illustrates the two basic needs of all types of waves, oscillations, or vibrations. There must be (1) some agency that acts to restore a system to a nominal zero condition and (2) residual energy or momentum in the system upon reaching zero. In the spring-mass system of Figure 5.1a, the linear spring applies a linear restoring force $F = -kx$, and the mass at $x = 0$ has linear momentum $m\dot{x}$. The defining equation is obtained using Newton's law ($F = dP/dt$), where ($P = m\dot{x}$). Combine $F = -kx$ and $F = d(m\dot{x})/dt = m\ddot{x}$ to get

$$\ddot{x} + \frac{k}{m}x = 0 \tag{5.1a}$$

with solution

$$x = X \sin(\omega t + A), \quad \omega = \sqrt{k/m} \tag{5.1b}$$

Here ω is the circular frequency, X is the maximum amplitude of the displacement, and A is a phase relationship. The motion is periodic and repeats with a period (τ) and frequency (f)

$$\tau = 2\pi/\omega, \quad f = \omega/2\pi \tag{5.1c}$$

A periodic motion that also is sinusoidal is called *harmonic*.

The critical elements in Figure 5.1a are (1) the restoring force of the spring and (2) the inertia property of the mass. Here both the force and the momentum are linear functions, with k and m constants

$$F = -kx, \quad P = mv$$

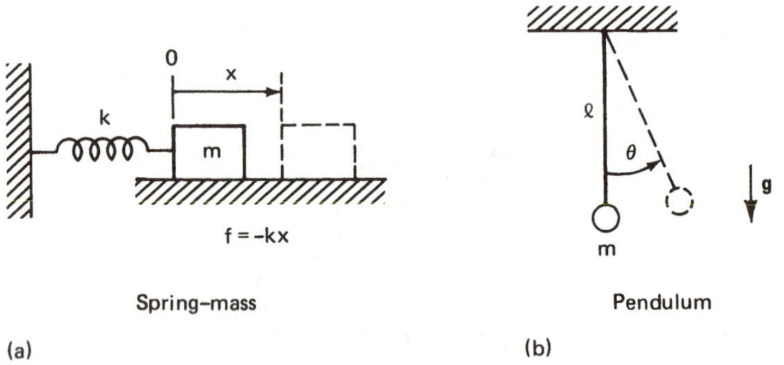

Figure 5.1: *A simple spring-mass system oscillates at a circular frequency of* $\omega = \sqrt{k/m}$. *The spring in (a) provides the restoring force. Gravity provides the restoring force for the pendulum in (b). It oscillates at* $\omega = \sqrt{g/\ell}$.

In many real situations the restoring force is not a linear function of displacement, and in relativistic mechanics, momentum is not a linear function of velocity. In all applications of waves and oscillations in this text, the inertia property of nonrelativistic mass is the agency that stores the residual momenta. Because the nature of the restoring force will determine the type of (and name of) the wave or oscillation, it is necessary to define carefully the physical basis for the restoring force and how that force is quantified as a function of displacement.

Figure 5.1b shows a pendulum that oscillates about the vertical if perturbed, i.e., given some initial momentum. In this example, gravity provides the restoring force and moment of force is a nonlinear function of angular displacement. Momentum here is in the form of moment of (i.e., angular) momentum, again derived from the inertia property of mass. The rotational equivalent of F = ma is M = Iα and the restoring moment is $M = -(mg)\ell \sin\theta$. Moment of inertia is $I = m\ell^2$. Equating these expressions for M gives

$$m \ell^2 \ddot{\theta} + mg \ell \sin\theta = 0$$

For small θ, the approximation $\sin \theta \approx \theta$ yields an equation similar to that of the linear spring-mass oscillator

$$\ddot{\theta} + \frac{g}{\ell}\theta = 0 \qquad (5.2a)$$

with solution

$$\theta = \theta_0 \sin(\omega t + A), \quad \omega = \sqrt{g/\ell} \tag{5.2b}$$

Here ω is the circular frequency of oscillation, which is a constant based only on g, the acceleration due to gravity, and ℓ, the length of the pendulum. The quantities θ, $\dot{\theta}$, and $\ddot{\theta}$ are functions of time. In this example the gravitational property of mass supplies the restoring force (potential energy) that interacts with the inertia property of mass (kinetic energy) to perpetuate the oscillation. The quantity m cancels out and does not appear in the solutions for displacement or frequency.

In a more general case of a linear oscillator, the equation of motion is

$$m\ddot{x} + c\dot{x} + kx = F(t) \tag{5.3}$$

where the term $c\dot{x}$ quantifies energy dissipated by the motion due to damping, and F(t) is an applied external force. Damping in most real applications is usually not a constant times velocity to the first power, except, e.g., in the case of damping caused by a simple Couette flow between adjacent parts. This form is useful however because it is a linear model and often is a reasonable approximation.

Oscillation of a compressible elastic material has additional complexities. The oscillation now can move as a wave through the material and can be either a longitudinal (compression) wave or a transverse (shear) wave. As in the case of a simple spring-mass oscillation, it is again the mass inertia property of the material that interacts with the restoring force to perpetuate the wave motion.

Figure 5.2a illustrates a longitudinal wave using a mechanical model consisting of discrete masses and springs. If mass n of the longitudinal wave model is suddenly displaced a distance Δx from its equilibrium position, it compresses spring $(n + 1)$ and stretches spring (n). The compressed spring accelerates mass $(n+1)$, which compresses the next spring, which accelerates mass $(n+2)$, etc. for succeeding springs and masses. This disturbance of each successive mass progresses at a finite speed called the *wave speed.* It is called a *longitudinal* wave because the restoring force and the displacement of the masses are in the same direction as the wave motion. Note that each mass must always remain in the near vicinity of its equilibrium position while the wave can continue indefinitely, or at least until its energy is dissipated by friction, viscosity, or material hysteresis.

Figure 5.2b illustrates a transverse wave using a similar mechanical model. In a transverse wave the amplitude of the motion and the restoring force are nominally perpendicular to the direction of wave propagation. Mass n, if suddenly displaced a distance Δy, begins to oscillate about the $y = 0$ axis. The adjacent springs transfer the motion successively to adjacent masses, which in turn begin to oscillate about $y = 0$ also. The propagation proceeds in both the +x and −x directions, as it does also in the longitudinal wave.

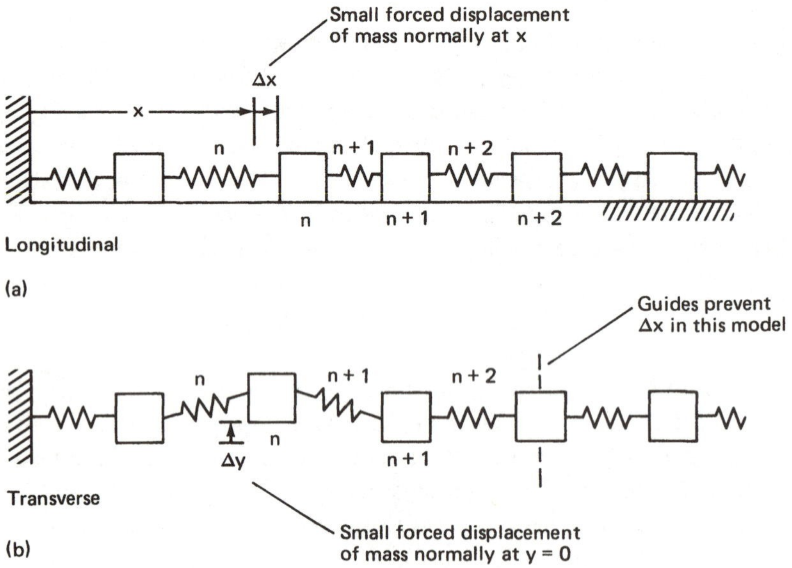

Figure 5.2: *Spring-mass systems as models for longitudinal (compression) waves (a) and for transverse (shear) waves (b).*

This section continues with a brief listing of wave terminology.

1. *Plane, circular, or spherical waves.* A description of the wave propagation direction. Waves in a continuous medium tend to progress in all directions (spherical--as in a midair explosion) but can be restricted by the geometry of an enclosure (or the assumptions of an analyst) to circular waves (ripples on a pond) or plane waves (along a narrow channel). At a large distance from a wave source, and over a small region of space, waves can usually be approximated as plane waves.

2. *Free vs. forced waves.* Free waves occur at the natural resonant frequency or frequencies of the system in the absence of (usually some time after the application of) a forcing function. When the forcing function has a frequency, this frequency is usually apparent in the wave motion during its application.

3. *Trapped, stationary, or standing waves.* Any wave within a physical container or limiting force field that is sized to a natural wavelength or multiple of that wavelength. A string between fixed supports and vibrating in its first mode shape (a half sine wave alternately above and then below the neutral axis) is an example,

as is simple sloshing of a liquid in a rectangular tank. Many applications are more subtle and more complicated.

4. *Dispersive wave.* The wave speed is a function of its frequency. As a result, a dispersive waveform representable as a sum of separate component frequencies will disperse or spread out because the higher frequencies will propagate at a different speed than the lower frequencies.

5. *Longitudinal wave.* The wave forcing function, fluid element motion, and stored momenta are aligned with the direction of wave propagation. Compression waves are a type of longitudinal wave; acoustic waves are small amplitude compression waves. They need not be in the audible region to be classified as acoustic waves.

6. *Transverse wave.* Forcing function, fluid element motion, and stored momenta are transverse (or mostly transverse) to the direction of wave propagation. Surface waves in an incompressible fluid are transverse waves, although in many cases longitudinal motion occurs also.

7. *Surface wave.* Transverse displacements of a liquid-gas surface, e.g., the ripples on a water surface. Gravity is usually the restoring force, although in some applications the restoring force is surface tension or centrifugal force.

8. *Internal wave.* Transverse displacements normal to a surface of constant density, angular momenta, or other fluid property within one or more fluids. A surface wave could be included as a subset because it is internal to a liquid-gas region, but it is conventional to restrict the term *internal* to regions where the densities are nearly the same or perhaps continuously varying, e.g., inside the atmosphere or inside the ocean.

9. *Gravity wave.* Any wave where gravity is the restoring force.

10. *Buoyancy wave.* Any wave where one portion of a fluid is buoyant or partially buoyant in another portion of that fluid or adjacent fluid; gravity plus density variation is the common mechanism, although density variations in any other force field related to mass (e.g., centripetal acceleration) produce the same phenomena.

11. *Inertial wave (oscillation).* Any wave where the inertia property of matter provides momenta to perpetuate the wave motion. All waves (oscillations) in this text are inertial waves in this sense. More often authors restrict the name *inertial* to waves where an acceleration of mass also provides the restoring force for the wave (see Chapter 13). Of course waves and oscillations exist in which inertia of a material substance has no role at all. An electrical oscillator circuit consisting of an inductance and a capacitor is a simple example. Electrical energy oscillating between an electrostatic field and a magnetic field define the oscillation.

Waves are related to studies of laminar vs. turbulent motion by stability theory. One stability analysis resulted in the Orr-Sommerfeld equation introduced in Section 4.5 that has been used successfully to predict transition from laminar to turbulent flow past a flat plate. The Orr-Sommerfeld technique investigates solutions of the equations of motion to determine whether an assumed disturbance ($\sim e^{\beta t}$) is amplified ($\beta > 0$) or is damped ($\beta < 0$). In more general analyses, an unstable laminar flow may make the transition to one or more other laminar flows (sometimes wavelike) before transitioning finally to turbulence. Although other analytic techniques are available, all ask the same basic question, *Will a given perturbation stay within specific limits or grow without limit?* Dissipative mechanisms (friction in a mechanical system, resistance in an electrical system, and viscosity in a fluid system) play an important role in such analyses.

Stability in a fluid system often can be analyzed by examining the force (pressure) resulting from an arbitrary displacement of a parcel of fluid. If the force returns the parcel to its initial position, the fluid is in stable equilibrium. If the force is zero after the perturbation, the flow is neutrally stable, and if the force is positive (drives the parcel further away from the initial position), the flow is unstable. In other fluid analyses, stability is determined using an energy criteria and, as will be shown in Chapter 13, sometimes using a vorticity criteria.

Figure 5.3 illustrates a stability analysis based on an arbitrary displacement of a parcel of fluid. The Brunt-Vaisala (buoyancy) frequency is the frequency of vertical oscillation of a parcel of stably stratified fluid (here air) adiabatically displaced vertically from a nominal equilibrium level. In a neutrally stable atmosphere the displaced parcel is neutrally buoyant after the displacement and remains at the new position. In an unstable atmosphere the displacement (perturbation) continues to grow, resulting in a vertical convection cell. In a stable atmosphere the parcel returns to its original position and begins to oscillate from above to below that position as follows.

In the earth's atmosphere, pressure (p) decreases with height. If a parcel of air is displaced vertically (Δz), its pressure must adjust to the ambient pressure at that level. If the displacement occurs quickly enough, no heat is added or lost to the parcel, and the parcel adjusts adiabatically to the new conditions. Its density can then be compared to the ambient density at that level as shown in Figure 5.3. If Δz is up, and the new adiabatic density is heavier than the ambient (point 1), the parcel will return toward its initial level. The atmosphere is then stable, and the parcel will oscillate about the initial level provided the perturbation energy is adequate to overcome dissipation. If Δz is up, and the new adiabatic density is less than ambient (point 2), the parcel is buoyant at that level and continues to rise. Such an atmosphere is unstable and will quickly develop convective cells. Similar arguments hold for a negative Δz as shown in Figure 5.3

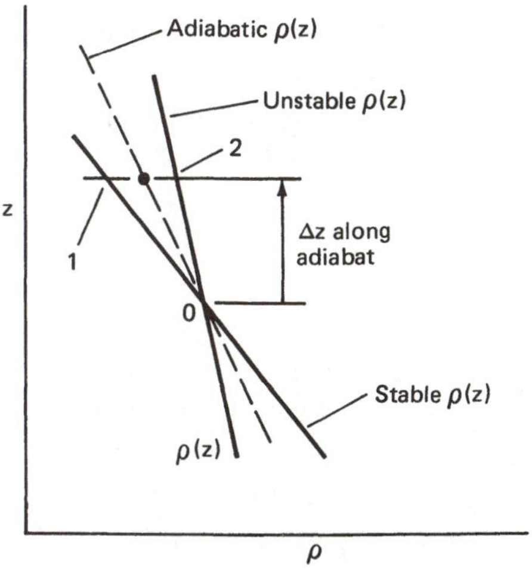

Figure 5.3: An atmosphere oscillates vertically at the Brunt-Vaisala frequency when it is stable against convection.

Density is a very difficult quantity to measure directly. It is more common to place the stability criterion in terms of pressure (p) and temperature (T), both of which can be measured easily and accurately. In these variables, an atmosphere is stable if

$$\left[-\left(1-\frac{1}{\gamma}\right)\frac{T}{p}\frac{\partial p}{\partial z}\right] > \left[-\frac{\partial T}{\partial z}\right] \tag{5.4a}$$

where γ is the ratio of the specific heats (c_p/c_v). Note that $\partial p/\partial z$ and $\partial T/\partial z$ are typically negative in the atmosphere so the bracketed quantities are usually positive quantities. The right-hand side is the actual temperature gradient and the left-hand side is the adiabatic temperature gradient. In a static atmosphere $\partial p/\partial z = -\rho g$. The Brunt-Vaisala frequency (n) for a static atmosphere becomes

$$n^2 = \frac{g}{T}\left[\frac{\partial T}{\partial z} + \frac{g}{c_p}\right] \tag{5.4b}$$

Passage of air that is stable against convection over a mountain range can provide a vertical disturbance leading to a set of stationary, transverse, internal lee waves oscillating at the Brunt-Vaisala frequency. The Sierra wave, made visible by condensation at the wave high points, just east of the main Sierra Nevada range in California provides an impressive example. The waves appear stationary relative to the earth by travelling upwind at the wind velocity.

Flow patterns can sometimes transform from one stable flow pattern to another stable flow pattern at some higher parameter region. As an example, a stability criterion might measure stability of a laminar flow against transformation to a wavelike flow; a further stability criterion might define a second or third transformation to more complex waveforms, or to convection or turbulence. Equation (5.4a) is a criterion that distinguishes a static atmosphere from a thermally convective atmosphere. When an atmosphere is stable in this sense, it can oscillate vertically at the Brunt-Vaisala frequency (5.4b).

5.2 Longitudinal Waves

In the classic longitudinal wave, the restoring force is the compressibility of the substance. Obviously longitudinal (compression) waves cannot exist in an incompressible fluid. The conflict between this and the often used approximation of water and sometimes air as being incompressible fluids is resolved by noting that the compressibility needed for production of a weak compression wave is very small. Therefore, a fluid such as air will often be analyzed as being incompressible relative to pressures of scale $p \sim \rho u^2$ but will still be able to support weak compression wave phenomena.

This section will derive compression (longitudinal) waves, using sound waves in air as an example and will lead to a wave equation having the general form

$$\frac{\partial^2 \xi}{\partial t^2} = \eta \frac{\partial^2 \xi}{\partial x^2} \tag{5.5}$$

where ξ is any one of several possible properties, and x is displacement along the direction of wave propagation. The variable η will be shown to be equal to the square of the speed of wave propagation. Note that η must have units of $(L/T)^2$ so that $(1/t^2)$, the left side, can be equated to $(1/x^2)$, the right side.

Equation (5.5), the wave equation, should be compared to the diffusion equation studied earlier (4.1) and repeated here in a format similar to (5.5),

$$\frac{\partial \xi}{\partial t} = \eta \frac{\partial^2 \xi}{\partial x^2} \tag{5.6}$$

Here ξ is again any one of several possible properties, and x is displacement along the direction of diffusive propagation. In this case, η is the coefficient of diffusivity and measures the areal rate at which the material property is dispersed in time. The units for η required to balance the units in (5.6) must be (L^2/T). The wave equation, involving $(\partial^2/\partial t^2)$, quantifies the acceleration of mass and oscillatory phenomena; the diffusion equation, involving $(\partial/\partial t)$, does not. Both involve a forcing function proportional to a dispersion over space $(\partial^2/\partial x^2)$. In the three-dimensional (3-D) case, the forcing function $(\partial^2\xi/\partial x^2)$ is replaced by the Laplacian $(\nabla^2\xi)$.

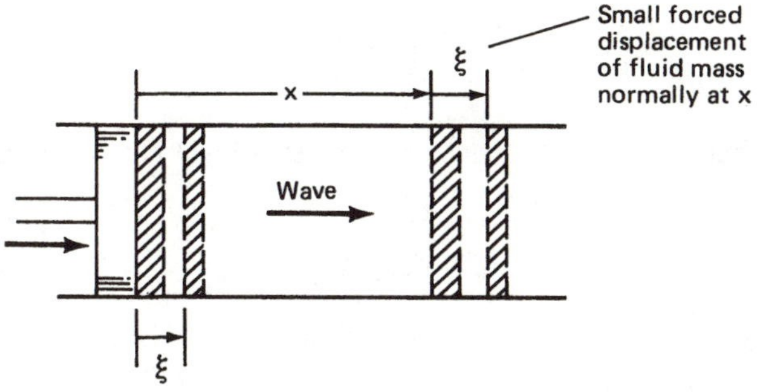

Figure 5.4: *Generation of a compression (acoustic) wave in a cylinder. The air mass provides inertia, and its compressibility provides the restoring force.*

Figure 5.4 shows a long circular pipe filled with air. The right end is open and the left end is fitted with a piston. If the piston is moved abruptly to the right, it accelerates the air in front of it, and the air, because of its mass inertia, resists this acceleration, causing it to be slightly compressed. A wave motion is initiated that propagates along the length of the pipe by interaction between the compressed region of air and the mass inertia property of the air itself. In the spring-mass systems of Figure 5.1a the springs and masses are discrete and separate items. In the case of a compressible continuous material both functions are supplied by the material; its compressibility provides the restoring force, and its mass inertia provides the residual momenta that perpetuates the motion.

In Figure 5.4, x is the equilibrium coordinate of an element of air, and ξ is its displacement from the equilibrium position. Assuming the air is motionless except for the wave motion, its velocity u(x, t) is

$$u = \partial\xi/\partial t$$

Net force acting on this element of air mass is due to the negative of the pressure gradient $(-dp/dx)$. This force causes an acceleration

$$-\frac{\partial p}{\partial x} = \rho_o \frac{\partial^2 \xi}{\partial t^2} \tag{5.7a}$$

where ρ_o is undisturbed density.

An expression is needed that quantifies pressure gradient as a function of the pressure-density characteristics of the compressible medium. Fluid condensation, i.e., the percentage increase in density, is defined as

$$s = \frac{\rho - \rho_o}{\rho_o} \quad \text{or} \quad \rho = \rho_o(1+s)$$

Within a restriction that the disturbance is very small, (s) can be compared to the relative displacement of the substance

$$s = -\Delta\xi/\Delta x$$

In acoustic waves in air, both s and $d\xi/dx$ (the limit of $\Delta\xi/\Delta x$) are less than 10^{-4}, and the assumption is valid. This includes even very loud, painful noises, but it does not include pressure waves caused by explosions, for example, or shock waves from supersonic aircraft.

These density changes must then be related to equivalent pressure changes using the characteristic $p(\rho)$ relationship for that fluid. In a typical acoustic wave the total internal energy is modelled as a constant; readers familiar with thermodynamics know this as an adiabatic process. The pressure-density relationship for an adiabatic compression or expansion of a gas can be stated as

$$p = k\rho^{\gamma} \qquad (5.7b)$$

where k is a constant and γ is the ratio of the specific heat at constant pressure to the specific heat at constant volume (C_p/C_v). Details are available in thermodynamics texts.

The general wave equation can then be derived using only that $p(\rho)$ must exist. After some manipulation the gradient of the acoustic pressure $p^*(= p - p_o)$ becomes

$$\frac{\partial p}{\partial x} = \frac{\partial p^*}{\partial x} = -\rho_o\left(\frac{dp}{d\rho}\right)_o \frac{\partial^2 \xi}{\partial x^2} \qquad (5.7c)$$

Equations (5.7a) and (5.7c) are combined to give the wave equation

$$\frac{\partial^2 \xi}{\partial t^2} = c^2 \frac{\partial^2 \xi}{\partial x^2} \qquad (5.8)$$

where c^2 has replaced $(dp/d\rho)_o$ because, in the solution of (5.8), the square root of $(dp/d\rho)_o$ is shown to be equal to the acoustic wave speed (c). Acoustic wave speed is sometimes represented by the letter (a), and surface wave speed by the letter (c), although there is no firm convention. This text uses (c) for both and also sometimes to represent an arbitrary constant. The uses can be distinguished one from the other by the context in which they are used.

The ultimate test of a theory is its applicability to reality. Computation of $(dp/d\rho)_o$ using (5.7b) yields for a gas

$$c = (\gamma p_o/\rho_o)^{1/2} \qquad (5.9a)$$

For dry air at 0°C and standard pressure, $\gamma = 1.402$, $p_o = 1.013 \times 105\ \text{N/m}^2$, and $\rho_o = 1.293\ \text{kg/m}^3$. Equation (5.9a) predicts that c = 331.6 m/s, which agrees precisely with measured values, supporting the validity of the approximations used in the derivation. However, regardless of (5.9a), the speed of sound is nearly independent of p_o or ρ_o in a given fluid because, for a given temperature, the ratio p_o/ρ_o is nearly constant. Because of this, the wave speed is more often related to temperature T using the perfect gas law $p = \rho RT$, with R a constant for a given gas

$$c = (\gamma RT)^{1/2} \qquad (5.9b)$$

The speed of sound is less at jet aircraft altitudes, for example, not because of reduced p or ρ, but because T is lower at those altitudes.

Equation (5.8) is equally valid if ξ is replaced by p^*, s, u, or ϕ, the velocity potential. The complete solution to (5.8), again using ξ as the variable of interest, is

$$\xi = f(ct - x) + g(ct + x) \qquad (5.10)$$

where f and g are any functions of the arguments $(ct - x)$ and $(ct + x)$. Sine and cosine functions are the most useful. Simple small amplitude waves are precisely modelled by ordinary sine or cosine functions, and more complicated shapes can be modelled as a sum of sine and cosine functions using Fourier analysis. Equation (5.10) is explained further in the analysis of surface waves.

The velocity of sound (low-amplitude pressure waves) in liquids and solids again depends on the compressibility of the substance. Because this quantity is more difficult to model in liquids and solids, the velocity of sound in these substances is more difficult to predict. Table 5.1 gives typical measured values of c for common gases, liquids, and solids.

5.3 Transverse Waves

The motion of a vibrating string, and waves on the free surface of a liquid, are examples of transverse waves. In liquid surface waves, as with the pendulum, the gravity property of mass supplies the restoring force and the inertia property of mass supplies the residual momenta that maintains the oscillation. It is expected then that mass (density) will cancel out of the solution as it did in the pendulum example.

Table 5.1: Velocity of Sound in Various Substances

Substance	Speed of Sound (m/s)	Conditions
Air (dry)	331.6	0°C Atmospheric pressure
Hydrogen (H_2)	1270	0°C Atmospheric pressure
Mercury	1450	20°C Atmospheric pressure
Water (pure)	1481	20°C Atmospheric pressure
Lead	2050	---
Concrete	3100	Varies with type
Granite	5000	Surface rocks (approximate)
Steel	6100	---
Earth mantle	8 to 13 × 10^3	Subsurface to core (mantle)

Transverse waves occur in the interior of elastic solids because shear elasticity provides a restoring force to transverse perturbations of the material. Transverse waves do not occur in the interior of typical fluid masses because fluids possess no shear elasticity. However, if a density stratification exists in the presence of gravity, surfaces of density variation can support transverse waves. These internal transverse waves are not shear waves but rather gravity waves, because gravity, not shear elasticity, provides the restoring force. Chapter 13 will introduce a transverse wave that occurs in the interior of a constant density rotating fluid. There the restoring force is caused, not by gravity or shear elasticity, but rather by angular momentum (vorticity) gradients.

Small amplitude surface waves are illustrated in Figure 5.5. The restoring force (gravity) is perpendicular to the direction of wave motion, but fluid displacement consists of both longitudinal and transverse components. Vertical displacement (shape) of the free surface from the equilibrium level (h) is $\xi(x, t)$. The maximum amplitude (ξ_m) is assumed small compared to the wavelength of the wave (λ)

$$\xi_m << \lambda$$

and the liquid is assumed inviscid so that a velocity potential (ϕ) exists. A typical derivation for ϕ, u, v, and ξ begins with the Bernoulli equation for a time-dependent flow with u the x component of velocity and v the y component of velocity at position (x, y)

$$-\frac{\partial\phi}{\partial t}+\frac{p}{\rho}+\frac{(u^2+v^2)}{2}+g\xi=0$$

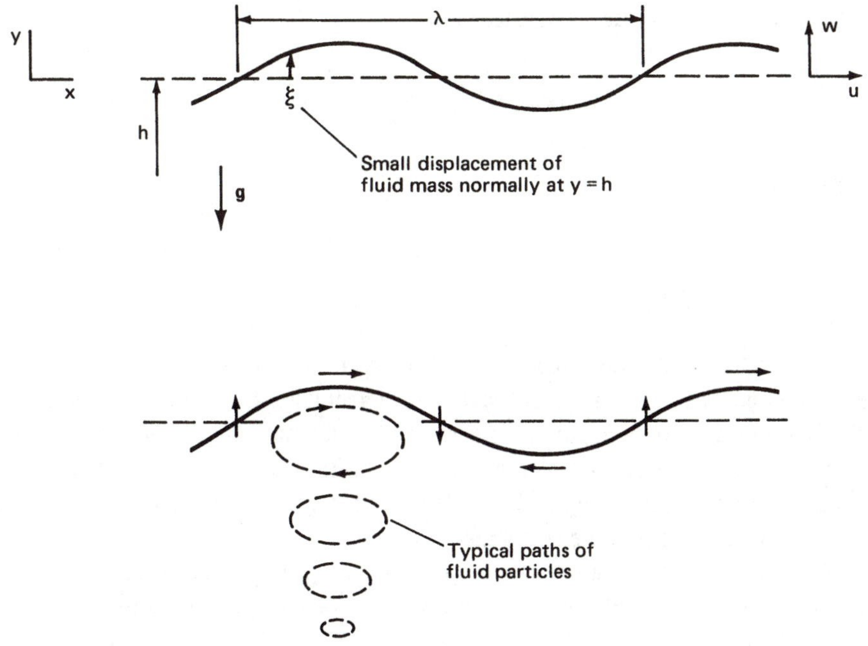

Figure 5.5: *Kinematics of liquid surface waves. Fluid motion at depth λ/2 is approximately 5% of surface motion. Gravity here provides the restoring force.*

After taking the time derivative of this equation, noting that at the surface p = 0 and the square of small velocities can be neglected, and substituting $\partial\xi/\partial t = v = -\partial\phi/\partial y$, it becomes

$$\frac{\partial^2\phi}{\partial t^2} + g\frac{\partial\phi}{\partial y} = 0$$

Laplace's equation for a 2-D incompressible inviscid fluid, $\nabla^2\phi = 0$, is introduced to satisfy continuity requirements. After some algebra, a wave equation is obtained, and the final solution is

$$\phi(x,y,t) = \frac{g\xi_m\lambda}{2\pi c}\,\frac{\cosh\frac{2\pi}{\lambda}(h-y)}{\cosh\frac{2\pi}{\lambda}h}\,\cos\frac{2\pi}{\lambda}(x-ct) \tag{5.11a}$$

which can be used to compute u and v. Here h is unperturbed fluid depth and c again is wave speed. Different authors use slightly different, but equivalent, versions of (5.11a). The equation of the free surface is

$$\xi(x,t) = \xi_m \sin \frac{2\pi}{\lambda}(x - ct) \tag{5.11b}$$

and the wave speed is

$$c = \left[\frac{g\lambda}{2\pi} \tanh\left(\frac{2\pi h}{\lambda}\right)\right]^{1/2} \tag{5.11c}$$

Figure 5.5 also illustrates the paths of fluid particles associated with a wave with the velocity potential (5.11a). Fluid particles (elements) move in circular to elliptical paths, depending on fluid depth. At a depth of $\lambda/2$, the maximum fluid displacement is approximately 5% of that at the surface. Motion is nearly zero at depth λ. Submarines or fish, for example, do not feel even large surface waves in a constant density fluid, provided they are submerged a depth $\geq \lambda$.

Many waves of practical interest occur in liquids (usually water) that are shallow compared to the wavelength, and where further, the wave amplitude is small compared to the liquid depth

$$\xi << h << \lambda$$

In the case of small amplitude surface waves in shallow liquid, $\tanh(2\pi h/\lambda) \rightarrow (2\pi h/\lambda)$, and

$$c = \sqrt{gh} \tag{5.12}$$

Near the shallow bottom the vertical component of flow velocity is suppressed, leaving only the horizontal component. The net effect is a significant longitudinal surging motion through the entire depth with scouring of the bottom surface.

Equation (5.10) represents a wave travelling in the +x direction $(x - ct)$, superimposed on a wave travelling in the $-x$ direction $(x + ct)$. Figure 5.6 illustrates this mathematical representation using an arbitrary wave and a sinusoidal periodic wave, both moving to the right. Arbitrary here implies any function f whose argument is $(x \pm ct)$, e.g., $\sin(x \pm ct)$, $(x \pm ct)^2$, $e^{(x \pm ct)}$, etc. As the arbitrary wave in the figure travels to the right, the argument $(x - ct)$ will always preserve the wave shape to be identical to the shape at t_o. The computation $(x_1 - ct_1)$ essentially transforms the displaced wave shape back to the t_o wave shape, point by point. In a sinusoidal wave representation, the displacement is measured as an angle with a full

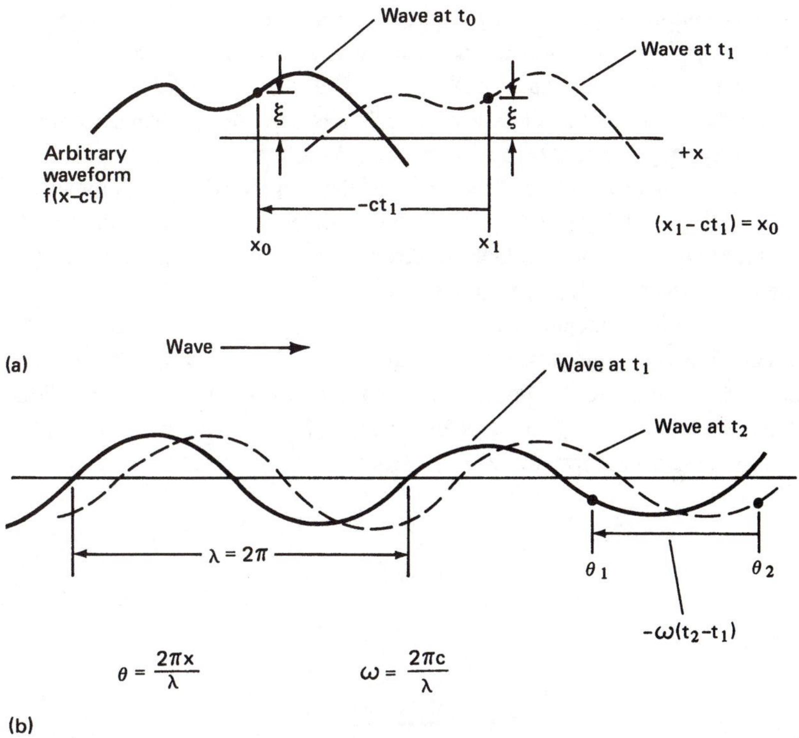

***Figure 5.6:** A wave of arbitrary shape (a) and a wave with sinusoidal shape (b). By representing the wave as some function $f(x - ct)$ or $g(x + ct)$, the shape of the wave is mathematically preserved at later times and positions. Note that each point ξ at $(x_1 - ct_1)$ must equal the original ξ at x_0.*

wave repeating each 2π radians. It is related to the actual displacement of the wave using

$$\theta = \frac{2\pi x}{\lambda} \quad \text{or} \quad x = \frac{\lambda\theta}{2\pi}$$

The time rate of change of θ is $\omega = \dot{\theta}$, which is related to c, the time rate of wave displacement (wave speed), by

$$\omega = \frac{2\pi c}{\lambda} \quad \text{or} \quad c = \frac{\lambda\omega}{2\pi}$$

If a wave (longitudinal or transverse) impacts on a nondissipative barrier, its energy is reflected in the reverse direction; the end that was leading in the original wave leads again in the reflected wave but in the reverse direction. If two barriers are placed a multiple wavelength apart, a wave trapped between the barriers appears as a *standing wave*. A steady-state oscillating violin string or a steady-state sloshing liquid in a rectangular channel can be regarded as examples of standing waves, but more properly they are analyzed as *travelling waves* that reflect between opposing barriers placed a multiple (or fractional) wavelength apart. End conditions determine the phase at the reflecting barriers. Node points exist at the barriers for the vibrating string. Mass conservation and lack of shear strength at the walls dictate that wave maxima must exist at the barriers for a sloshing liquid.

Wave speed is a function of wavelength for deep water waves, compare (5.11c) and (5.12). When two progressive waveforms of different wavelengths interact, they introduce waves with frequencies, i.e., wavelengths, equal to the sum and difference of their basic wavelengths ($\lambda_1 + \lambda_2$ and $\lambda_1 - \lambda_2$). If λ_1 and λ_2 are nearly the same, a new modulated waveform develops. This *group* waveform has an effective wavelength of

$$\lambda^* = 2\lambda_1\lambda_2 / (\lambda_1 - \lambda_2)$$

and a group speed of

$$c^* = \frac{c_2\lambda_1 - c_1\lambda_2}{\lambda_1 - \lambda_2}$$

where c_1 is the speed of wave λ_1, and c_2 the speed of λ_2. In the limit of $\lambda_1 \approx \lambda_2$ and $c_1 \approx c_2$, $c^* \approx c_1/2 \approx c_2/2$, and the group pattern travels half as fast as the individual waves. The individual waves continuously overtake the group waveforms and pass through them. The phenomena is common in the ocean and helps to explain *wave sets* with, say, every seventh wave being the largest. For finite amplitude waves, the crest will move faster than the trough, fall forward into the trough, and produce a breaking wave. Large breaking waves at a beach illustrate this and also illustrate the concentration of total wave energy into a thinner and thinner layer of water.

Internal waves can occur at any density interface. Note that the air-water surface is also a density interface, but one where air density is so small compared to water density that it has been assumed to be zero. When the densities are comparable, the top fluid must be included in the analysis also. The top fluid participates because of (1) its mass inertia property and (2) its mass gravitational property. Internal waves are more sluggish (have reduced speed) because of the added inertia of the upper fluid. Also, the bottom fluid is more nearly the density of the upper fluid (than, say, for water and air) so that small energy disturbances create large amplitude waves. In many applications the density variation occurs continuously over depth (height), resulting in more complicated waveforms and more difficult analyses.

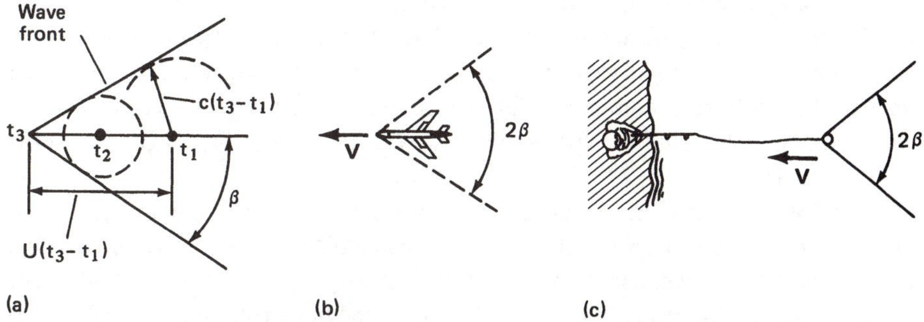

Figure 5.7: *Mach cones in a compressible medium, and ripples on a liquid surface. (a) When the speed of a perturbing object exceeds the speed of the generated waves (U > c), the fluid cannot flow smoothly past the object, and a discontinuity forms with angle $\beta = \sin^{-1}(c/U)$. (b) When an aircraft flies faster than the speed of sound (U > c), a three-dimensional conical shock wave forms. (c) When a fisherman pulls a float across the surface of a shallow lake at a speed greater than the speed of a surface wave (U > c), a similar phenomenon occurs as a two-dimensional V-shaped ripple.*

5.4 Mach Cones

Figure 5.7a illustrates a Mach cone. An object is moving at a velocity U in a stationary fluid. It produces a wave moving at velocity c relative to the fluid. Assume U > c. At time t_1 the object is at point 1; at time t_3 it is at point 3 having travelled a distance $U(t_3 - t_1)$, while the wave it produced at point 1 has travelled a shorter distance $c(t_3 - t_1)$. At intervening times the waves travel distances in the same proportion (i.e., U/c) as that travelled by the object.

The locus of wave fronts caused by an aircraft moving faster than the ambient speed of sound describes a 3-D cone as shown in Figure 5.7b. The wave speed here is the acoustic speed given by (5.9b). Texts on supersonic flow theory detail the pressure, temperature, and density differences that occur at the *shock wave* conical surface. The differences are so extreme and so abrupt that the fluid cannot be considered a *continuum* at that surface; thermodynamic considerations dominate the problem. In Figure 5.7c a similar 2-D shape is produced on the surface of a shallow lake by a fisherman reeling in a float. Here the wave speed in the shallow lake (5.12) is less than the speed of the float moving through the water. Similar patterns are produced in rotating fluids but are based on different wave mechanisms.

The following numerical analysis quantifies Figures 5.7b and 5.7c; it makes use of the Froude number (2.31) and the Mach number (2.32) introduced first in Section 2.5. A fisherman pulls a small float across still water that is 0.2 m deep. The fisherman observes a surface wave pattern following the float with an included angle of 80°. *How fast is the float moving, and what is the Froude number?* At the same time an aircraft flying overhead at an altitude of 10,000 m is producing a conical shock wave, also with an included angle of 80°. *How fast is the aircraft flying, and what is its Mach number?*

The speed of a small amplitude surface wave in a shallow liquid is $c = \sqrt{gh}$ from (5.12). Any movement of the float causes a wave disturbance on the surface that propagates at this speed. As the float moves from t_1 to t_3, it generates a continuous series of waves. The ratio of the speed of the float and the waves determines the locus of the wave front as shown. For example, after the float has moved a distance $U(t_3 - t_1)$ from position 1, the wave front has moved $c(t_3 - t_1)$. The relationship between the two is

$$\sin\beta = c\Delta t/U\Delta t = c/U$$

With c and β known

$$U = \sqrt{gh}/\sin\beta = \frac{\sqrt{9.81 \times 0.2}}{\sin 40^\circ} = 2.18 \text{ m/s}$$

The Froude number is

$$F_r = U/\sqrt{gh} = U/c = 2.18/1.40 = 1.56$$

The aircraft's shock wave is produced because the aircraft velocity (U) is faster than the speed of an acoustic wave with velocity c. During time Δt, a supersonic aircraft flies a distance $U\Delta t$, while its disturbance travels through the air a distance $c\Delta t$. The angle β is related to these by

$$\sin\beta = c\Delta t/U\Delta t = c/U$$

Appendix E.6, for an altitude of 10^4 m, specifies $T = -49.7\,^\circ\text{C} = 223.5$ K, and then $c = \sqrt{\gamma RT} = \sqrt{1.4 \times 287 \times 223.5} = 300$ m/s. With c and β known

$$U = c/\sin\beta = 300/\sin 40^\circ = 467 \text{ m/s}$$

The aircraft's Mach number is

$$M_a = U/c = 467/300 = 1.56$$

The two phenomena are similar in that they each depend on the velocity of an object and the velocity of a wave propagation created by the object. When the wave patterns have the same included angles, the ratios U/c must be the same, and the Froude number and Mach number also are equal.

PART II

Rotating Fluid Theory

Chapter 6

Rotating Coordinate Systems

6.1 Intermediate Reference Frames

Use of an intermediate reference frame, with a coordinate system fixed in that frame, is often advantageous in the analysis of particle and rigid body dynamics. Sometimes several frames are desirable, as for example, a helicopter attempting to land on an ocean-going ship. The pilot must adjust his reference frame (the helicopter) to the ship's reference frame and also to a reference frame fixed in the moving wind in that region. Further, if the approach is instrument aided, and the instruments use $\mathbf{F} = m\mathbf{A}$, an inertial frame is a part of the problem.

Remember that there are an infinity of inertial frames relative to translation, but there is one and only one inertial frame relative to rotation. Observers in distant galaxies who establish themselves in *the* nonrotating frame (as verified by Newton's law) would find that they are not rotating relative to the nonrotating frame selected by earth observers; however, they might be translating relative to the earth frame. It is correct to write *an* inertial frame and *the* nonrotating frame.

In an analysis of rotating fluids based on an Eulerian formulation of mechanics (refer to Chapter 2), the use of an intermediate frame is more than just advantageous--it is often essential to a successful analysis. Consider in Figure 6.1 a contained quantity of fluid defined as the flow region of interest. The container may be translating and rotating relative to inertial space. If an Eulerian grid fixed in the inertial frame (n) is used, the Navier-Stokes equations (again refer to Chapter 2) are valid, but the container moves into or out of the grid points, and analysis is nearly impossible. Assume next that an Eulerian grid is fixed in a moving container reference frame that is not inertial. Now the fluid region boundaries are easily defined, but the Navier-Stokes equations cannot be used in the noninertial reference frame.

The difficulty is resolved by using an Eulerian grid fixed to b, the container, and then including terms in the Navier-Stokes equation to compensate for the fact that the grid is noninertial. Figure 6.2 shows a simplified two-dimensional (2-D) sketch. The included terms are easily understood in terms of relative acceleration of a point. The

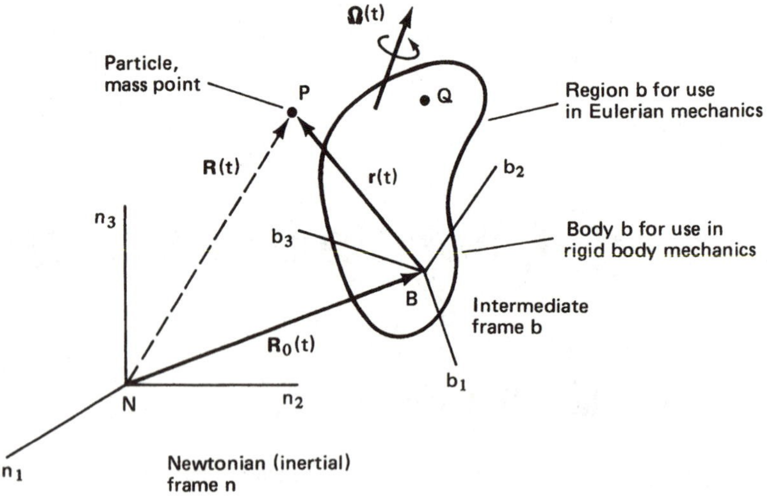

Figure 6.1: *Reference frame b has an arbitrary time-varying position and orientation relative to frame n. Although the illustration is valid for any two reference frames, applications in this text are limited to a nonaccelerating (Newtonian) frame n, and an accelerating (e.g., rotating) reference frame b.*

position of point P in Figure 6.1 relative to origin point N of an inertial frame n is given by

$$\mathbf{R} = \mathbf{r} + \mathbf{R}_o \tag{6.1}$$

Many of the difficulties commonly encountered in the analysis of typical real motions, whether of particles, rigid objects, or fluids, are due to use of inadequate notation. For example, when several reference frames are needed to define a problem, confusion can arise because a vector quantity might be defined relative to one frame, its derivative taken relative to a second frame, and then that placed in components in yet another frame. The notation must adequately distinguish these and yet be easy to recognize and remember.

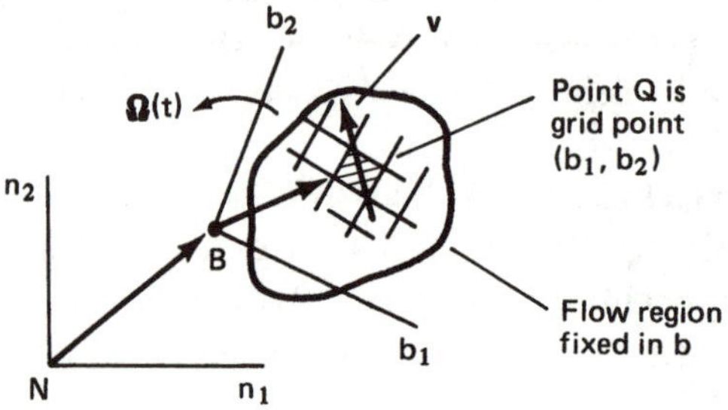

Figure 6.2: *A 2-D simplification of Figure 6.1 illustrating an Eulerian grid fixed in frame b. Fluid motion is defined relative to the accelerating grid.*

The notation used here is adequate for use with the two reference frames n and b. In the following equations, **R**, **V**, and **A** are position, velocity, and acceleration of point P relative to the inertial frame, respectively, while **r**, **v**, and **a** are equivalent terms but relative to the b frame. $\mathbf{R}_o$, $\mathbf{V}_o$, and $\mathbf{A}_o$ define position, velocity, and acceleration of the origin of the b frame relative to the n frame, respectively. The derivative of **R** relative to the n frame gives **V** relative to the n frame. In the same way the b frame derivative of **r** gives **v** relative to the b frame. This convention, combined with (6.1), gives

$$\mathbf{V} = \frac{{}^{n}d}{dt}\mathbf{R} = \frac{{}^{n}d}{dt}\mathbf{r} + \frac{{}^{n}d}{dt}\mathbf{R}_o$$

The term ${}^{n}d\mathbf{r}/dt$ is an n-frame derivative of a b-frame vector. The complication is resolved by using a general operator identity

$$\frac{{}^{n}d(\,)}{dt} = \frac{{}^{b}d(\,)}{dt} + \mathbf{\Omega}\times(\,)$$

where $\mathbf{\Omega}$ is the angular velocity of frame b relative to frame n. This identity is valid for any vector and any two reference frames. Applied to **r** and frames n and b, the term becomes

$$\frac{{}^{n}d}{dt}\mathbf{r} = \frac{{}^{b}d}{dt}\mathbf{r} + \mathbf{\Omega}\times\mathbf{r}$$

After use of the previous conventions and rearrangement, **V** becomes

$$\begin{array}{c} \mathbf{V} \;=\; \mathbf{v} \;+\; [\mathbf{V}_o \;+\; \mathbf{\Omega}\times\mathbf{r}] \\ \text{Inertial} = \text{Local} + \text{Origin} + \text{Tangential} \end{array} \tag{6.2}$$

Note in Figure 6.1, **V** is the velocity of point P relative to n, **v** is the velocity of point P relative to b, $\mathbf{V}_o$ is the velocity of origin point B of the b frame relative to n, and the cross product $\mathbf{\Omega}\times\mathbf{r}$ is the tangential velocity of the location of point P in the b frame due to the rotation of b relative to n.

The n-frame derivative of **V** (relative to n) yields **A** (relative to n). After use of (6.2), **A** becomes

$$\mathbf{A} = \frac{{}^{n}d}{dt}\mathbf{V} = \frac{{}^{n}d}{dt}\mathbf{v} + \frac{{}^{n}d}{dt}\mathbf{V}_o + \frac{{}^{n}d}{dt}(\mathbf{\Omega}\times\mathbf{r})$$

After repeated use of the rules for derivatives and simplification, the expression for **A** reduces to

$$\begin{array}{c} \mathbf{A} \;=\; \mathbf{a} \;+\; \left[\mathbf{A}_o \;+\; \mathbf{\Omega}\times(\mathbf{\Omega}\times\mathbf{r}) \;+\; 2\mathbf{\Omega}\times\mathbf{v} \;+\; \boldsymbol{\alpha}\times\mathbf{r}\right] \\ \text{Inertial} = \text{Local} + \text{Origin} + \text{Centripetal} + \text{Coriolis} + \text{Tangential} \end{array} \tag{6.3}$$

with $\boldsymbol{\alpha} = {}^b d\boldsymbol{\Omega}/dt = {}^n d\boldsymbol{\Omega}/dt$. Here ${}^b d/dt = {}^n d/dt$ because $\boldsymbol{\Omega} \times \boldsymbol{\Omega} = 0$. The terms in brackets in (6.2) and (6.3) are called *transport* velocities and accelerations, respectively, because they define the transport of point P due to motion of frame b.

6.2 Fluids in a Rotating Frame

In rigid body dynamics, all points Q of the body are fixed relative to frame b, and therefore $\mathbf{v}^Q$ and $\mathbf{a}^Q$ must equal zero. In the same way in fluid mechanics, points Q define Eulerian grid points fixed relative to b, and $\mathbf{r}^Q$ locates these grid points at (b_1, b_2, b_3) as Figure 6.2 shows in 2-D. The velocity and acceleration of the fluid relative to b at points Q are $\mathbf{v}^f$ and $\mathbf{a}^f$, respectively. Equation (6.3) can be rewritten as

$$\mathbf{A}^f = \mathbf{a}^f + \left[\mathbf{A}_o + \boldsymbol{\Omega} \times \left(\boldsymbol{\Omega} \times \mathbf{r}^Q\right) + 2\boldsymbol{\Omega} \times \mathbf{v}^f + \boldsymbol{\alpha} \times \mathbf{r}^Q\right] \tag{6.4}$$

In the Eulerian formulation, fluid acceleration relative to the b frame is simply the substantive derivative (see Chapter 2) taken in the b frame

$$\mathbf{a}^f = \frac{{}^b D\mathbf{v}^f}{Dt} = \frac{{}^b\partial \mathbf{v}^f}{\partial t} + (\mathbf{v}^f \cdot {}^b\nabla)\,\mathbf{v}^f \tag{6.5}$$

In (6.5), all variables and space and time derivatives are relative to the b frame.

Equation (6.4), with (6.5) defining $\mathbf{a}^f$, is then used in computing $\mathbf{A}^f$ in the Navier-Stokes equations. It is no longer necessary to retain the identifying superscripts and $\mathbf{A}^f$ can be written simply as $\mathbf{A}$

$$\mathbf{A} = \frac{\partial \mathbf{v}}{\partial t} + (\mathbf{v} \cdot \nabla)\,\mathbf{v} + \mathbf{A}_o + \boldsymbol{\Omega} \times (\boldsymbol{\Omega} \times \mathbf{r}) + 2\boldsymbol{\Omega} \times \mathbf{v} + \boldsymbol{\alpha} \times \mathbf{r} \tag{6.6}$$

where $\mathbf{A}_o$, $\boldsymbol{\Omega}$, and $\boldsymbol{\alpha}$ define motion of the b frame; $\mathbf{r}$ represents positions of Eulerian grid points in the b frame; and v, ∇, and $\partial/\partial t$ are all relative to the b frame. Velocity v is the dependent variable, and $\mathbf{A}_o$, $\boldsymbol{\Omega}$, $\boldsymbol{\alpha}$, and boundary conditions on $\mathbf{r}$ are generally known.

The remaining task in deriving the Navier-Stokes equations relative to the b frame is to place forces, originally derived as force components in the n frame, into force components in the b frame. Consider the term $(-\nabla p)$. Surfaces of constant pressure in the fluid, at time t, exist in the fluid independent of who observes them or which reference frame is used. If pressure at some instant is given as $p(n_1, n_2, n_3)$, and the del operator as $(\partial/\partial n_1, \partial/\partial n_2, \partial/\partial n_3)$, then ${}^n\nabla p$ provides the gradient (real-world direction and magnitude) of p as shown in Figure 6.3. But for the same fluid, if $p(b_1, b_2, b_3)$ is given at the same instant and $(\partial/\partial b_1, \partial/\partial b_2, \partial/\partial b_3)$ is used, then ${}^b\nabla p$ must provide the same real world vector gradient of p. Therefore

$${}^n\nabla p = {}^b\nabla p$$

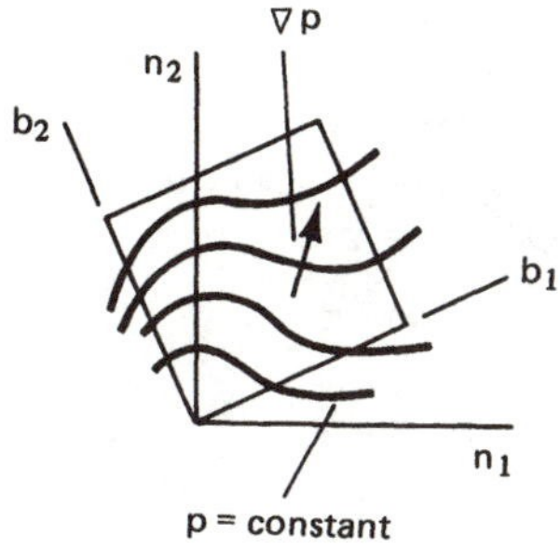

Figure 6.3: *The isobars (lines of constant pressure) can be defined at any instant in coordinates* (n_1, n_2, n_3) *or in* (b_1, b_2, b_3)*, and gradients can be taken as partials relative to the n axes* $({}^n\nabla)$ *or relative to the b axes* $({}^b\nabla)$*. Either* ${}^n\nabla p_n$ *or* ${}^b\nabla p_b$ *must yield the same absolute spatial rate of change of the pressure field, and therefore* $({}^n\nabla p_n = {}^b\nabla p_b)$*.*

Refer to Chapter 3 for details on the various mechanisms that contribute to the pressure field.

The vector Laplacian (see Appendix A.6) ${}^n\nabla^2 \mathbf{v}^f$ is also invariant to coordinate transformations so that

$$^{n}\nabla^2 \mathbf{v}^f = {}^b\nabla^2 \mathbf{v}^f$$

In many applications of interest, $\mathbf{A}_o$ and $\boldsymbol{\alpha}$ are zero or negligible so that the (ρ = c, μ = c) Navier-Stokes vector equation written for use in a reference frame rotating at constant angular velocity becomes simply

$$\frac{\partial \mathbf{v}}{\partial t} + (\mathbf{v} \cdot \nabla)\mathbf{v} + \boldsymbol{\Omega} \times (\boldsymbol{\Omega} \times \mathbf{r}) + 2\boldsymbol{\Omega} \times \mathbf{v} = \frac{1}{\rho}\mathbf{f} - \frac{1}{\rho}\nabla p + \nu \nabla^2 \mathbf{v} \qquad (6.7)$$

All quantities are now defined in, and expressed as components in, the rotating frame, with centripetal acceleration, $\boldsymbol{\Omega} \times (\boldsymbol{\Omega} \times \mathbf{r})$, and Coriolis acceleration, $2\boldsymbol{\Omega} \times \mathbf{v}$, the only additional complications. The continuity equation is also invariant to the given coordinate transformation. It becomes $(\nabla \cdot \mathbf{v} = 0)$, with all quantities now relative to, and expressed as components in, the rotating frame.

If, in (6.7), ρ is placed on the left-hand side of the equation, each term has units of force. A question arises whether terms such as $(2\rho\boldsymbol{\Omega} \times \mathbf{v})$ are "real" forces. Each term on the right-hand side is a component of the total applied real force represented by solid line components in Figure 6.4. A vector equation states that both sides are equal, in magnitude, in direction, and in physical dimension. The sum of terms $(\rho\, \partial \mathbf{v}/\partial t + \ldots 2\rho\boldsymbol{\Omega} \times \mathbf{v})$ therefore also equals total applied real force. Each separate term, e.g., $\rho\, \partial \mathbf{v}/\partial t$, can be considered as one of a different set of components of the total applied real force as shown by the dashed line components in Figure 6.4. The risk in considering them individually as separate real forces is that none of them typically correspond to any one of the actual applied forces. The critical question to

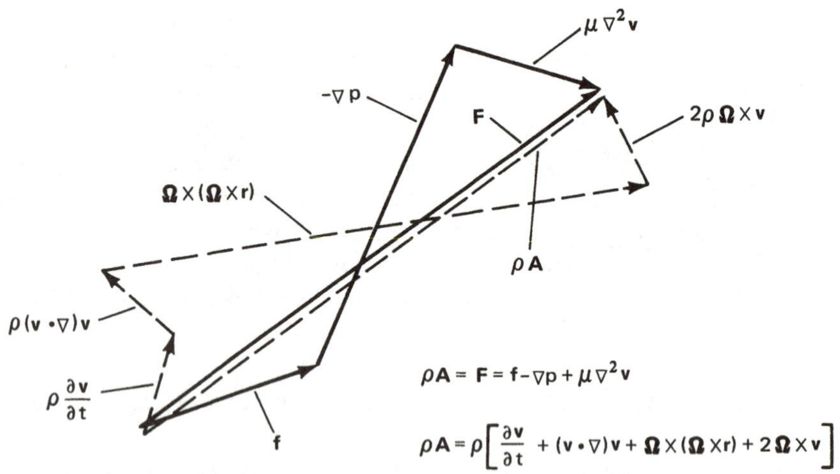

Figure 6.4: *A vector equation, such as* $\mathbf{F} = p\mathbf{A}$*, states both sides have the same magnitude, direction, and units. Either side can be defined as sums of component vectors, and each complete sum adds up to either side.*

ask is this, *Does the entire equation properly represent* $\mathbf{F} = m\mathbf{A}$*, where* $\mathbf{A}$ *is taken relative to inertial space*? The derivation of this section proves the answer is, *Yes*, for (6.7).

6.3 Ekman and Rossby Numbers

When the Navier-Stokes equations are reduced to dimensionless form (see Chapter 2) with no gravity forces, the four parameters (ρ, μ, L, U) reduce to one parameter, the Reynolds number ($R_e = \rho LU/\mu$). In inviscid flow with ρg as a gravity force, the four parameters (ρ, g, L, U) reduce to one dimensionless parameter, the Froude number ($F_r = U/\sqrt{gL}$). More generally, the Buckingham pi theorem states that conversion from dimensional to dimensionless form reduces the number of parameters by the number of independent physical quantities. In the study of mechanics (omitting electric and magnetic phenomena), the physical universe is described by three quantities: mass, distance, and time (SI units); or force, distance, and time (US units). Accordingly, if a problem involves all five parameters (ρ, g, μ, L, U), conversion to dimensionless form yields $5 - 3 = 2$ dimensionless parameters, R_e and F_r.

Equation (6.7) with $\mathbf{f} = \rho \mathbf{g}$ involves six parameters (ρ, g, μ, Ω, L, U), and conversion to dimensionless form should lead to $6 - 3 = 3$ parameters. An alternative dimensionless form involving only two dimensionless parameters is more relevant for motions dominated by rotation. This form uses Ω^{-1} as characteristic time, assumes $\mathbf{f}$

is a conservative force (e.g., gravity) so that it can be expressed as the gradient of some potential function (Φ), and notes further that $\boldsymbol{\Omega}\times(\boldsymbol{\Omega}\times\mathbf{r})=-\frac{1}{2}\nabla[(\boldsymbol{\Omega}\times\mathbf{r})\cdot(\boldsymbol{\Omega}\times\mathbf{r})]$. In this simplification, where p^* represents all pressures other than those caused by gravity and/or centrifugal effects

$$\nabla p=\nabla\left[p^*+\rho\Phi-\tfrac{1}{2}\rho(\boldsymbol{\Omega}\times\mathbf{r})\cdot(\boldsymbol{\Omega}\times\mathbf{r})\right] \tag{6.8}$$

Pressure now is made dimensionless by ($\rho\Omega UL$) rather than (ρU^2). The dimensionless equation becomes

$$\frac{\partial \mathbf{v}}{\partial t}+R_o(\mathbf{v}\cdot\nabla)\mathbf{v}+2\hat{\mathbf{k}}\times\mathbf{v}=-\nabla p+E_k\nabla^2\mathbf{v} \tag{6.9}$$

where E_k is the Ekman number

$$E_k=\nu/\Omega L^2 \tag{6.10}$$

and R_o is the Rossby number

$$R_o=U/\Omega L \tag{6.11}$$

The Ekman number is the ratio of viscous to Coriolis accelerations, and the Rossby number is the ratio of convective to Coriolis accelerations. Use of Ω^{-1}, rather than L/U as characteristic time, presupposes that the equation will be used when rotational phenomena dominate the application. Engineers, analyzing liquids in spinning containers, sometimes refer to a Reynolds number computed as E_k^{-1} with $L\sim R$ and $U\sim\Omega R$. This practice is misleading. It might appear also that the Reynolds number is the ratio of R_o/E_k as follows

$$R_e=\frac{R_o}{E_k}=\frac{U}{\Omega L}\frac{\Omega L^2}{\nu}=\frac{UL}{\nu}$$

This is not correct, however, because the proper characteristic length for R_o is rarely the same characteristic length appropriate for E_k as shown in Example 6.4. When using dimensionless ratios it is important to remember their origin and to select characteristic L, U, Ω, and ν appropriate to each specific application.

Another dimensionless number, the *frequency ratio*, is useful when an oscillation is imposed on a rotating fluid. The ratio of $\partial\mathbf{v}/\partial t$ and $2\boldsymbol{\Omega}\times\mathbf{v}$ can then be represented as

$$\text{Frequency ratio}=\frac{\omega U}{\Omega U}=\frac{\omega}{\Omega} \tag{6.12}$$

where ω is the frequency of the oscillatory phenomenon. Applications are given in Chapter 13.

6.4 Application Examples

This section places the Navier-Stokes and continuity equations, with ρ and μ constant, in a coordinate system rotating at an angular velocity $\boldsymbol{\Omega}$. The equations are then transformed to dimensionless variables, with the Ekman number (E_k) and the Rossby number (R_o) replacing the Reynolds number. Examples 6.1 through 6.4 illustrate details associated with the actual computation of relative positions, velocities, accelerations, and pressure fields when more than one coordinate system is necessary.

Example 6.1. A drop of dark dye is used to observe flow in a tank spinning at 3 rpm. Position and velocity of the drop relative to the laboratory (assumed inertial) are measured at a given instant in laboratory coordinates as $\mathbf{R} = 8\hat{\mathbf{n}}_2 + 3\hat{\mathbf{n}}_3$, $\mathbf{V} = -2\hat{\mathbf{n}}_1 + \hat{\mathbf{n}}_2 + \hat{\mathbf{n}}_3$. What is the position and velocity of the drop relative to the tank in tank fixed coordinates? Is information available to deduce all or any of $\mathbf{A}$ or $\mathbf{a}$?

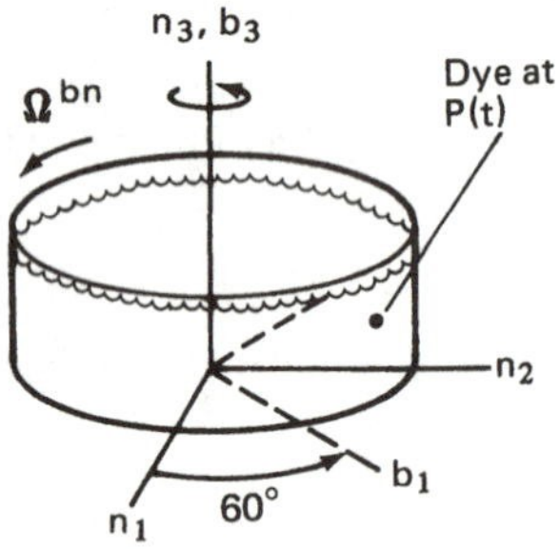

Discussion. The direction cosine matrix of b relative to n is $[C^{bn}]$ (see Appendix A.3)

$$\begin{bmatrix} \hat{\mathbf{b}}_1 \\ \hat{\mathbf{b}}_2 \\ \hat{\mathbf{b}}_3 \end{bmatrix} = \begin{bmatrix} C\theta & S\theta & 0 \\ -S\theta & C\theta & 0 \\ 0 & 0 & 1 \end{bmatrix} \begin{bmatrix} \hat{\mathbf{n}}_1 \\ \hat{\mathbf{n}}_2 \\ \hat{\mathbf{n}}_3 \end{bmatrix} \qquad \begin{matrix} \theta = 60^\circ \\ \mathbf{R} = \mathbf{r} + \mathbf{R}_o \end{matrix}$$

$$\mathbf{r} = \begin{bmatrix} C\theta & S\theta & 0 \\ -S\theta & C\theta & 0 \\ 0 & 0 & 1 \end{bmatrix} \begin{bmatrix} 0 \\ 8 \\ 3 \end{bmatrix}_n = \begin{bmatrix} 8\sin 60^\circ \\ 8\cos 60^\circ \\ 3 \end{bmatrix} = \begin{bmatrix} 6.93 \\ 4 \\ 3 \end{bmatrix}_b$$

$$\mathbf{r} = 6.93\hat{\mathbf{b}}_1 + 4\hat{\mathbf{b}}_2 + 3\hat{\mathbf{b}}_3$$

$$\mathbf{V} = \mathbf{v} + \boldsymbol{\Omega} \times \mathbf{r} \qquad \boldsymbol{\Omega} = (3\pi/30 \text{ rad/s})\hat{\mathbf{b}}_3$$

$$\mathbf{v} = \mathbf{V} - \boldsymbol{\Omega} \times \mathbf{r} \qquad \mathbf{V} = \begin{bmatrix} C\theta & S\theta & 0 \\ -S\theta & C\theta & 0 \\ 0 & 0 & 1 \end{bmatrix} \begin{bmatrix} -2 \\ 1 \\ 1 \end{bmatrix}_n = \begin{bmatrix} -0.13 \\ 2.23 \\ 1 \end{bmatrix}_b$$

$$\boldsymbol{\Omega} \times \mathbf{r} = \begin{bmatrix} 0 & -\pi/10 & 0 \\ \pi/10 & 0 & 0 \\ 0 & 0 & 0 \end{bmatrix} \begin{bmatrix} 6.93 \\ 4 \\ 3 \end{bmatrix}_n = \begin{bmatrix} -1.26 \\ 2.18 \\ 0 \end{bmatrix}_b = -1.26\hat{\mathbf{b}}_1 + 2.18\hat{\mathbf{b}}_2$$

See Appendix A.2 for representation of $(\boldsymbol{\Omega}\times)$ as a skew-symmetric matrix.

$$\mathbf{v} = -0.13\hat{\mathbf{b}}_1 + 2.23\hat{\mathbf{b}}_2 + 1\hat{\mathbf{b}}_3 - \left(-1.26\hat{\mathbf{b}}_1 + 2.18\hat{\mathbf{b}}_2\right) = 1.13\hat{\mathbf{b}}_1 + 0.05\hat{\mathbf{b}}_2 + \hat{\mathbf{b}}_3$$

$$\mathbf{A} = \underset{\text{Zero}}{\mathbf{a} + \mathbf{A}_o} + \underset{\text{Zero}}{\boldsymbol{\alpha} \times \mathbf{r}} + \underset{\text{Known}}{\boldsymbol{\Omega} \times (\boldsymbol{\Omega} \times \mathbf{r})} + \underset{\text{Known}}{2\boldsymbol{\Omega} \times \mathbf{v}}$$

The data available gives **R** and **V** at some instant of time, not as functions of time. Two or more measurements of **R** and **V** over short intervals of time would be needed to estimate **a** and **A**. Neither **a** nor **A** are available here, but if, as in many applications, **a** ~ 0, then **A** is the sum of the computed centripetal and Coriolis accelerations.

Example 6.2. A satellite is ejected spinning from a space platform (assumed inertial) with its spin axis (b_3) parallel to the n_3 axis of the platform. The b and n frames coincide at ejection. In SI units, $\boldsymbol{\Omega} = 2\pi\hat{\mathbf{b}}_3$, $\mathbf{r} = -\hat{\mathbf{b}}_2 + 2\hat{\mathbf{b}}_3$, $\mathbf{R}_0 = 4t\hat{\mathbf{n}}_1 + 3t\hat{\mathbf{n}}_2 + 5t\hat{\mathbf{n}}_3$. Compute **R**, **V**, and **A** in n-frame components at t = 6.125 s.

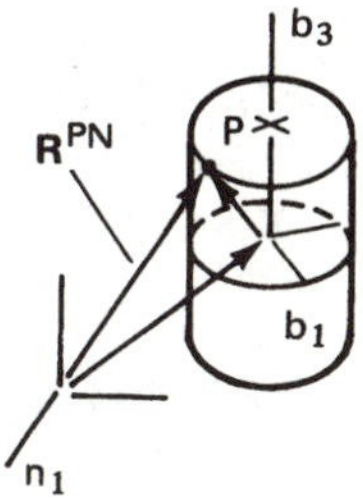

Discussion.

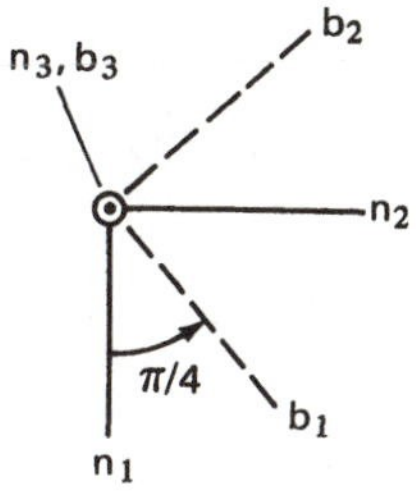

$$\boldsymbol{\alpha} = \frac{{}^{n}d}{dt}\boldsymbol{\Omega} = 0, \qquad \mathbf{V}_o = \frac{{}^{n}d}{dt}\mathbf{R}_o = 4\hat{\mathbf{n}}_1 + 3\hat{\mathbf{n}}_2 + 5\hat{\mathbf{n}}_3$$

$$\mathbf{A}_o = \frac{{}^{n}d}{dt}\mathbf{V}_o = 0, \qquad \mathbf{v} = \frac{{}^{b}d}{dt}\mathbf{r} = 0 \quad \text{and} \quad \mathbf{a} = 0$$

$\mathbf{R} = \mathbf{r} + \mathbf{R}_o$ (at t = 6.125 s)

$\mathbf{R} = -\hat{\mathbf{b}}_2 + 2\hat{\mathbf{b}}_3 + 24.5\hat{\mathbf{n}}_1 + 18.38\hat{\mathbf{n}}_2 + 30.63\hat{\mathbf{n}}_3.$

Convert b_i to n_i. In one-dimensional (1-D) rotation

$\theta = \Omega t = 2\pi\, 6.125$, b has rotated 6.125 times.

$$\begin{bmatrix} \hat{\mathbf{b}}_1 \\ \hat{\mathbf{b}}_2 \\ \hat{\mathbf{b}}_3 \end{bmatrix} = \begin{bmatrix} C45 & S45 & 0 \\ -S45 & C45 & 0 \\ 0 & 0 & 1 \end{bmatrix} \begin{bmatrix} \hat{\mathbf{n}}_1 \\ \hat{\mathbf{n}}_2 \\ \hat{\mathbf{n}}_3 \end{bmatrix} = \begin{bmatrix} 0.707\hat{\mathbf{n}}_1 + 0.707\hat{\mathbf{n}}_2 \\ -0.707\hat{\mathbf{n}}_1 + 0.707\hat{\mathbf{n}}_2 \\ \hat{\mathbf{n}}_3 \end{bmatrix}$$

$\mathbf{R} = -(-0.71\hat{\mathbf{n}}_1 + 0.71\hat{\mathbf{n}}_2) + 2\hat{\mathbf{n}}_3 + 24.5\hat{\mathbf{n}}_1 + 18.38\hat{\mathbf{n}}_2 + 30.36\hat{\mathbf{n}}_3$

$\mathbf{R} = 25.2\hat{\mathbf{n}}_1 + 17.7\hat{\mathbf{n}}_2 + 32.6\hat{\mathbf{n}}_3$ (m)

$\mathbf{V} = \mathbf{v} + \mathbf{V}_o + \boldsymbol{\Omega} \times \mathbf{r} = 0 + (4\hat{\mathbf{n}}_1 + 3\hat{\mathbf{n}}_2 + 5\hat{\mathbf{n}}_3) + 2\pi\hat{\mathbf{b}}_3 \times (-\hat{\mathbf{b}}_2 + 2\hat{\mathbf{b}}_3)$

$\mathbf{V} = (4\hat{\mathbf{n}}_1 + 3\hat{\mathbf{n}}_2 + 5\hat{\mathbf{n}}_3) + 2\pi\hat{\mathbf{b}}_1$, but $\hat{\mathbf{b}}_1 = 0.707\hat{\mathbf{n}}_1 + 0.707\hat{\mathbf{n}}_2$, so

$\mathbf{V} = 8.44\hat{\mathbf{n}}_1 + 7.44\hat{\mathbf{n}}_2 + 5\hat{\mathbf{n}}_3$ (m/s)

$\mathbf{A} = \mathbf{a} + \mathbf{A}_o + \boldsymbol{\alpha} \times \mathbf{r} + \boldsymbol{\Omega} \times (\boldsymbol{\Omega} \times \mathbf{r}) + 2\boldsymbol{\Omega} \times \mathbf{v}$

$\mathbf{A} = 0 + 0 + 0 + 2\pi\hat{\mathbf{b}}_3 \times [2\pi\hat{\mathbf{b}}_3 \times (-\hat{\mathbf{b}}_2 + 2\hat{\mathbf{b}}_3)] + 0$

$\mathbf{A} = 2\pi\hat{\mathbf{b}}_3 \times [2\pi\hat{\mathbf{b}}_1] = 4\pi^2\hat{\mathbf{b}}_2$, but $\hat{\mathbf{b}}_2 = -0.707\hat{\mathbf{n}}_1 + 0.707\hat{\mathbf{n}}_2$

$\mathbf{A} = -27.9\hat{\mathbf{n}}_1 + 27.9\hat{\mathbf{n}}_2$ (m/s^2)

Example 6.3. The satellite contains two 1 m diameter spherical tanks centered on the spin axis. One is 80% filled with liquid fuel and the other is 80% filled with oxidizer (ρ and $\mu \approx$ water). Assuming the liquids have the same angular velocity as the satellite, compute each term in (6.9), then compute ∇p using (6.8), and finally deduce the surfaces of constant pressure and the shape of the liquid surface.

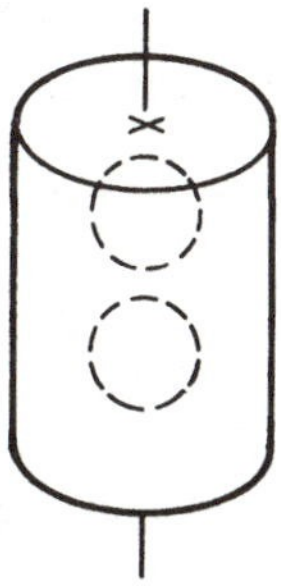

Discussion.

$$\frac{\partial \mathbf{v}}{\partial t} + R_o(\mathbf{v} \cdot \nabla)\,\mathbf{v} + 2\hat{\mathbf{k}} \times \mathbf{v} = -\nabla p + E_k \nabla^2 \mathbf{v} \tag{6.9}$$

Because $\mathbf{v}$ here equals zero for fluid in rigid rotation (Ω = constant), each term with a $\mathbf{v}$ is zero, resulting in $\nabla p = 0$. This is the reduced pressure given by

$$\nabla p = \nabla[p^* + \rho\Phi - \tfrac{1}{2}\rho(\mathbf{\Omega} \times \mathbf{r}) \cdot (\mathbf{\Omega} \times \mathbf{r})] = 0 \tag{6.8}$$

Because $\mathbf{V}_o$ = constant, and because ($g = 0$) in a free-space orbit, $\rho\Phi = 0$.

$$\nabla p^* = \tfrac{1}{2}\rho\nabla[(\mathbf{\Omega} \times \mathbf{r}) \cdot (\mathbf{\Omega} \times \mathbf{r})] = \tfrac{1}{2}\rho\nabla[\Omega r \sin\alpha\,\hat{\mathbf{e}} \cdot \Omega r \sin\alpha\,\hat{\mathbf{e}}]$$

For notational simplicity let $s = r \sin\alpha$, the distance to the spin axis, and then $\nabla p^* = \frac{1}{2}\rho\Omega^2\nabla s^2$. Using cylindrical coordinates (s, θ, z), only $\partial/\partial s$ has a value.

$$\frac{\partial p^*}{\partial s} = \rho\Omega^2 s \quad \text{or} \quad dp^* = \rho\Omega^2 s\, ds$$

Integrate to get

$$p = p_o + \tfrac{1}{2}\rho\Omega^2(s^2 - s_o^2)$$

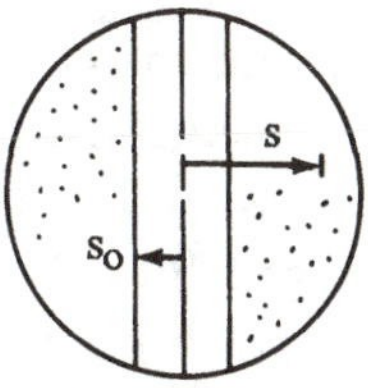

P_o = pressure in vapor

where the asterisk is omitted, and with p now actual pressure in the liquid. Surfaces of constant pressure are cylinders, and the free surface at $p = p_o$ is a cylinder centered about the axis with $s = s_o$.

Note that the answer here is obtained more easily using an n-frame free-body diagram with $\rho\mathbf{\Omega}\times(\mathbf{\Omega}\times\mathbf{r})$ as force. The benefit of often including $\mathbf{\Omega}\times(\mathbf{\Omega}\times\mathbf{r})$ in the pressure term will be more apparent in later sections.

Example 6.4. Estimate the importance of rotation relative to gravity, convection, and viscosity in the following flow situations.

(a). A square auditorium, 150 ft on each side by 20 ft high, has air inlets along one wall and outlets along the opposite wall. Air in the room is completely replaced every 2 min.

(b). A spherical tank partially filled with water (as in Example 6.3) is tested in a laboratory in an attempt to simulate conditions in a spinning satellite. Discuss use of R_o, E_k, and the Froude number, F_r.

Discussion.

(a). The ventilation air traverses the 150 ft wide room in 2 min, so its velocity is 150/120 = 1.25 ft/s. The room, as part of the earth, rotates once every 24 hr, so $\Omega \approx 1$ revolution/24 hr $= 7.27 \times 10^{-5}$ rad/s. Kinematic viscosity of air at sea level and 68°F is approximately $\nu = 1.6 \times 10^{-4}$ ft²/s.

The Rossby number measures the ratio of convective to Coriolis phenomena. It can easily be rederived by comparing the convective term in the Navier-Stokes equation $[\rho(\mathbf{v}\cdot\nabla)\mathbf{v}]$ to the Coriolis term $(2\rho\mathbf{\Omega}\times\mathbf{v})$

$$R_o = \left|\frac{\rho(\mathbf{v}\cdot\nabla)\mathbf{v}}{2\rho\mathbf{\Omega}\times\mathbf{v}}\right| = \frac{\rho\, U\, L^{-1}\, U}{\rho\, \Omega\, U} = \frac{U}{\Omega L}$$

The L is obtained from the x in $(u\, \partial u/\partial x)$, so the proper characteristic length is the width of the room, L = 150 ft, and

$$R_o = \frac{1.25}{7.27\times 10^{-5}\times 150} = 115$$

which indicates convective phenomena due to velocity of the air relative to the room is much more important then the rotation rate of the room.

Note that if it took 12 hr for the air to cross the room, the room would have rotated 180°, and Coriolis phenomena then would not be negligible. Chapters 20 and 21 apply this concept to atmospheric and oceanic flows that move large distances; in those flows Coriolis phenomena dominate the flow patterns completely.

The Ekman number measures the ratio of viscous to Coriolis phenomena. It can easily be rederived by comparing the viscous term in the Navier-Stokes equation

($\mu\, \nabla^2 \mathbf{v}$) to the Coriolis term

$$E_k = \left| \frac{\mu \nabla^2 \mathbf{v}}{2\rho \boldsymbol{\Omega} \times \mathbf{v}} \right| = \frac{\mu\, L^{-2}\, U}{\rho\, \Omega\, U} = \frac{\nu}{\Omega\, L^2}$$

This L is obtained from $\mu \nabla^2 \mathbf{v}$ with terms like $\mu\ \partial^2 u/\partial y^2$, where the y is measured perpendicular to the flow direction. The proper characteristic length now is the height of the room, so L = 20 ft, and

$$E_k = \frac{1.6 \times 10^{-4}}{7.27 \times 10^{-5} \times 20^2} = 5.50 \times 10^{-3}$$

which indicates viscous effects are even less important than Coriolis effects. Note that this analysis does not apply to the ventilation ducts and fans.

The Reynolds number is important for this problem to assess whether flow in the room is laminar or turbulent.

$$R_e = \frac{U\,L}{\nu} = \frac{1.25 \times 20}{1.6 \times 10^{-4}} = 1.56 \times 10^5$$

and the flow is definitely turbulent (pipe flow is usually turbulent for R_e over 2300, with L = diameter). This compromises the validity of the solution for E_k, because turbulent stresses ($\sim \overline{u'v'}$) are more important now than viscous stresses ($\sim \mu$). A turbulent Ekman number can be defined as $E_{kt} = \overline{u'v'} / \Omega L U$, although its use is compromised by the usual difficulty in estimating a value for $(\overline{u'v'})$.

(b). After the water is spun-up to the angular velocity of the tank, it is not moving relative to the rotating coordinate system fixed in the tank, and **v** and **a** are zero. The convective term $[\rho(\mathbf{v} \cdot \nabla)\mathbf{v}]$ is zero, and the Coriolis term ($2\rho\ \boldsymbol{\Omega} \times \mathbf{v}$) is zero. In addition, the viscous term ($\mu \nabla^2 \mathbf{v}$) also is zero. The Rossby number and Ekman number have no meaning in this application. If the magnitude or direction of the angular velocity of the tank is changed, then **v** would exist, and both R_o and E_k would provide useful information. Later chapters discuss these cases in detail.

In the present application a Froude number is relevant, where the Froude number compares the ratio of centripetal acceleration to gravity. In this use of F_r

$$F_r = \left| \frac{\rho\, \boldsymbol{\Omega} \times (\boldsymbol{\Omega} \times \mathbf{r})}{\rho g} \right| = \frac{\Omega^2 L}{g}$$

where, as with all dimensionless ratios, only absolute values are used. The appropriate L here is distance from the spin axis.

The shape of the free surface is determined by F_r. When $\Omega = 0$, the surface is horizontal; as Ω is increased, the free surface is a paraboloid; and when $\Omega^2 L >> g$, the free surface is approximately a cylinder symmetric about the spin axis. This latter case can be achieved exactly when $g = 0$, as in an orbiting space vehicle. In most laboratory (earth-based) experiments, earth gravity is neglected when $F_r \gtrsim 10$. Nonspherical, partly liquid-filled tanks operated at $F_r = 40$ are used to model equivalent fuel tanks in orbiting space vehicles (see Chapter 18).

Chapter 7

Coriolis Phenomena

7.1 Coriolis Forces vs. Accelerations

The centripetal and Coriolis terms quantify the major phenomena that distinguish fluids with large-scale rotations. In a rotating Eulerian coordinate system, the centripetal term, $\mathbf{\Omega} \times (\mathbf{\Omega} \times \mathbf{r})$, is computed using the angular velocity of the rotating coordinate system ($\mathbf{\Omega}$) and fixed locations (r) in the rotating Eulerian grid. It can be represented by a radial pressure distribution imposed on that region, independent of fluid motion relative to the rotating grid, and computed as the gradient of a scalar function. Because of this, the term can be included in the pressure term and disappears from the typical rotating fluid computation. It is recovered when details of the reduced pressure term (6.8) are computed, as in Example 6.3.

The Coriolis term, $2\mathbf{\Omega} \times \mathbf{v}$, requires in its computation the velocity of the fluid (v) relative to the rotating grid. This term never disappears unless $\mathbf{\Omega}$ or v are zero or are parallel; in dimensionless form it remains as $2\hat{\mathbf{k}} \times \mathbf{v}$. Understanding Coriolis phenomena, both in the mathematical sense and in a physically intuitive sense, is a necessary prerequisite to understanding rotating fluid phenomena.

Figure 7.1 illustrates Coriolis phenomena computed for motions along axes of a cylindrical coordinate system (r, θ, z). The coordinate system coincides with a circular wheel having four radial spokes. Its axle is the z axis. The wheel rotates at constant angular velocity, $\mathbf{\Omega} = \Omega\hat{\mathbf{z}}$, relative to inertial space. A rod parallel to the z axis is attached to the rim at one point. Small masses are located on the rod, on one spoke, and on the rim. Each can slide freely as shown.

The mass on the vertical rod can only move in the z direction. Assume it is given a constant velocity v_z. Application of the term for Coriolis acceleration yields

$$2\mathbf{\Omega} \times \mathbf{v} = 2\Omega v_z \hat{\mathbf{z}} \times \hat{\mathbf{z}} = 0$$

The conclusion is, *Motion parallel to the instantaneous axis of rotation does not involve Coriolis phenomena.* Although Figure 7.1 illustrates the case for $\mathbf{\Omega}$ and v

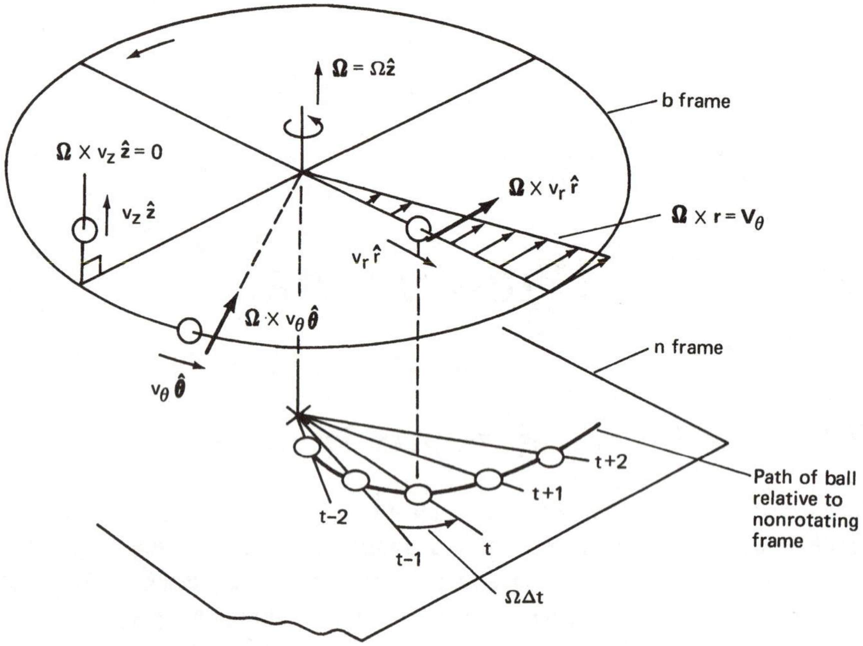

Figure 7.1: Beads sliding on a wire wheel assembly illustrate motions and forces in cylindrical coordinates (r, θ, z). Here the b frame rotates with constant angular velocity about the inertially fixed z axis. Motions v_r and v_θ involve real (relative to inertial space) accelerations and real forces.

constant, the conclusion is the same for $\mathbf{\Omega}$(t) and v(t), since $\mathbf{\Omega} \times \mathbf{v}$ is computed using the instantaneous values of the two terms.

Analyze next the motion of the mass sliding on the radial spoke. Assume, as shown, that the mass is given some constant velocity v_r so that it slides outward on the spoke. Two phenomena occur simultaneously; each has magnitude and direction $\mathbf{\Omega} \times \mathbf{v}$.

1. At all times the mass is constrained to have the same tangential velocity, relative to the nonrotating (inertial) frame, as the rotating spoke. Therefore, as the mass moves outward at rate v_r, its tangential velocity must continually increase. The mass is being convected at rate v_r to a region of increased tangential velocity, equivalent to a term such as $u\partial v/\partial x$. This change of tangential velocity is an acceleration relative to inertial space, and the spoke must be pressing sideways against the mass to cause it. This transverse force of the spoke on the mass is a real force, and is computed as $\mathbf{F} = m\Omega v_r \, \hat{\mathbf{z}} \times \hat{\mathbf{r}} = m\Omega v_r \, \hat{\boldsymbol{\theta}}$. Because each action must have an equal and opposite reaction, the mass presses in the opposite direction against the spoke. If the mass moves inward, the phenomena is reversed

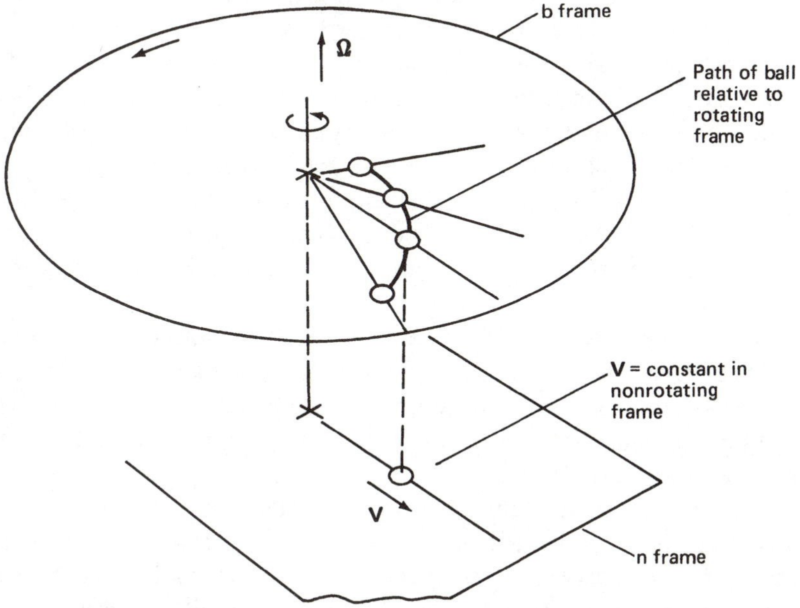

Figure 7.2: *The bead moves with constant velocity relative to an inertial frame n. Acceleration relative to inertial space (and force) is zero. The bead accelerates relative to the rotating frame b, but acceleration relative to a noninertial frame is not part of Newton's law, **F** = mA.*

because the transverse component of velocity is continually being reduced; the mass experiences a transverse deceleration relative to inertial space.

2. The velocity $v_r\hat{\mathbf{r}}$ is a vector, and is constrained by the spoke to remain radially directed as the wheel rotates. Therefore, $\hat{\mathbf{r}}$ is constantly changing direction at the rate $\mathbf{\Omega}$. This also is an acceleration with magnitude and direction $\mathbf{\Omega}\times\mathbf{v}$. The two phenomena combine to give $\mathbf{\Omega}\times\mathbf{v}+\mathbf{\Omega}\times\mathbf{v}=2\mathbf{\Omega}\times\mathbf{v}$.

Note that an observer in the rotating frame only sees the mass moving in a straight line at constant speed. Computed in the rotating frame, the mass has zero acceleration--yet the observer would measure a real transverse force and wonder why. An observer in the nonrotating (inertial) frame sees an accelerated motion of the mass and computes the transverse force simply by using $\mathbf{F} = m\mathbf{A}$. Use of the rotating coordinate system equations of Chapter 6 permits the rotating observer to get the same answer as the inertial observer, provided the observer knows s/he is rotating and adds the necessary Coriolis term.

Consider now the alternative situation shown in Figure 7.2. In this case, the mass is moving along a straight line with constant speed in the inertial frame. Inertial acceleration is zero, and Newton's law states that net force is zero. However, an observer who measures this motion while fixed to a rotating surface will see the mass accelerate (the spiral path shown in Figure 7.2). The observer in the rotating frame might incorrectly assume a force has been applied to the mass. However, the acceleration is caused by measurements (observations) in a noninertial frame, and Newton's law applies only to *accelerations relative to an inertial frame.* If the observer knows s/he is rotating, s/he can again correct for the error by applying the negative of the Coriolis term. The term ${}^{b}d\mathbf{v}/dt$ will be cancelled exactly by $-2\boldsymbol{\Omega}\times\mathbf{v}$. Figure 7.2 shows a radial path only for convenience. The concept is the same for any constant **V** path.

Some authors use the term *Coriolis* to describe the phenomena shown in Figure 7.1, while others use the term specifically for the phenomena of Figure 7.2. Either can be correct depending upon the physical situation being analyzed. Care must be used to distinguish motions observed or computed in an inertial frame from those in a rotating or other noninertial frame.

Consider next the mass sliding along the rim of the wheel in Figure 7.1. If the mass is glued to the rim, it has the same tangential velocity (relative to inertial space) as the rim, and it experiences the same centripetal acceleration as the rim. The mass is motionless to an observer on the wheel, and the discrepancy in its acceleration between the inertially fixed observer and the rotating observer is corrected by the term $\boldsymbol{\Omega}\times(\boldsymbol{\Omega}\times\mathbf{r})$.

Coriolis phenomena arises when the mass begins to slide along the rim at which time $\boldsymbol{\Omega}\times\mathbf{v}=\Omega v_\theta\hat{\mathbf{z}}\times\hat{\boldsymbol{\theta}}=\Omega v_\theta(-\hat{\mathbf{r}})$ as shown in Figure 7.1. The physical nature of this component of the Coriolis acceleration (force) is more difficult to show on a drawing than that of the mass sliding on the spoke; however, its explanation is equally easy. When the mass moves forward as shown, it has a larger tangential velocity relative to inertial space than a mass stationary on the rim. Its centripetal acceleration is therefore greater than the rim--the net difference between centripetal acceleration of the rim and the sliding mass requires an additional inward force. This *increase in centripetal acceleration,* caused by $v_\theta\hat{\boldsymbol{\theta}}$ of the mass, *is computed by the Coriolis acceleration term.* In the case shown, the rim provides the additional inward (centripetal) force, with the mass providing a negative reaction (outward centrifugal force) to the rim. The other half of $2\boldsymbol{\Omega}\times\mathbf{v}$ again arises due to the fact that $v_\theta\hat{\boldsymbol{\theta}}$ changes direction at the rate $\boldsymbol{\Omega}$ as the wheel rotates. This motion involves an inertial acceleration and requires the application of real forces that, for example, could do work by deforming the rim.

Reverse motion of the mass would reduce its centripetal acceleration, reversing the phenomena. Note that if the mass slides on the rim in the reverse direction as that shown, and with the same speed as the inertial tangential velocity as the rim, the mass is then stationary relative to the inertial frame. The inertial observer sees the mass stationary and computes $\mathbf{F}=0$. A rotating observer, familar with rotating coordinate

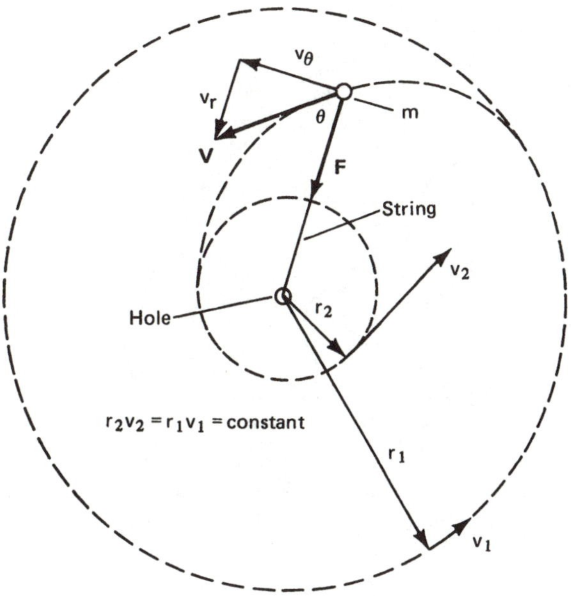

Figure 7.3: *Conservation of angular momentum is due to Coriolis force/acceleration. A string attached to a mass (m), is pulled down through a central hole. Angular momentum of the mass is maintained constant by Coriolis effects, so that v_θ increases for negative v_r. Rotational kinetic energy of the mass is not conserved. It is increased by the amount of work done by pulling the string inward ($E = \int \mathbf{F} \cdot d\mathbf{R}$).*

systems, will also deduce that force between the rim and mass is zero. In each case, reversing the direction of $\mathbf{\Omega}$ or $\mathbf{v}$, or the frame in which the computation is made (inertial or rotating), reverses the direction of $2\mathbf{\Omega} \times \mathbf{v}$. Each case should be analyzed separately; a physical sketch is always helpful.

7.2 Coriolis and Angular Momentum

Figure 7.3 shows a third case similar to the dynamics of many important flow fields. In this example, one end of a string is tied to a mass m, and the other end passes down through a small hole. The mass, while at r_1, is given a tangential velocity v_1 so that it has angular momentum per unit mass, $H/m = r_1 v_1$. Then the string is pulled down through the hole, causing the mass to move inward on the frictionless surface at a velocity v_r. Because the string force can only be radial, applied moment ($\mathbf{M} = \mathbf{r} \times \mathbf{F}$) is zero. Therefore, angular momentum does not change ($\mathbf{M} = \mathbf{H} = 0$), and $\mathbf{H}$ = constant

$$h = \frac{H}{m} = rv_\theta = \text{constant} = r_1 v_1 = r_2 v_2 \tag{7.1}$$

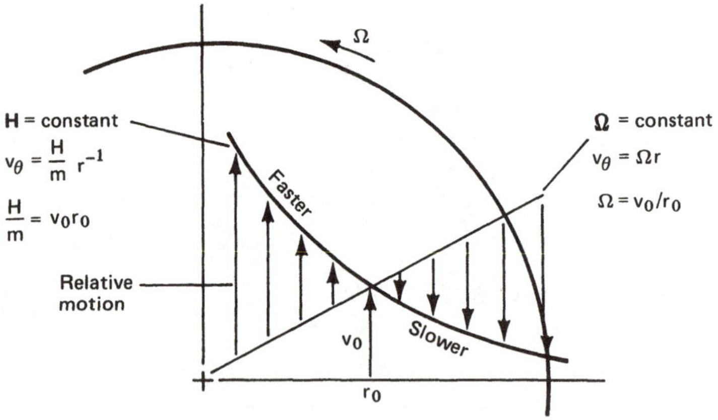

Figure 7.4: *Select as angular velocity ($\Omega = v_o/r_o$), with r_o the present or a nominal position of m (or a fluid element), when analyzing motion with constant angular momentum.*

Kinetic energy per unit mass does change. Consider the spinning mass at a radius (r) with v_r brought to zero

$$\frac{E}{m} = \frac{1}{2} v_\theta^2 = \frac{1}{2} \frac{(rv)^2}{r^2} = \frac{1}{2} \frac{h^2}{r^2} \tag{7.2}$$

With each decrease in r, E increases. The energy change can also be computed as the product of force required to pull the string through the hole times displacement, $\int \mathbf{F} \cdot \mathbf{dr}$. The corresponding fluid application has a radial pressure gradient (replacing the string) that acts on a fluid mass having some initial angular motion.

Coriolis accelerations again are computed as $2\boldsymbol{\Omega} \times \mathbf{v}$, with $\dot{\mathbf{v}}$ the velocity of m relative to a frame rotating at $\boldsymbol{\Omega}$. In Figure 7.1 angular velocity of the spoke remains constant and provides a useful $\boldsymbol{\Omega}$. Angular velocity of the string-mass in Figure 7.3 varies as r changes, complicating the matter. A typical analysis selects a coordinate system rotating at $\Omega = v_o/r_o$, where r_o is the nominal or present position of m, and v_o represents its present tangential velocity. As the mass moves inward from r_o, its velocity increases relative to the rotating coordinate system as shown in Figure 7.4, and as it moves outward, it slows as shown. In Figure 7.3 the Coriolis term $2\boldsymbol{\Omega} \times \mathbf{v}$ defines the tangential acceleration associated with an increase (decrease) in v_θ.

Motion, as in Figure 7.3, is similar to flows called *geostrophic* while that in Figure 7.1 is similar to flows called *ageostrophic* or *not geostrophic*. More specifically, a geostrophic flow will later be defined as a flow where a pressure gradient induces only Coriolis phenomena. Coriolis phenomena applied to real fluid motions

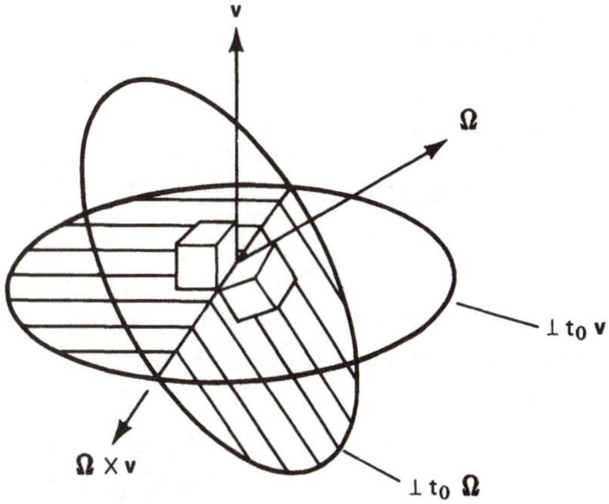

Figure 7.5: *In some 3-D analyses it is best to keep **Ω** and v as vectors. This shows **Ω** × v as a vector perpendicular to both **Ω** and v.*

usually involve a combination of Figures 7.1, 7.2, and/or 7.3. For example, fluid in the vicinity of a rotating surface normally will be attempting to follow a force-free or torque-free path but will be disturbed by viscosity or other mechanisms to follow partly the motion of the rotating surface. This motion is sometimes called *quasi-geostrophic.*

An arbitrary vector can always be analyzed component by component in any convenient coordinate system but need not be. Figure 7.5 shows a three-dimensional (3-D) application of $2\mathbf{\Omega} \times \mathbf{v}$ and illustrates the well-known result that a cross product must be perpendicular to both vectors comprising the cross product. The fact that $\mathbf{\Omega} \times \mathbf{v}$ is perpendicular to $\mathbf{\Omega}$ means $\mathbf{\Omega} \times \mathbf{v}$ can have no component in the $\mathbf{\Omega}$ direction, normally taken as the z direction. The fact that $\mathbf{\Omega} \times \mathbf{v}$ is also perpendicular to $\mathbf{v}$ means $\mathbf{\Omega} \times \mathbf{v}$ has no component in the direction of $\mathbf{v}$ and provides additional insight in specific applications.

7.3 Coriolis Force-Acceleration Criteria

Coriolis *acceleration* is generally significant in a particular application if the Rossby number, the ratio of convective acceleration to Coriolis acceleration, is less than about one. The Rossby number ($R_o = U/\Omega L$) was obtained by placing the Navier-Stokes equation in a rotating coordinate system into dimensionless form. A more physically intuitive explanation is obtained by asking the question, *Under what conditions will a particle, translating relative to a rotating surface, appear to be significantly influenced by the rotation of the surface?* The answer is, *Coriolis acceleration, representing the influence of rotation, is significant when the time required for the particle to complete its translational motion is long enough for the surface to have rotated a significant angle*. In general, the characteristic time of translation is (L/U), and the time of rotation is Ω^{-1}. The criterion states that

$$\frac{L}{U} \gtrsim \Omega^{-1} \quad \text{or} \quad R_o = U/\Omega L \lesssim 1 \tag{7.3}$$

for Coriolis phenomena to be relevant. Air on the rotating earth, accelerated by a pressure gradient but free to move in a transverse direction, is a classic example, as will be shown in Chapter 20.

Coriolis *force*, as a result of an inertial acceleration as in Figure 7.1, is relevant when it is large relative to other forces. The Rossby number compares only U and Ω -- not other forces. As a consequence, it is now not necessarily relevant. A fluid constrained to move radially in a rotating machine, or perhaps the flow of water through a long narrow channel on the rotating earth are examples where Coriolis forces must be compared with other forces, e.g., gravity.

Sometimes a comment is made, without condition, that Coriolis force and acceleration exist only because a rotating coordinate system was chosen for the analysis. The comment is correct when discussing Figure 7.2. Sometimes it is correct for Figure 7.3. It is not correct for Figure 7.1.

To appreciate this further, consider the following. When analyzing the motion of a falling object, if a coordinate system is selected at a skewed angle to the vertical, or especially if the coordinate system is rotating and accelerating relative to the earth, gravitational attraction will appear in complex ways in many terms. Choice of an earth stationary coordinate system with one axis vertical causes gravity to appear in only one term. In the same way, choice of an appropriate rotating coordinate system permits Coriolis force and/or acceleration to appear in only one term. The forces in Figure 7.1 do not disappear simply by choosing a different set of coordinates. The mass sliding outward on the spoke in Figure 7.1 experiences a transverse force due to Coriolis acceleration, and/or a downward force due to gravity, regardless of what coordinate system a person chooses for the analysis. Only when motion is that of Figure 7.2 are centripetal and/or Coriolis forces purely fictional, i.e., exist only because a rotating coordinate system was chosen for the analysis.

There are no applied forces shown in Figure 7.2 so momentum ($\rho \mathbf{V}$) and energy ($\rho V^2/2$), relative to inertial space, are constant. In Figure 7.3, angular momentum ($\mathbf{r} \times \rho \mathbf{V}$) is constant because applied torque is zero, but because there is an applied radial force, translational momentum and energy relative to inertial space change with changes in r. In Figure 7.1, there is an applied radial force and an applied torque; all three, translational momentum, angular momentum, and energy relative to inertial space, change with changes in r.

7.4 Application Examples

This section examines use of the Coriolis acceleration term $2(\mathbf{\Omega} \times \mathbf{v})$ and discusses its use in several specific cases. In some cases, the term represents acceleration relative to inertial space and can be associated with a specific force. In other cases, the acceleration is relative to a rotating (noninertial) space and does not represent a specific force. Criteria are presented for determining whether Coriolis acceleration or force will be significant in a problem involving rotation. Examples 7.1 through 7.4 apply the theory to simple situations involving objects and fluids. An ability to visualize these concepts physically, and to represent them mathematically, is critical to understanding fluid rotation. *Always* take the time to draw one or more 3-D perspective sketches of the physical layout, the motions, and the forces and accelerations.

Example 7.1. A train of mass m at latitude 50°S is travelling due north at 300 km/hr. What forward force, if any, above that necessary to overcome wheel resistance and air drag, is needed to maintain a constant velocity? What is its transverse (east-west) acceleration relative to the earth and relative to inertial space? What, if any, transverse force is exerted on the train wheel flanges by the rails? How does the train's kinetic energy relative to the earth, and relative to inertial space, change, if at all?

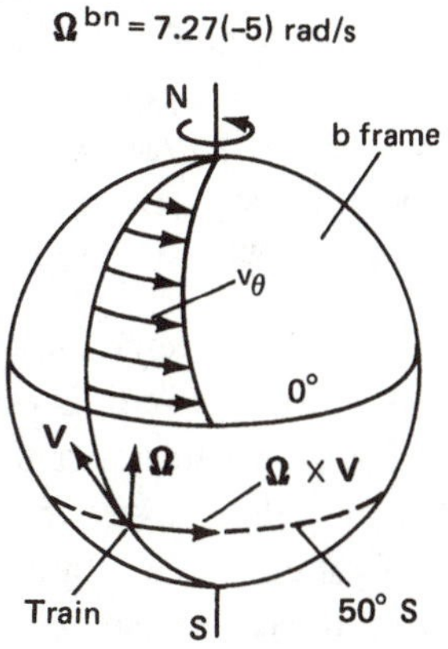

Discussion. The first step in solving a problem involving 3-D motion is to make a 3-D perspective drawing and possibly one or two auxiliary projection drawings. Skill in visualizing and representing 3-D objects and motions is as critical as skill in basic mathematics.

As the train T moves toward the equator, its east-west velocity relative to inertial space increases (accelerates) eastward at the rate $2\mathbf{\Omega} \times \mathbf{v}$, with $\mathbf{\Omega}$ the angular velocity of the earth relative to inertial space. $\Omega = 7.27(-5)$ rad/s.

The earth's oblateness balances gravitational and centrifugal forces. Wheel resistance and drag are the only forces in the north-south (N-S) direction.

Transverse east-west (E-W) acceleration, relative to the earth's surface, is zero if T stays on a N-S track. The magnitude of the transverse acceleration relative to inertial space at the speed and latitude shown is $2|\mathbf{\Omega} \times \mathbf{v}| = 2\Omega v \sin 50° = 0.00928\ \mathrm{m/s^2}$.

The rails must exert an eastward force $(2m\ \mathbf{\Omega} \times \mathbf{v})$ on the wheel flanges to cause the acceleration $2\mathbf{\Omega} \times \mathbf{v}$. The train resists this E-W acceleration with a westward reaction.

Since E = $mv^2/2$, kinetic energy relative to the earth is constant, but relative to inertial space it is increasing because the E-W component of its velocity is increasing. What is the change in E-W velocity between a latitude near a pole and near the equator? (Answer ~ 1000 mph.) Where could a train be at 50°S latitude?

Example 7.2. A 2 m diameter flood drain in Alaska (at 70°N) is half-filled with water flowing due north at 40 km/hr. What is the angle α? Which way is north in the sketch (into or out of the page)?

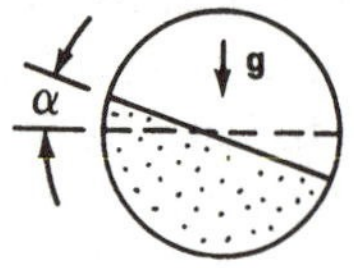

Discussion. A cross section through the earth and a view from above the North Pole are drawn. The earth rotates counterclockwise, viewed from the north, at a rate relative to inertial space of 7.27×10^{-5} rad/s. The distance, drain pipe to earth spin axis, is 6360 sin 20° = 2175 km. Centripetal acceleration is rarely needed for earth rotation phenomena because the earth long ago adjusted its shape to balance **g** with $\mathbf{\Omega} \times (\mathbf{\Omega} \times \mathbf{r})$. See details in Chapters 19, 20, and 21. In any event, the problem specifies the north-south velocity, so only transverse phenomena remain. The flow shown is being moved to regions having smaller transverse velocities $(\mathbf{\Omega} \times \mathbf{r})$, so the pipe must be forcing the fluid westward by retarding its eastward velocity. Action-reaction indicates that the water presses against the eastward side of the pipe. Angle α is the ratio of $2\mathbf{\Omega} \times \mathbf{v}$ to **g**.

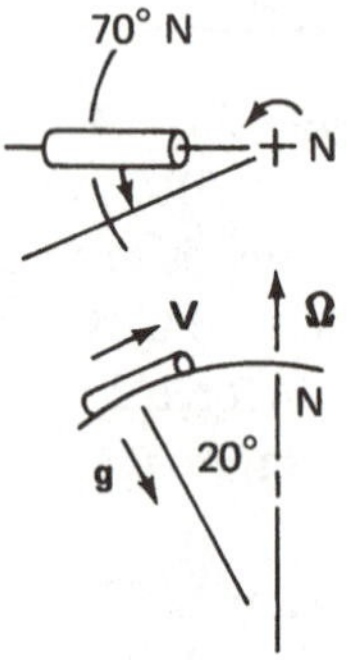

$\alpha = \tan^{-1}(2\Omega v \sin 70°/g) = \tan^{-1}(1.55 \times 10^{-4}) = 32$ arc sec.

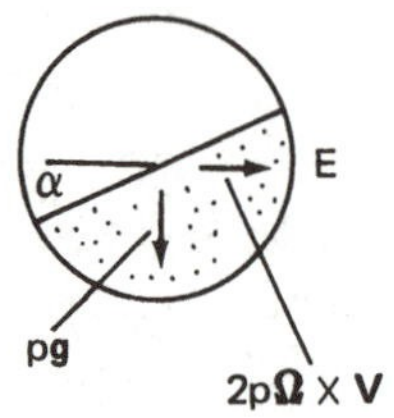

The original sketch shows east to the left, so that is a view looking south (north is out of the page for it). Note that earth radius, pipe diameter, and fill ratio are not used. The tilt angle is too small to measure, but a 1 km wide river, flowing north or south at 70° latitude and 40 km/hr, would show a lateral height difference of 15.5 cm.

Note that this problem is similar to Figure 7.1 of the text. The water is constrained to follow the pipe, just as the small mass was constrained to follow the radial spoke.

Example 7.3. A person stands near the flood drain of Example 7.2 and fires a bullet (500 m/s) at a target 30 m due north. Next, she throws a stone (40 km/hr) at the target. At the same time water slowly flows (1 cm/s) down a field gently sloping north near the pipe. What is the Coriolis deflection in each case? Neglect air and water viscosity, and assume flat trajectories.

Discussion. The distance of 30 m is so small relative to the earth that it can be represented by a flat horizontal surface. Computations such as $\mathbf{\Omega} \times \mathbf{v}$ will make use of the component of earth angular velocity at that latitude (Ω sin 70°) rather than Ω.

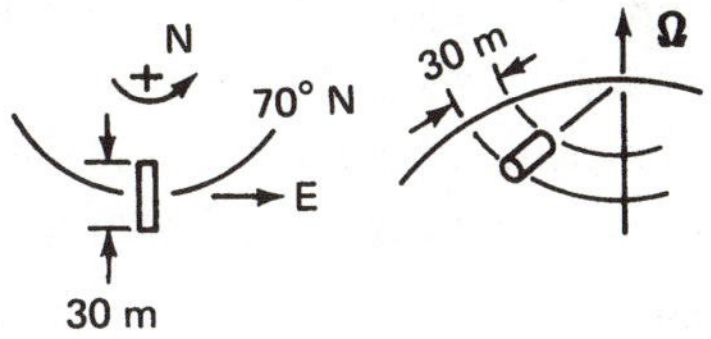

Rossby numbers (U/LΩ)

Bullet = $500/(30 \times 7.27 \times 10^{-5}) = 2.29 \times 10^5$
Stone = $11.11/(30 \times 7.27 \times 10^{-5}) = 5.09 \times 10^3$
Water = $0.01/(30 \times 7.27 \times 10^{-5}) = 4.59$

The Rossby numbers indicate that only the slowly flowing water will be significantly affected by Coriolis phenomena. Deflections at the target are obtained by integrating the Coriolis acceleration over the time of motion (or distance).

Time to target

Bullet = 30/500 = 0.06 s
Stone = 30/11.11 = 2.70 s
Water = 30/0.01 = 3,000 s

Transverse (Coriolis) acceleration = $2\Omega v$ sin 70°

Bullet = 6.83×10^{-2} m/s^2
Stone = 1.52×10^{-3} m/s^2
Water = 1.37×10^{-6} m/s^2

Transverse eastward velocity relative to earth at target, v = at

Bullet = Stone = Water = 4.10×10^{-3} m/s

Eastward deflection (average velocity is 1/2 of the transverse velocity)

Bullet = 0.12 mm
Stone = 0.55 cm
Water = 6.15 m

The bullet travels only 30 m at 1.5 times the speed of sound, but even so has a measurable deflection. Is it a coincidence that all have the same transverse velocity of 4.10×10^{-3} m/s after travelling 30 m? No, because all moved from the same starting point to the same finish line. It is the difference in transverse velocities of the starting point and finish line that was computed in all three cases. Note that this problem is similar to Figure 7.2 of the text.

Example 7.4. Design a satellite that is capable of reorienting itself in free space without ejecting mass or using external torques. Discuss its changes of attitude in terms of energy, momenta, Coriolis forces, and centripetal acceleration. Knowledge of rigid body motion is useful in the study of rotating fluids because

1. It provides a review of rotational dynamics.
2. It helps to understand what fluids might or might not do.
3. It provides the flow boundary conditions in some applications, e.g., communication satellites.

Likins (1973) includes a review of rigid body dynamics appropriate to this problem and its solution.

Discussion. Conservation laws apply to mass, energy, and momentum, but not to mass orientation. The drawing below shows a simplified version of an object that can change its attitude about its z axis using only internal motions. The top and bottom portions spin on a common axis, and an internal mechanism can cause them to rotate relative to each other. Both portions also have rods that can be extended or retracted, thereby changing their moments of inertia about their z axis.

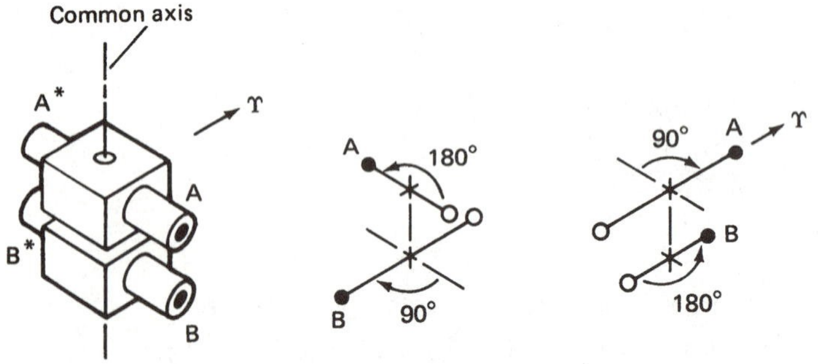

Orientation (attitude) is initially as shown in the drawing, and the system is not rotating. Rotational kinetic energy (E) is zero, and moment of momentum (**H**) is zero. Since relative rotation is only about the z axis, $H_z = I_A\omega_A + I_B\omega_B = 0$. Any relative rotation between the top and bottom portions, caused by the internal mechanism, will result in

$$\omega_A = -(I_B / I_A)\,\omega_B$$

If moment of inertia with the rods extended is twice that with the rods retracted, then with A-A* in and B-B* out, $\omega_A = -2\omega_B$. In a given interval of time, A-A* will rotate twice as far as B-B* and in the opposite direction, as shown in Step 1. Each part has equal and opposite angular momentum during the rotations shown in Step 1, but H_z for the total system remains zero at all times. In Step 2, rods A-A* are out and B-B* are in, and the relative rotation shown results in rods A and B now pointing 90° counterclockwise relative to inertial space from their original position.

No external torques are employed, so $d\mathbf{H}/dt = \mathbf{M} = 0$, and no energy is dissipated in a perfect system, e.g., comparable to a reversible thermodynamic process. There are no Coriolis forces because **v** of the rods, in or out, occur when ω_A and ω_B are zero. There is centripetal acceleration during the rotations, but it changes neither E nor **H**.

Three such devices, each aligned along one of three orthogonal axes (say x, y, z) would permit reorientation of an object to completely arbitrary directions in 3-D space, although a more complex and facile mechanism, such as a falling cat or an astronaut in free space, can achieve the same result by movement of legs, arms, and waist.

A less complicated device also permits 3-D reorientations, but requires some initial moment of momentum **H**. This value of **H** remains constant, while rotational kinetic energy is changed by an internal energy source and an energy sink.

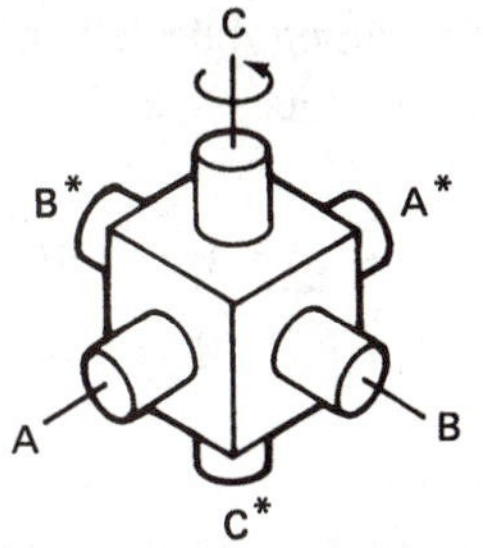

Its operation is based on several facts.

1. The inertia property of any rigid object is equivalent to some solid, constant density, ellipsoid (its inertia tensor) with mutually orthogonal maximum, minimum, and intermediate moments of inertia.
2. For a rigid object, spin about the maximum or minimum moments of inertia is stable, but spin about the intermediate axis is violently unstable, as can be illustrated by attempting to spin a book (held closed with a rubber band) about its principal axes.
3. For a given moment of momentum (**H**), the object will have more rotational kinetic energy (E) for spin about the minimum axis of inertia than for spin about any other axis.

Spin about the maximum axis of inertia requires the least energy for a given **H**. In the simple proof that follows, if I_1 is minimum and I_2 is maximum, then energy about axis 2 (i.e., E_2) is minimum and E_1 is maximum

$$\frac{E_1}{E_2} = \frac{\frac{1}{2} I_1 \omega_1^2}{\frac{1}{2} I_2 \omega_2^2} = \frac{I_1 \omega_1^2}{I_2 \omega_2^2} \frac{(I_1 I_2)}{(I_1 I_2)} = \frac{(I_1 \omega_1)^2 I_2}{(I_2 \omega_2)^2 I_1} = \frac{H_1^2 I_2}{H_2^2 I_1} = \frac{I_2}{I_1}$$

As noted previously, an energy source is necessary for retracting the rods against centrifugal forces, when necessary. An energy dissipator, driven by "wobbling" motion of the device, is also needed. The energy dissipator can be a tank of viscous liquid (e.g., fuel).

Assume rods C-C* are in, and A-A* and B-B* are out, and spin about the z axis (the C-C* rods) defines **H**. Adjust the rods so the C-C* axis is the intermediate axis of inertia. The device will then begin to gyrate wildly. Next, adjust the rods so that the maximum moment of inertia coincides with the desired spin axis, say B-B*. This requires A-A* and C-C* out, and B-B* in. The internal energy sink driven by the wobbling motion will cause the device to spin about I_{max}, i.e., the minimum energy orientation. The magnitude and direction of **H** relative to inertial space does not change; it is the orientation of the device, and its angular velocity, that changes.

Energy is added when rods are pulled in. Coriolis forces, caused by retracting or extending rods during rotation ($2m\,\mathbf{\Omega} \times \mathbf{v}$), transfer momentum from the rods to the central mass and vice-versa—but all transfers are internal.

$$I_{total} = I_c + 2mr_1^2$$

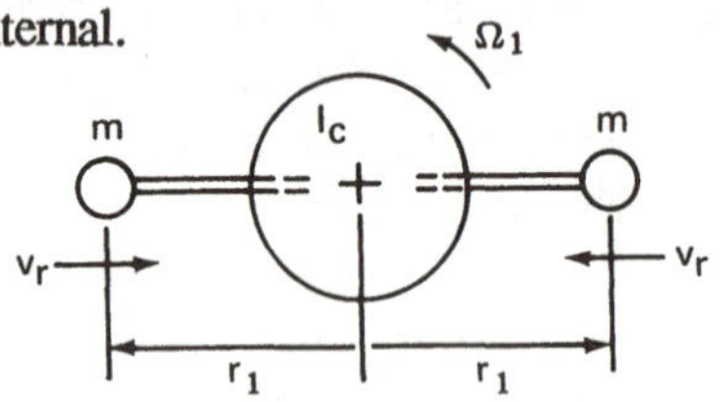

During movement from r_1 to a smaller radius r_2, masses experience Coriolis forces ($2m\mathbf{\Omega} \times \mathbf{v}_r$). This force, times r, applies a torque to I_c and accelerates it, while I_c retards the tangential velocity of the m below the value $v_\theta = \kappa/r$ they would have had, had they not been constrained by I_c. Total **H** for the system is conserved during the motion. Rotational kinetic energy is increased by the energy required to retract the masses against centrifugal force, but the energy is dissipated again during the wobbling motion as spin converges to the new axis of maximum moment of inertia. Note that by proper manipulation of the rods, any axis of the assembly can be caused to be the maximum axis of inertia.

Chapter 18 applies this to a problem sometimes encountered with wobbling space vehicles, and Chapter 19 applies it to the earth. In those chapters, such rigid-body motion provides the necessary boundary conditions for solution of the enclosed fluid problem. Coriolis and centripetal phenomena obviously can become very complicated for arbitrary 3-D rotations and changes in rotation. Tracking details of the transition during the wobbling motion requires use of tensor analysis; the analysis given here uses only the end limits that can be described using scalars.

Chapter 8

Rotation, Vorticity, and Circulation

8.1 Rotation

Figures 8.1a, 8.1b, and 8.1c show three possible motions for a fluid element (only the x-y view is shown). Pure rotation is shown in (a), (b) shows pure distortion, and (c) shows a combination of rotation and distortion. In these figures, quantities such as $\partial v/\partial x$ applied to a line (such as line A) are simply alternative expressions for the angular velocity of the line. Expanding and rearranging $\partial v/\partial x$, for example, gives

$$\frac{\partial v}{\partial x} = \frac{\partial}{\partial x}\left(\frac{\partial y}{\partial t}\right) \approx \frac{\partial}{\partial t}\left(\frac{\Delta y}{\Delta x}\right) \approx \frac{\partial}{\partial t}(\theta) = \omega_z \tag{8.1}$$

Another way to visualize this is to have line A coincide with the x axis, and then have it rotate counterclockwise at a rate ω_z. For an arbitrary point (x) on line A, $v = \omega_z x$, so that $dv/dx = \omega_z$.

When analyzing angular velocity (ω_z) of a rigid body, rotation of a single line fixed to the body defines rotation of the body. Although it would be redundant, one could take ω_z of any two lines (say A and B), perpendicular or not, and say

$$\omega_z = \frac{1}{2}\left(\omega_z^A + \omega_z^B\right) \tag{8.2}$$

since $\omega_z = \omega_z^A = \omega_z^B$. When analyzing the angular velocity of a nonrigid body, an equation like (8.2), with two perpendicular lines (A and B), is essential because the body may be undergoing distortion (strain) as it rotates. Angular velocity about the z axis of a rigid body, distorting body, or fluid mass can be defined as

$$\omega_z = \frac{1}{2}\left(\omega_z^A + \omega_z^B\right) = \frac{1}{2}\left(\frac{\partial v}{\partial x} - \frac{\partial u}{\partial y}\right) \tag{8.3}$$

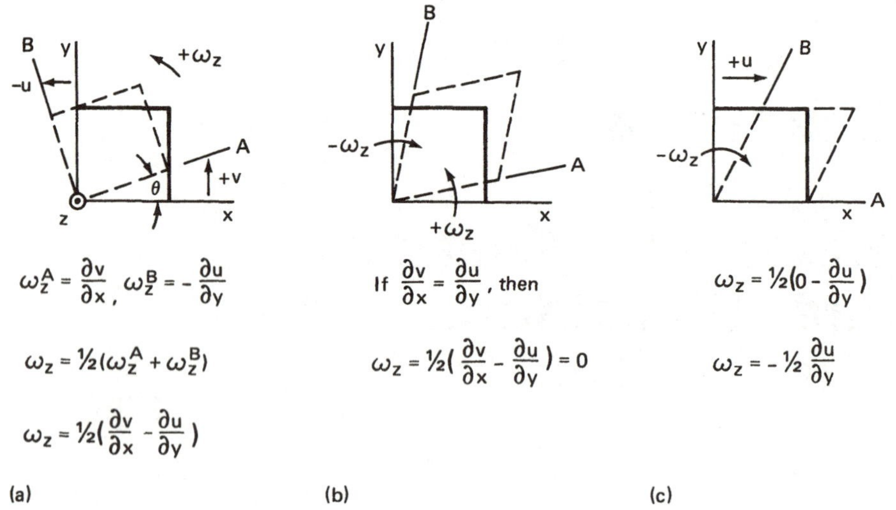

Figure 8.1: *(a) Pure rotation of a fluid element without distortion, (b) pure distortion of a fluid element without net rotation, and c) a deformation that involves both rotation and distortion.*

as shown in Figures 8.1a, 8.1b, and 8.1c. In three-dimensions (3-D), the components of $\boldsymbol{\omega}$ are

$$\omega_x = \frac{1}{2}\left(\frac{\partial w}{\partial y} - \frac{\partial v}{\partial z}\right),\ \omega_y = \frac{1}{2}\left(\frac{\partial u}{\partial z} - \frac{\partial w}{\partial x}\right),\ \omega_z = \frac{1}{2}\left(\frac{\partial v}{\partial x} - \frac{\partial u}{\partial y}\right) \tag{8.4}$$

Chapter 2 develops the same expression, see (2.20), by separating a velocity gradient tensor into its symmetric distortion part (2.21) and its skew-symmetric rotation part (2.22). The factor of 1/2 is seen to be essential in that derivation.

8.2 Vorticity

Vorticity, except for a factor of 1/2, is identical to the angular velocity of fluid elements. It is computed as the curl of the velocity field and is a measure of how much *curl* or rotation the flow has at each point. The mathematical expression for the curl of the velocity field, called *vorticity*, is

$$\boldsymbol{\zeta} = \nabla \times \mathbf{V} \tag{8.5}$$

For Cartesian (rectangular) coordinates, this can be written in matrix form as

$$\begin{bmatrix} \zeta_x \\ \zeta_y \\ \zeta_z \end{bmatrix} = \begin{bmatrix} 0 & -\partial/\partial z & \partial/\partial y \\ \partial/\partial z & 0 & -\partial/\partial x \\ -\partial/\partial y & \partial/\partial x & 0 \end{bmatrix} \begin{bmatrix} u \\ v \\ w \end{bmatrix} \tag{8.6}$$

which, expanded to scalar form, becomes

$$\zeta_x = \frac{\partial w}{\partial y} - \frac{\partial v}{\partial z}, \quad \zeta_y = \frac{\partial u}{\partial z} - \frac{\partial w}{\partial x}, \quad \zeta_z = \frac{\partial v}{\partial x} - \frac{\partial u}{\partial y} \tag{8.7}$$

Comparison of (8.7) for $\boldsymbol{\zeta}$, with (8.4) for angular velocity, shows

$$\boldsymbol{\zeta} = 2\boldsymbol{\omega} \tag{8.8}$$

Many of the applications in this text are for flows best expressed in polar, cylindrical, or spherical coordinates. Curl **V** in cylindrical coordinates (r, θ, z) is

$$\zeta_r = \left(\frac{1}{r}\frac{\partial v_z}{\partial \theta} - \frac{\partial v_\theta}{\partial z}\right) \tag{8.9a}$$

$$\zeta_\theta = \left(\frac{\partial v_r}{\partial z} - \frac{\partial v_z}{\partial r}\right) \tag{8.9b}$$

$$\zeta_z = \frac{1}{r}\left[\frac{\partial}{\partial r}(r v_\theta) - \frac{\partial v_r}{\partial \theta}\right] \tag{8.9c}$$

Organized vorticity, as opposed to disorganized vorticity (turbulence), can be observed by placing small objects in a flow field. If the objects rotate, the flow has vorticity; if they do not rotate, there is no (organized) vorticity. Figures 8.2 through 8.4 illustrate several 2-D flow fields. Two are steady-state viscous flows between parallel walls, and the third is an inviscid vortex. Small corks floating on the surfaces are used as indicators. They rotate in the parallel flows, but not in the inviscid vortex flow. In the parallel flows the corks translate in straight lines as they rotate; in the vortex flow the corks translate in circular paths but do not rotate. Figures 1.1 and 1.2 illustrate the same flow fields. Real fluids modify these results slightly.

In 2-D flow analyses, rectangular (x, y) or polar (r, θ) coordinates, only the z component of $\boldsymbol{\zeta}$ exists. Application of (8.7) to the velocity field of Couette flow shown in Figure 8.2a gives

$$\mathbf{V} = u\hat{\mathbf{i}} + v\hat{\mathbf{j}} = \frac{Uy}{2b}\hat{\mathbf{i}} + 0\hat{\mathbf{j}}$$

$$\zeta_z = \frac{\partial v}{\partial x} - \frac{\partial u}{\partial y} = 0 - \frac{U}{2b} = -\frac{U}{2b}$$

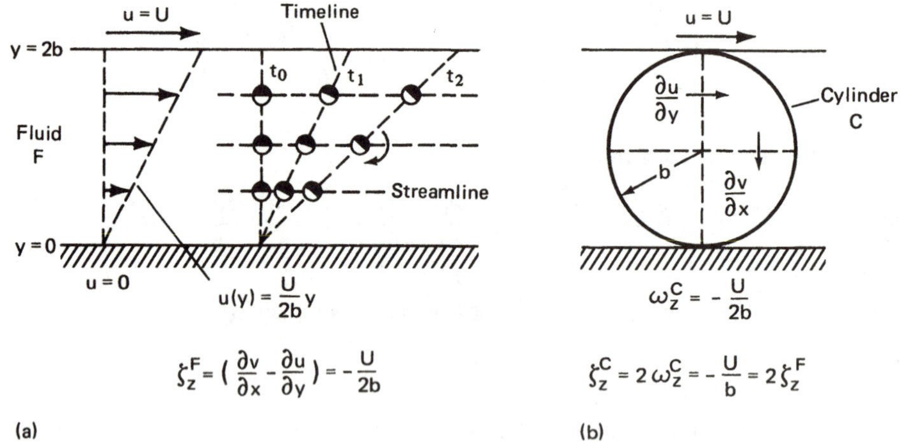

Figure 8.2: *(a) Couette flow driven by motion of the upper wall. Vorticity is constant everywhere in the flow, and small objects suspended in, or on, the fluid all rotate at the same rate. (b) A cylinder rotating between the two walls with the same $\partial u/\partial y$ as the Couette flow. Note that ($\partial v/\partial x$ = 0) for Couette flow, but not for the cylinder.*

This ζ_z is a negative (clockwise) rotation of the fluid elements. The corks spin because the flow velocity at their tops is higher than that at their bottoms.

It is instructive to compare the vorticity of the flow field with the angular velocity and vorticity of a cork (Figure 8.2b) with the same flow differential, top to bottom. Assume a cork at midstream, and to eliminate some arithmetic, let its diameter be equal to the plate spacing (diameter = 2b). Its *angular velocity* (ω_z^c), computed as the angular rate of change of *one line*, or the *average of any two lines*, $(\omega^A + \omega^B)/2$, is

$$\omega_z^c = -U/2b$$

Vorticity is computed as the algebraic *sum* of the angular rates of *two perpendicular lines.* Vorticity of the cork (not the fluid) is

$$\zeta_z^c = \frac{\partial v}{\partial x} - \frac{\partial u}{\partial y} = -\frac{U}{2b} - \frac{U}{2b} = -\frac{U}{b}$$

yielding the correct relationship $\zeta_z^c = 2\omega_z^c$.

Couette flow has vorticity (−U/2b), but the cork with the same flow differential has vorticity (−U/b). The difference is that the horizontal axis of the rigid cork

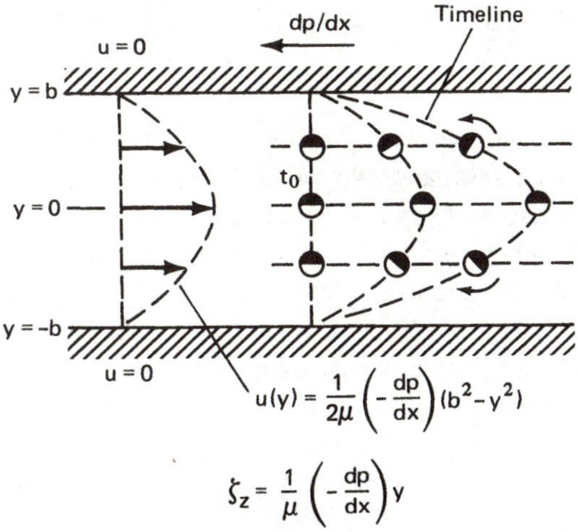

Figure 8.3: *Flow between fixed parallel walls, driven by dp/dx. Vorticity, indicated by rotation of the small objects, is related to viscous shear rate and varies from zero at the centerline to a maximum at each wall. See also Figure 1.1.*

rotates, but the horizontal axis of the fluid element does not. The cork only rotates; the fluid element rotates and distorts at a rate

$$\dot{\varepsilon}_{xy} = \frac{1}{2}\left(\frac{\partial v}{\partial x} + \frac{\partial u}{\partial y}\right) \tag{8.10}$$

It is possible to rotate a fluid volume as if it were a rigid body by filling a container, and then spinning the container until all the fluid is rotating at the same angular rate as the container. This fluid motion is called *rigid rotation* flow. The flow has both $\partial v/\partial x$ and $\partial u/\partial y$, and its vorticity, just like the cork of Figure 8.2b, is twice that of Couette flow. Because any two perpendicular axes rotate at the same rate, there is no distortion, as can be verified using (8.10)

$$\dot{\varepsilon}_{xy} = \frac{1}{2}\left(\frac{U}{2b} - \frac{U}{2b}\right) = 0$$

Figure 8.3 depicts a $\rho = c$, $\mu = c$ fluid in steady-state laminar flow between parallel walls. The x axis, assumed horizontal, is selected to be in the flow direction with the z axis projecting out of the page and the y axis perpendicular to the walls. Flow in the y and z directions are assumed zero. Because of the assumptions, only the x component of the Navier-Stokes equations and the continuity equation need to be used, and most terms in them are zero. For simplicity, assume also that the fluid extends to infinity in both the $-z$ and $+z$ directions.

$$\frac{\partial u}{\partial t}+u\frac{\partial u}{\partial x}+v\frac{\partial u}{\partial y}+w\frac{\partial u}{\partial z}=-\frac{1}{\rho}\frac{\partial p}{\partial x}+\nu\left(\frac{\partial^2 u}{\partial x^2}+\frac{\partial^2 u}{\partial y^2}+\frac{\partial^2 u}{\partial z^2}\right)$$

Terms equal to zero are first removed. With $v = w = 0$, two of the convective terms ($v\ \partial u/\partial y$ and $w\ \partial u/\partial z$) are zero, and two terms ($\partial v/\partial y$ and $\partial w/\partial z$) in the continuity equation are zero. Therefore, $\partial u/\partial x$ and, of course, $\partial^2 u/\partial x^2$ must be zero also. The steady-state condition infers $\partial u/\partial t = 0$, and $\partial^2 u/\partial z^2$ is zero because the flow is assumed uniform in the z direction. The driving force that provides the energy dissipated by viscosity is supplied by a known pressure gradient. Because the driving force is only a function of x, and u is only a function of y, the partials are no longer needed. The final equation to be solved is

$$\frac{dp}{dx}=\mu\frac{d^2 u}{dy^2} \tag{8.11}$$

Solution of (8.11), subject to the no-slip condition at the walls of $u = 0$ at $y = \pm b$, is

$$u(y)=\frac{1}{2\mu}\left(-\frac{dp}{dx}\right)\left(b^2-y^2\right) \tag{8.12}$$

with flow (u) from high pressure to low pressure. The pressure gradient $(-dp/dx)$ is assumed known.

Vorticity is

$$\zeta_z=\frac{1}{\mu}\left(-\frac{dp}{dx}\right)y \tag{8.13}$$

For y positive (negative), vorticity is positive (negative). At $y = 0$, vorticity is zero as illustrated in Figure 8.3.

The value of μ does not appear in the solution for Couette flow (Figure 8.2a). Motion is maintained by forces applied to the walls. This force (per unit area) can be deduced from the solution for u(y) as $\tau_w = \mu\ (du/dy)_o = \mu(U/2b)$. Shear stress on the walls in channel flow is computed in the same way.

The inviscid vortex of Figure 8.4 can be used as an approximate model for a hurricane. Although air is not inviscid, the value of $\tau \sim \mu\ \partial v_\theta/\partial r$ is negligible over much of the flow because both μ and $\partial v_\theta/\partial r$ are small. The flow field of an inviscid vortex is $rv_\theta = \kappa$, with κ a constant called *vortex strength*. Vorticity is

$$\zeta_z=\frac{1}{r}\left[\frac{\partial}{\partial r}(rv_\theta)-\frac{\partial v_r}{\partial \theta}\right]=\frac{1}{r}\left[\frac{\partial}{\partial r}(\kappa)-0\right]=0 \tag{8.14}$$

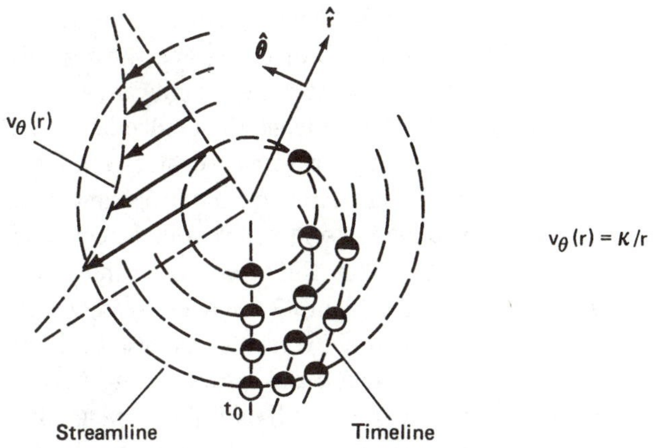

Figure 8.4: *An inviscid vortex (μ = 0) dissipates no energy and requires no driving force. The flow is irrotational (ζ = 0); fluid elements and infinitesimally small objects do not rotate in this flow. Finite-sized objects in a nearly inviscid vortex will rotate slightly. (See also Figure 1.2.)*

Because $\mu = 0$, the inviscid vortex is conservative; no energy is dissipated, and no driving force exists for the steady-state flow.

Also, because the flow is irrotational ($\zeta_z = 0$), a cork immersed in the inviscid vortex will not rotate. Compare this with a cork immersed in a fluid rotating as a rigid body where the cork rotates at the same rate as the fluid. In the case of the inviscid vortex, the relative velocity of the flow field across the diameter of the cork is the differential needed to cancel the rotational effect caused by a displacement $\Delta\theta$. This is made more explicit by reexamining (8.14)

$$\zeta_z = \frac{1}{r}\frac{\partial}{\partial r}(rv_\theta) = \frac{1}{r}\left(\frac{\partial r}{\partial r}v_\theta\right) + \frac{1}{r}\left(\frac{\partial v_\theta}{\partial r}r\right) = \frac{v_\theta}{r} + \frac{\partial v_\theta}{\partial r} \tag{8.15}$$

The term (v_θ/r) measures the rotational effect of a change $(\Delta\theta)$, and $(\partial v_\theta/\partial r)$ is the relative velocity of the flow field. Together they equal ζ_z, which is zero when $\kappa = rv_\theta$ = constant, as seen in (8.14). The relative velocity is associated with distortion rate of the fluid, but as long as $\mu = 0$, the distortion rate causes no stress or fluid interaction--even if $r \to 0$, and $\partial v_\theta/\partial r \to \infty$.

Note in Figure 8.4 that a cork placed with its center at the origin of the vortex will rotate if it shares the velocity of the fluid touching its periphery, and the smaller the cork, the faster it will rotate. As the radius of the cork approaches zero, its angular velocity approaches infinity, but subject to the limitation

$$rv_\theta = r^2\omega = \kappa$$

The matter is resolved mathematically by defining a *line vortex* of strength κ at the vortex center. In a real fluid, with μ small, there is some value of $\partial v_\theta/\partial r$ near $r = 0$ where the product $\mu\, \partial v_\theta/\partial r$ is large enough to induce significant viscous effects. In a hurricane this is called the *eye*. Viscous effects, the production of turbulence, and inflow of nonrotating air, cause the eye to rotate approximately as if it were a rigid body. In this case, vorticity is not concentrated in a line but is distributed throughout the finite volume of the eye. Compare the inviscid vortex with the equivalent model in Figure 7.3 that uses a mass m rather than fluid mass of a unit volume. The inviscid vortex uses constant κ, while the Figure 7.3 model uses constant angular momentum H, but $\kappa = H/\rho = rv_\theta$ for the inviscid vortex, so the result is identical.

Chapter 3 discusses the velocity potential ϕ and shows that it exists only when fluid element rotation is zero. This section identifies fluid element rotation with vorticity ($\zeta = \nabla \times \mathbf{V} = 2\omega$). A more direct relationship between ϕ and absence of ζ is obtained by noting, if $\mathbf{V} = -\nabla\phi$, then $\zeta = \nabla \times \mathbf{V} = \nabla \times (-\nabla\phi) = -\nabla \times \nabla\phi$, which must equal zero because of the basic identity that states, *The curl of the gradient of any scalar field always equals zero.*

Often fluid motions are best understood by analyzing the production, transport, and dissipation of fluid vorticity. Succeeding sections will explore theorems that define vorticity and its attributes more precisely, including the use of vorticity (ζ) rather than pressure (p) as the variable of interest in a modified form of the Navier-Stokes equations. When ($\nabla\times$) is applied to the Navier-Stokes vector equations, $\nabla \times \mathbf{V}$ yields ζ, but $\nabla \times \nabla p$, as the curl of a gradient, yields zero.

8.3 Circulation and Stokes Theorem

Circulation is defined as the integral of the scalar (dot) product of vector velocity times vector displacement around a closed curve at some instant, as shown in Figure 8.5. Note that it is not the same as the integral of the velocity of a point (e.g., a race car) times the distance it travels in traversing the same closed curve during some finite time. Circulation (Γ) is defined mathematically as

$$\Gamma = \int_c \mathbf{V} \cdot d\boldsymbol{\ell} \tag{8.16}$$

with units of L^2/T.

Circulation has two major applications in the study of fluids. One is its application to aerodynamics. Chapter 16 discusses the concept of lift (L), a force transverse to the flow direction, and derives the surprisingly simple result that ($L = -\rho U\Gamma$). The more basic application is the direct relationship of circulation to vorticity. A theorem by Stokes states, *Vorticity integrated over a closed surface is equal to circulation around its periphery*, or

$$\int_S \zeta \cdot dS = \int_C \mathbf{V} \cdot d\boldsymbol{\ell} = \Gamma \tag{8.17}$$

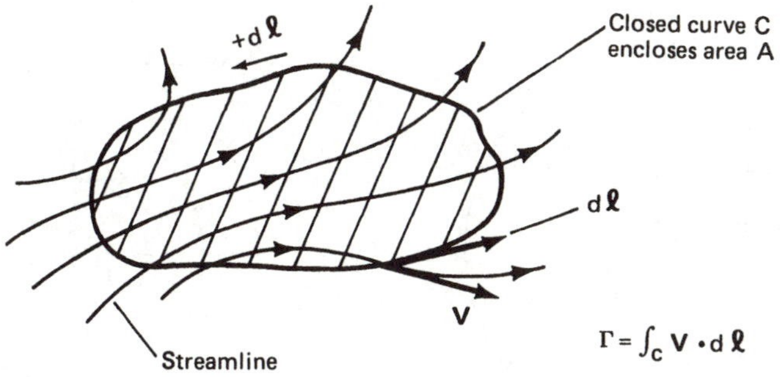

Figure 8.5: *Circulation around the closed curve C is computed as the integral of* $\mathbf{V} \cdot d\boldsymbol{\ell}$ *counterclockwise around the curve at a specific instant of time.*

There are restrictions and conditions on C, S, and **V** that modify application of Stokes's theorem to both real and theoretical flow fields. These are detailed in the following paragraphs and sections.

Figure 8.6 illustrates Stokes's theorem and its applicability to an arbitrary (but piecewise smooth) 3-D surface (S), and its enclosing curve (C), in the region of some continuous flow field **V**(x, y, z, t) having continuous first partial derivatives in that region. The relationship between vorticity and circulation can be visualized by subdividing the surface into small area elements as shown. The net effect of summing vorticity over the surface is to cancel out flows at element interfaces, leaving only those flows bounded by the curve C, whose sum is then the circulation about the surface, exactly as specified by Stokes's theorem.

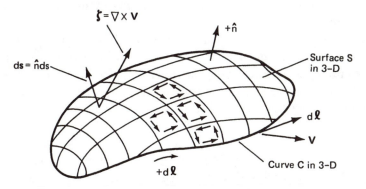

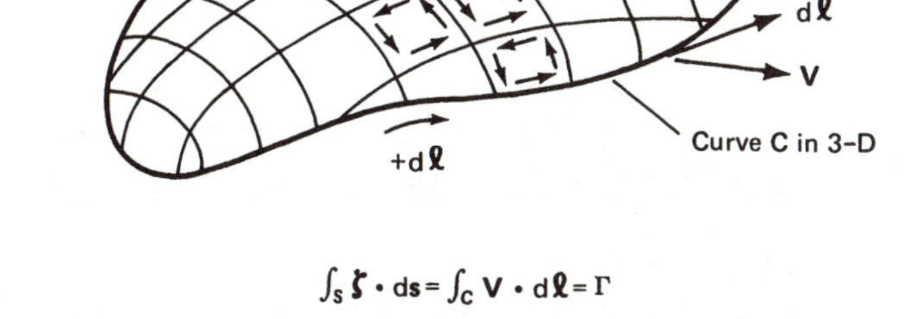

$$\int_S \boldsymbol{\zeta} \cdot d\mathbf{s} = \int_C \mathbf{V} \cdot d\boldsymbol{\ell} = \Gamma$$

Figure 8.6: *The 3-D generalization for the computation of circulation, and Stokes's theorem, which relates vorticity over a 3-D surface to circulation around an enclosing 3-D curve.*

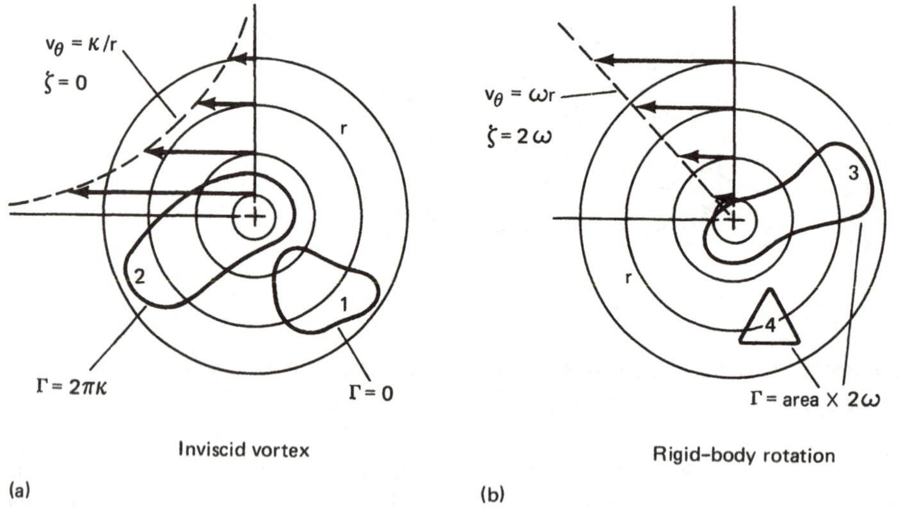

Figure 8.7: *Stokes's theorem applied to an inviscid vortex (a) and to a rigid rotation flow (b) illustrates the advantages of sometimes using Stokes's theorem to calculate circulation.*

Figure 8.7 applies Stokes's theorem to an inviscid vortex and to a fluid rotating as if it were a rigid body. Both are in 2-D, so for these examples the areas are flat, and polar coordinates are most appropriate for phenomena descriptions. In some applications it is necessary to compute vorticity enclosed within a region, and in other applications, circulation about the region. Stokes's theorem sometimes provides a trivial solution to what otherwise could be an extremely difficult mathematical exercise.

Suppose, for example, circulation about region 1 of Figure 8.7 is needed. Stokes's theorem notes that the inviscid vortex has $\boldsymbol{\zeta} = 0$ everywhere, except at its center; therefore, any curve not enclosing the center has $\int_S \boldsymbol{\zeta} \cdot d\mathbf{S} = \Gamma = 0$. If, like for region 2, it encloses the center, then it encloses the line vortex at the center with magnitude (κ), resulting in

$$\Gamma = 2\pi\kappa \tag{8.18}$$

To compute $\int_C \mathbf{V} \cdot d\boldsymbol{\ell}$ would require a precise definition of the curve C and an evaluation of $\mathbf{V}$ at each point on C, a tedious and often difficult task as illustrated in the examples.

In the case of rigid body rotation, $\zeta = 2\omega$ is uniform, and the area enclosed by a given curve times 2ω is equal to Γ. The center is not a singular point, as it is in the inviscid vortex, so whether the curve encloses the center, as for region 3, or not, as for region 4, is immaterial. Note that for the triangle, region 4, the calculation of Γ is trivial because the location or orientation of the triangle in the flow field does not matter. While $\int_C \mathbf{V} \cdot d\boldsymbol{\ell}$ would be difficult to compute, Stokes's theorem states that circulation, for this flow and this shape, must always be equal to the integral of $\boldsymbol{\zeta} \cdot dS$ over that area

$$\Gamma = 2\omega \times \frac{1}{2} ab = ab\omega$$

As applications where Stokes's theorem cannot be used, consider first a column of fluid, in rigid rotation Ω out to radius r_o, immersed in a larger body of fluid at rest. Curve C lies entirely in the fluid at rest but encloses the rotating column. The integral of enclosed vorticity is $(2\Omega\pi r^2)$ but circulation computed around C is zero. The difficulty of course is that **V** as defined has a discontinuity at r_o. Consider next a region where **V** of the rotating fluid transitions smoothly into the fluid at rest. The flow can be represented as $v_\theta = \Omega r$ out to r_o, then some function that varies smoothly from $v_\theta = \Omega r_o$ at r_o to $v_\theta = 0$ at $(r_o + \Delta r)$, with $v_\theta = 0$ at larger r. The discrepancy still remains, because *continuous* here does not mean *smoothly varying* but rather *the same function over the entire region of interest.* In this last example of a flow that transitions smoothly, there are three separate functions, each with its own region of validity, not one function valid for the entire region. However it is possible to construct a flow field, consisting of more than one function, for which Stokes's theorem is valid, e.g., Rankine's combined vortex described in Chapter 10.

Applications with several different velocity fields within a region of interest are common in many real flows. For example, a jet engine's exhaust has swirl (rotates about the flow axis) and an axial component of vorticity, yet measurements around a curve enclosing the jet's exhaust, but at a larger radii, might detect nothing. Tornadoes are observed sometimes to drop down as a funnel from an overhead cloud. They represent large accumulations of vorticity, yet the velocity field fairly close to the funnel can be nearly zero. Use of Stokes's theorem is inappropriate in both cases.

Even more common are time-dependent generations of vorticity by viscous shear or other mechanisms. The generation of vorticity may occur quickly within the region of the operating mechanism, and a measurement of circulation somewhat outside that region at the same instant may detect a different velocity field. Chapter 10 introduces additional theorems on the production, diffusion, and dissipation of vorticity and its relationship to circulation.

8.4 Application Examples

Vorticity is introduced and computed for several flows in both Cartesian coordinates (x, y, z) and cylindrical coordinates (r, θ, z). Motion in a circular path is carefully distinguished from fluid element rotation and vorticity. Example 8.1 examines a mechanical model that includes small gears that move in circular paths, while not rotating, similar to fluid elements in an inviscid vortex. Examples 8.2 and 8.3 compute and illustrate vorticity in more complex 3-D flows. Section 8.3 presents

Stokes's theorem equating total vorticity over a surface to circulation around the enclosing curve. The theorem provides insight into the nature of vorticity and also provides a practical method for computing circulation. Example 8.4 illustrates the computation of circulation, and Example 8.5 investigates application of the theorem to viscous and time-dependent flows.

Example 8.1. The sketch shows a planetary gear system. Parts (a), (b), and (c) rotate about the center on concentric axles. Parts (d) roll without slipping on (a) and (b), and are carried by part (c). If the carrier rotates at 120 rpm, how fast must (a) and (b) rotate for the vorticity of (d) to remain zero?

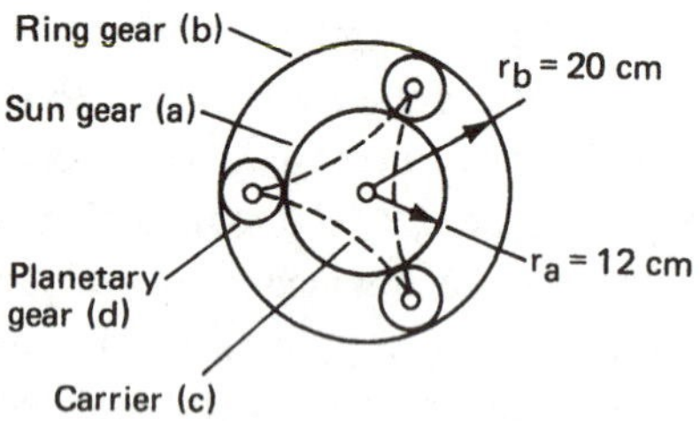

Discussion. Texts and handbooks on mechanisms give formal procedures for solving planetary (epicyclic) gear systems. Gear (a) is often the power input, and (b) or (c) is the output. If (b) is the output, power is engaged simply by braking (c), or vice versa. No gears are shifted, and no clutch is used. Automatic transmissions use epicyclic gear combinations.

In this 1-D problem, $\zeta = 2\omega$, so if $\zeta = 0$, then $\omega = 0$ also. The axles of each (d) must have a translational tangential velocity $v_c = \omega_c r_c = (120\ \pi/30)\ 16 = 64\pi$ cm/s. If the gears (d) do not rotate, but only translate in circular paths, then all points on each (d) must have the same velocity at any instant. Next, because the (d) do not slip on (a) or (b), the peripheries of (a) and (b) must also have tangential speeds of 64π cm/s. The angular velocity of (a) is $\omega_a = v_c/r_a = 64\pi/12$

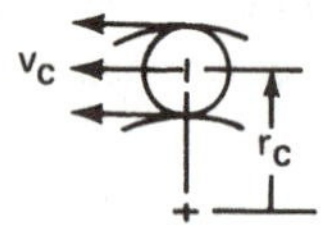

$$\omega_a = \frac{v_c}{r_a} = \frac{64\pi}{12} \cdot \frac{30}{\pi} = 160 \text{ rpm}$$

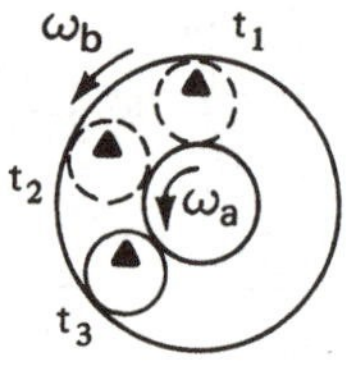

and the angular velocity of (b) is

$$\omega_b = \frac{v_c}{r_b} = \frac{64\pi}{20} \cdot \frac{30}{\pi} = 96 \text{ rpm}$$

For these angular velocities, gears (d) translate in circular paths but do not rotate.

The irrotational motion of each (d), and the motion of the adjacent gears (a) and (b), are similar to the irrotational motion of an inviscid vortex only if $r_d \to 0$. For finite sized (d), ω of the adjacent gears is $\sim r^{-1}$, while for the inviscid vortex, it is v_θ that is $\sim r^{-1}$.

Example 8.2. The annulus between two concentric circular cylinders is filled with a $\nu = 1000$ cs fluid ($\rho \approx 1$ g/cm^3). Cylinder (a) is stationary and (b) rotates at 6 rpm, producing circular Couette flow as shown in the end view. At the same time, an axial flow is induced by an axial pressure gradient of 10^4 dynes/cm^2 over a length of 10 cm. Assume a linear relationship for $v_\theta(r)$ and a parabolic relationship for $v_z(r)$. Compute ζ, and draw it on a sketch at r = 5.1, 5.25, and 5.4 cm.

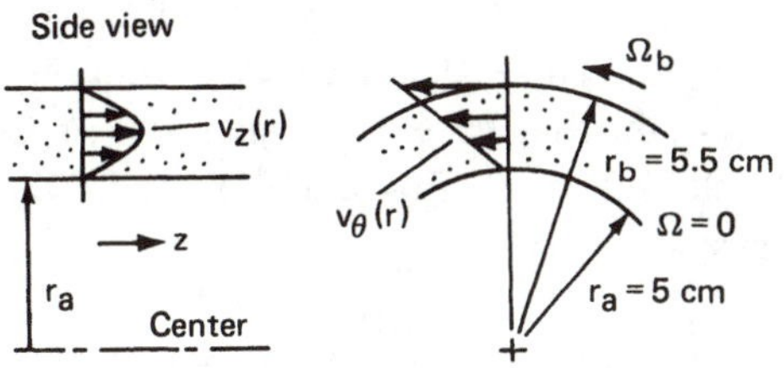

Discussion. Vector velocity of the fluid is $\mathbf{V} = v_r\,\hat{\mathbf{r}} + v_\theta\,\hat{\boldsymbol{\theta}} + v_z\,\hat{\mathbf{z}}$, with $v_r = 0$, and v_θ and v_z specified as f(r) only. Placing the origin at r = 0

$$v_\theta = \frac{r_b\Omega_b - r_a\Omega_a}{r_b - r_a}(r - r_a) = \frac{5.5\times 6\times \pi/30 - 0}{5.5 - 5}(r-5) = 2.2\pi(r-5)$$

$$v_z = \frac{1}{2\mu}\left(-\frac{dp}{dx}\right)\left(b^2 - y^2\right) \text{ from (8.12), or } = \frac{1}{2\mu}\left(-\frac{dp}{dx}\right)\left[0.25^2 - (r-5.25)^2\right]$$

$$\mu = \nu\rho = 10 \times 1 = 10 \text{ poise}$$

$$v_z = \frac{1}{20}\left(\frac{10^4}{10}\right)\left(-r^2 + 10.5r - 27.5\right) = 50\left(-r^2 + 10.5r - 27.5\right)$$

The components of ζ in cylindrical coordinates are given in (8.9a,b,c), indicating $\zeta_r = 0$, and

$$\zeta_\theta = -\frac{\partial v_z}{\partial r} = 100r - 525$$

$$\zeta_z = \frac{1}{r}\frac{\partial}{\partial r}(rv_\theta) = \frac{1}{r}\frac{\partial}{\partial r}\left[2.2\pi\left(r^2 - 5r\right)\right] = \left(4.4 - \frac{11}{r}\right)\pi$$

At r	=	5.1	5.25	5.4
ζ_θ	=	−15	0	15
ζ_z	=	7.05	7.24	7.42

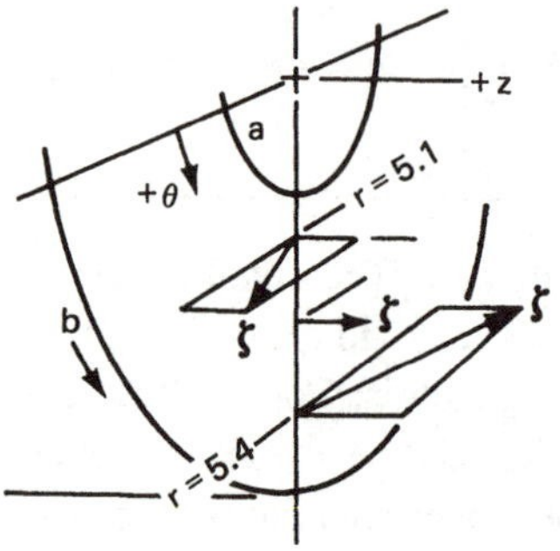

Example 8.3. Flow within a 3 ft diameter by 3 ft tall control volume is given by $v_r = 0$, $v_\theta = 2r(z-2)$, $v_z = 0$ (ft/s). Draw the flow with the origin at the bottom center of the volume. Does the fluid have $\rho = c$, and is $\mu = 0$? Compute $\boldsymbol{\zeta}$, and show on the drawing for points A(1, 0, 2), B(0.5, π, 0.5), and C(1, π, 2.5) defined in a cylindrical coordinate system (r, θ, z).

Discussion. If density is constant, then $\nabla \cdot \mathbf{V}$ must equal zero. In cylindrical coordinates

$$\nabla \cdot \mathbf{V} = \frac{1}{r}\frac{\partial}{\partial r}(rv_r) + \frac{1}{r}\frac{\partial v_\theta}{\partial \theta} + \frac{\partial v_z}{\partial z} = 0 + 0 + 0$$

and therefore, $\rho = c$ for this flow. In cylindrical coordinates

$$\boldsymbol{\zeta} = \nabla \times \mathbf{V}$$

$$= \left(\frac{1}{r}\frac{\partial v_z}{\partial \theta} - \frac{\partial v_\theta}{\partial z}\right)\hat{\mathbf{r}} + \left(\frac{\partial v_r}{\partial z} - \frac{\partial v_z}{\partial r}\right)\hat{\boldsymbol{\theta}} + \frac{1}{r}\left[\frac{\partial}{\partial r}(rv_\theta) - \frac{\partial v_r}{\partial \theta}\right]\hat{\mathbf{z}}$$

$$= -2r\hat{\mathbf{r}} + 0\hat{\boldsymbol{\theta}} + 4(z-2)\hat{\mathbf{z}}$$

Point	$\boldsymbol{\zeta}$	Units
A(1,0,2)	$-2\hat{\mathbf{r}}$	s^{-1}
B(0.5,π,0.5)	$-1\hat{\mathbf{r}} - 6\hat{\mathbf{z}}$	s^{-1}
C(1,π,2.5)	$-2\hat{\mathbf{r}} + 2\hat{\mathbf{z}}$	s^{-1}

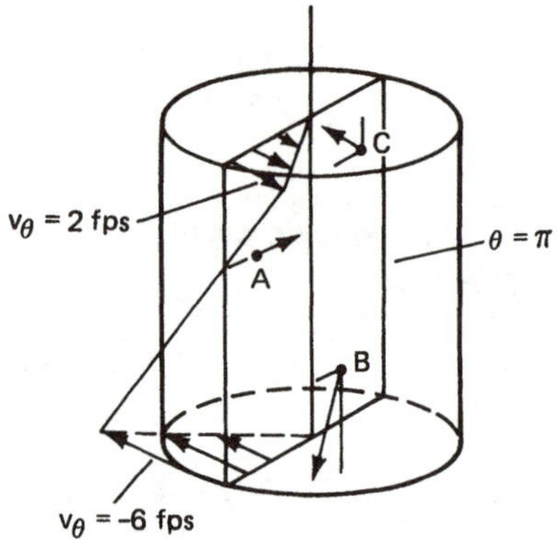

Next, is viscosity zero for this flow? That $\zeta \neq 0$ indicates the flow is rotational (not irrotational). The flow field cannot be derived as the gradient of some velocity potential (ϕ), i.e., it is not a potential flow field. Nonzero vorticity usually indicates the presence of viscosity; however, there are other mechanisms to be covered in Chapter 10 that also produce vorticity. Because of that, not enough information is available here to answer with certainty whether the given flow field does or does not represent an inviscid flow.

Example 8.4. Assume a flow field $\mathbf{V} = 3(x+y)\,\hat{\mathbf{i}} + (x^2 - 3y)\,\hat{\mathbf{j}}$. Compute Γ about a rectangle with corners at $(-1,-1)$, $(2,-1)$, $(2,3)$, $(-1,3)$, both by integrating $\mathbf{V}$ around the boundary and $\boldsymbol{\zeta}$ over the area. Is one approach significantly easier than the other?

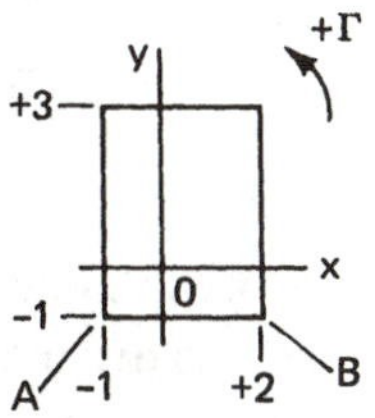

Discussion. First, integrate $\int_C \mathbf{V} \cdot d\boldsymbol{\ell}$ going counterclockwise. Line AB is given in detail.

$$\int_{-1}^{2} \left[3(x+y)\hat{\mathbf{i}} + (x^2 - 3y)\hat{\mathbf{j}}\right] \cdot \hat{\mathbf{i}}\, dx \text{, over the line } y = -1 \text{ is } \int_{-1}^{2} 3(x-1)\, dx$$

The computation for circulation sums the effect of all four sides as follows

$$\Gamma = \int_{-1}^{2} 3(x-1)\,dx + \int_{-1}^{3} (4-3y)\,dy + \int_{2}^{-1} 3(x+3)\,dx + \int_{3}^{-1} (1-3y)\,dy$$

$$= 3\left[\frac{x^2}{2} - x\right]_{-1}^{2} + \left[4y - \frac{3y^2}{2}\right]_{-1}^{3} + 3\left[\frac{x^2}{2} + 3x\right]_{2}^{-1} + \left[y - \frac{3y^2}{2}\right]_{3}^{-1}$$

$$= 0 - 4.5 - 1.5 + 5.5 - 7.5 - 24 - 2.5 + 10.5 = -24$$

Circulation $\Gamma = -24\ (L^2 T^{-1})$

Next,

$$\nabla \times \mathbf{V} = \begin{bmatrix} 0 & -\partial/\partial z & \partial/\partial y \\ \partial/\partial z & 0 & -\partial/\partial x \\ -\partial/\partial y & \partial/\partial x & 0 \end{bmatrix} \begin{bmatrix} 3x+3y \\ x^2 - 3y \\ 0 \end{bmatrix} = \begin{bmatrix} 0 \\ 0 \\ 2x - 3 \end{bmatrix}$$

Therefore, $\boldsymbol{\zeta} = \nabla \times \mathbf{V} = \zeta_z\, \hat{\mathbf{k}}$ only.

Integrate $\int_S \boldsymbol{\zeta}\cdot d\mathbf{S} = \int_S \zeta_z dS = \int_{-1}^{3}\int_{-1}^{2} (2x-3)\,dx\,dy$

$$= \int_{-1}^{3} \left[x^2 - 3x\right]_{-1}^{2} dy = \int_{-1}^{3} (-6)\,dy = \left[-6y\right]_{-1}^{3} = -24$$

Circulation $\int_S \boldsymbol{\zeta}\cdot d\mathbf{S} = -24\left(L^2T^{-1}\right)$

This is consistent with Stokes theorem, which says $\int_S \boldsymbol{\zeta}\cdot d\mathbf{S} = \Gamma = \int_c \mathbf{V}\cdot d\boldsymbol{\ell}$. Either approach is moderately easy for this problem.

Example 8.5. Apply Stokes's theorem to a solution for a viscous flow (Couette flow) and also to a time-dependent flow, $v_\theta = tr + \frac{1}{r}e^{-tr}$, $v_r = v_z = 0$. In each case, independently evaluate and compare $\int_c \mathbf{V}\cdot d\boldsymbol{\ell}$ and $\int_S \boldsymbol{\zeta}\cdot d\mathbf{S}$ for convenient c, say a rectangle for Couette flow and a circle for the time-dependent flow. Does Stokes's theorem hold for these two flows, one viscous and the other time dependent?

Discussion. $u = \frac{U}{h}y$, $v = w = 0$ for Couette flow.

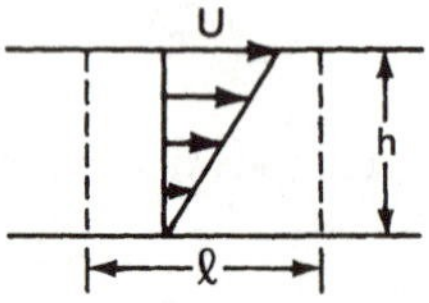

Use a rectangle ($\ell \times h$) for the control region.

$$\int_c \mathbf{V}\cdot d\boldsymbol{\ell} = 0\cdot\ell + 0\cdot h + U(-1)\ell + 0\cdot h = -U\ell$$

$$\boldsymbol{\zeta} = \nabla\times\mathbf{V} = \begin{bmatrix} 0 & -\partial/\partial z & \partial/\partial y \\ \partial/\partial z & 0 & -\partial/\partial x \\ -\partial/\partial y & \partial/\partial x & 0 \end{bmatrix}\begin{bmatrix} Uy/h \\ 0 \\ 0 \end{bmatrix} = \begin{bmatrix} 0 \\ 0 \\ -U/h \end{bmatrix}$$

$\int_s \boldsymbol{\zeta}\cdot d\mathbf{S} = (-U/h)\cdot \ell h = -U\ell$, and therefore, $\int_c \mathbf{V}\cdot d\boldsymbol{\ell} = \int_s \boldsymbol{\zeta}\cdot d\mathbf{S}$

The matrix formulation for $\nabla\times\mathbf{V}$ is only valid for Cartesian coordinates. Curl $\mathbf{V}$ in cylindrical coordinates is derived from the basic definition as

$$\boldsymbol{\zeta} = \left(\frac{1}{r}\frac{\partial v_z}{\partial\theta} - \frac{\partial v_\theta}{\partial z}\right)\hat{\mathbf{r}} + \left(\frac{\partial v_r}{\partial z} - \frac{\partial v_z}{\partial r}\right)\hat{\boldsymbol{\theta}} + \frac{1}{r}\left[\frac{\partial}{\partial z}(rv_\theta) - \frac{\partial v_r}{\partial\theta}\right]\hat{\mathbf{z}}$$

Use a circle with radius r centered at the origin for the control region for the time-dependent flow.

$$\int_c \mathbf{V} \cdot d\boldsymbol{\ell} = \left(tr + \frac{1}{r} e^{-tr} \right) \cdot (2\pi r) = 2\pi tr^2 + 2\pi e^{-tr}$$

$$\boldsymbol{\zeta} = \nabla \times \mathbf{V} = (0+0)\hat{\mathbf{r}} + (0+0)\hat{\boldsymbol{\theta}} + \frac{1}{r}\left[\frac{\partial}{\partial r}\left(tr^2 + e^{-tr} \right) - 0 \right]\hat{\mathbf{z}}$$

$$\boldsymbol{\zeta} = \frac{1}{r}\left(2tr - te^{-tr} \right)\hat{\mathbf{z}} = \left(2t - \frac{t}{r} e^{-tr} \right)\hat{\mathbf{z}}$$

$$\int_s \boldsymbol{\zeta} \cdot d\mathbf{S} = \int_0^r \left(2t - \frac{t}{r} e^{-tr} \right) 2\pi r dr = 2\pi t \int_0^r \left(2r - e^{-tr} \right) dr$$

$$= 2\pi t \left[r^2 + \frac{1}{t} e^{-tr} \right]_o^r = 2\pi t \left[\left(r^2 + \frac{1}{t} e^{-tr} \right) - \left(0 + \frac{1}{t} \right) \right]$$

$$\int_s \boldsymbol{\zeta} \cdot d\mathbf{S} = 2\pi tr^2 + 2\pi e^{-tr} - 2\pi$$

This differs from $\int_c \mathbf{V} \cdot d\boldsymbol{\ell}$ by a factor of (-2π). In this case, $\int_s \boldsymbol{\zeta} \cdot d\mathbf{S} \neq \int_c \mathbf{V} \cdot d\boldsymbol{\ell}$ because a line vortex of value $\kappa = 1$ exists at the origin. It contributes a value $\Gamma = 2\pi\kappa = 2\pi$ to the circulation, which, when added to $\int_s \boldsymbol{\zeta} \cdot d\mathbf{S}$, brings the integral of vorticity into equality with the integral of velocity.

Note the similar computation for an inviscid vortex, where $\boldsymbol{\zeta} = 0$ over every finite area. A line vortex at the origin of value κ, i.e., $\Gamma = 2\pi\kappa$, must be added to the integral of vorticity to validate Stokes's theorem. When using complex variables this difficulty does not arise because the singularity at the origin routinely yields the necessary term.

Chapter 9

Vorticity as the Variable

9.1 Vorticity in Navier-Stokes Equations

The Navier-Stokes equations for a constant density and constant viscosity fluid are shown in Table 9.1, with (**V**, p) as the dependent variables in an inertial (nonrotating) frame as (2.29a), and in a frame rotating at constant angular velocity as (6.7). Both equations can be transformed to dimensionless form (2.30) and (6.9). Each of these four equations can also be expressed as functions of the variable $\boldsymbol{\zeta}$. The use of vorticity as the variable of interest yields additional insight into rotating fluid phenomena, and in particular, makes more explicit the relationship between viscosity and vorticity.

After application of ($\nabla\times$) to each term of (2.29a), use of vector identities and some algebra, the equation in dimensional form in a nonrotating frame similar to (2.29a), but in the variables ($\boldsymbol{\zeta}$, **V**), is

$$\frac{\partial \boldsymbol{\zeta}}{\partial t} + (\mathbf{V}\cdot\nabla)\boldsymbol{\zeta} - (\boldsymbol{\zeta}\cdot\nabla)\,\mathbf{V} = \nu\nabla^2\boldsymbol{\zeta} \tag{9.1}$$

Equation (9.1) and all the following equations in $\boldsymbol{\zeta}$ are augmented with

$$\boldsymbol{\zeta} = \nabla\times\mathbf{V}$$

and

$$\nabla\cdot\boldsymbol{\zeta} = 0 \tag{9.2}$$

which follows from the vector identity, *The divergence of a curl always equals zero.* The pressure term disappears because the curl of a gradient ($\nabla\times\nabla p$) also equals zero. After using the same scaling factors as for (2.30), e.g., L/U as characteristic time, (9.1) becomes the dimensionless equation

$$\frac{\partial \boldsymbol{\zeta}}{\partial t} + (\mathbf{V}\cdot\nabla)\boldsymbol{\zeta} - (\boldsymbol{\zeta}\cdot\nabla)\,\mathbf{V} = \frac{1}{R_e}\nabla^2\boldsymbol{\zeta} \tag{9.3}$$

After application of $(\nabla\times)$ to each term of (6.7), the equation in $\boldsymbol{\zeta}$ in dimensional form in a rotating frame is

$$\frac{\partial\boldsymbol{\zeta}}{\partial t}+(\mathbf{V}\cdot\nabla)\boldsymbol{\zeta}-(\boldsymbol{\zeta}\cdot\nabla)\,\mathbf{V}-2\Omega\frac{\partial\mathbf{V}}{\partial z}=\nu\nabla^2\boldsymbol{\zeta} \tag{9.4}$$

Remember that all quantities, including $\nabla\times$, are now relative to the rotating frame. The convention of Chapter 6 to use lowercase letters (**a**, **v**, **r**) for the rotating frame is less useful now and is dropped. The centripetal term has been included with the pressure, and then removed by the $(\nabla\times)$ operation, leaving the Coriolis term as the only effective change from the nonrotating case. The Coriolis term is derived using $\boldsymbol{\Omega}=\Omega\hat{\mathbf{k}}$ as follows. The quantity $\nabla\times(2\boldsymbol{\Omega}\times\mathbf{V}) = 2[(\mathbf{V}\cdot\nabla)\,\boldsymbol{\Omega}-\mathbf{V}(\nabla\cdot\boldsymbol{\Omega})-(\boldsymbol{\Omega}\cdot\nabla)\,\mathbf{V}+\boldsymbol{\Omega}(\nabla\cdot\mathbf{V})]$ has only one nonzero term

$$-2(\boldsymbol{\Omega}\cdot\nabla)\,\mathbf{V}=-2\Omega(\hat{\mathbf{k}}\cdot\nabla)\,\mathbf{V}=-2\Omega\frac{\partial\mathbf{V}}{\partial z}$$

Equation (9.4) is made dimensionless using the scaling parameters used for (6.9), including Ω^{-1} for characteristic time.

$$\frac{\partial\boldsymbol{\zeta}}{\partial t}+R_o\left[(\mathbf{V}\cdot\nabla)\boldsymbol{\zeta}-(\boldsymbol{\zeta}\cdot\nabla)\,\mathbf{V}\right]-2\frac{\partial\mathbf{V}}{\partial z}=E_k\,\nabla^2\boldsymbol{\zeta} \tag{9.5}$$

The Rossby number again is a ratio of the convective acceleration to the Coriolis acceleration. This includes the convection of $\boldsymbol{\zeta}$ by the velocity field $(\mathbf{V}\cdot\nabla)\boldsymbol{\zeta}$ and the convection of $\mathbf{V}$ by the vorticity $(\boldsymbol{\zeta}\cdot\nabla)\mathbf{V}$.

In summary, eight different versions of the Navier-Stokes equations for a constant density, constant viscosity fluid have been presented in each combination of

Variables	Units	Frame
$(\mathbf{V}, p)$	Dimensional	Inertial frame
$(\boldsymbol{\zeta}, \mathbf{V})$	Dimensionless	Rotating frame

The eight equations are shown in Table 9.1 and will be used now and in later sections to seek exact and/or approximate solutions to rotating fluid phenomena.

9.2 Viscous Production of Vorticity

The viscous generation, dissipation, diffusion, and convection of vorticity can be illustrated using (9.1), ($\boldsymbol{\zeta}$, dimensional, inertial). This equation, if restricted to two-dimensional (2-D) flow in the x, y plane, has only ζ_z

$$\frac{D\zeta_z}{Dt}=\nu\nabla^2\zeta_z \tag{9.6}$$

Table 9.1: The Navier-Stokes Equation in Various Forms

	Inertial Frame	Rotating Frame	
(2.29a)	$\frac{\partial \mathbf{V}}{\partial t}+(\mathbf{V}\cdot\nabla)\mathbf{V} = -\frac{1}{\rho}\nabla p+\nu\nabla^2\mathbf{V}$	$\frac{\partial \mathbf{V}}{\partial t}+(\mathbf{V}\cdot\nabla)\mathbf{V}+\mathbf{\Omega}\times(\mathbf{\Omega}\times\mathbf{R})+2\mathbf{\Omega}\times\mathbf{V}=-\frac{1}{\rho}\nabla p+\nu\nabla^2\mathbf{V}$	(6.7)
(2.30)	$\frac{\partial \mathbf{V}}{\partial t}+(\mathbf{V}\cdot\nabla)\mathbf{V} = -\nabla p+\frac{1}{R_e}\nabla^2\mathbf{V}$	$\frac{\partial \mathbf{V}}{\partial t}+R_o(\mathbf{V}\cdot\nabla)\mathbf{V}+2\hat{\mathbf{k}}\times\mathbf{V} = -\nabla p+E_k\nabla^2\mathbf{V}$	(6.9)
(9.1)	$\frac{\partial \boldsymbol{\zeta}}{\partial t}+(\mathbf{V}\cdot\nabla)\boldsymbol{\zeta}-(\boldsymbol{\zeta}\cdot\nabla)\mathbf{V}=\nu\nabla^2\boldsymbol{\zeta}$	$\frac{\partial \boldsymbol{\zeta}}{\partial t}+(\mathbf{V}\cdot\nabla)\boldsymbol{\zeta}-(\boldsymbol{\zeta}\cdot\nabla)\mathbf{V}-2\Omega\frac{\partial \mathbf{V}}{\partial z} = \nu\nabla^2\boldsymbol{\zeta}$	(9.4)
(9.3)	$\frac{\partial \boldsymbol{\zeta}}{\partial t}+(\mathbf{V}\cdot\nabla)\boldsymbol{\zeta}-(\boldsymbol{\zeta}\cdot\nabla)\mathbf{V}=\frac{1}{R_e}\nabla^2\boldsymbol{\zeta}$	$\frac{\partial \boldsymbol{\zeta}}{\partial t}+R_o[(\mathbf{V}\cdot\nabla)\boldsymbol{\zeta}-(\boldsymbol{\zeta}\cdot\nabla)\mathbf{V}]-2\frac{\partial \mathbf{V}}{\partial z} = E_k\nabla^2\boldsymbol{\zeta}$	(9.5)

1. All quantities in the rotating frame equations are relative to the rotating frame. The convention of Chapter 6 to use **A**, **V**, **R** for inertial and **a**, **v**, **r** for rotating frames is less useful now and has been dropped.
2. The dimensionless form of each equation follows the same equation in physical variables.
3. Equations shown assume only pressure and viscous forces and are for constant density and constant viscosity fluids.
4. Velocity equations are augmented with $(\nabla\cdot\mathbf{V}=0)$.
5. Vorticity equations are augmented with $(\boldsymbol{\zeta}=\nabla\times\mathbf{V})$ and $(\nabla\cdot\boldsymbol{\zeta}=0)$.
6. All body forces are assumed to be conservative $(\mathbf{f}=-\nabla\phi)$ and are included in ∇p.
7. The $[\mathbf{\Omega}\times(\mathbf{\Omega}\times\mathbf{R})]$ term is included with the ∇p term in (6.9).
8. Pressure does not appear in any vorticity equation because $(\nabla\times\nabla p=0)$.
9. Angular velocity is constant in (6.7), and a specific constant $(\Omega\hat{\mathbf{k}})$ in (6.9), (9.4), and (9.5).

With no variation of V in the z direction, convection of the velocity field by vorticity has disappeared because

$$(\zeta \cdot \nabla)\,\mathbf{V} = \zeta_z(\hat{\mathbf{k}} \cdot \nabla)\mathbf{V} = \zeta_z \frac{\partial}{\partial z}\mathbf{V} = 0$$

Expansion of (9.6) gives

$$\frac{\partial \zeta_z}{\partial t} + u\frac{\partial \zeta_z}{\partial x} + v\frac{\partial \zeta_z}{\partial y} = \nu\left(\frac{\partial^2 \zeta_z}{\partial x^2} + \frac{\partial^2 \zeta_z}{\partial y^2}\right) \tag{9.7}$$

If $\nu = 0$, this equation equates changes in vorticity per unit time to the convective acceleration terms. With few exceptions, viscosity is required for the initial generation, dissipation, and/or direct diffusion of vorticity.

Solution of (9.7) is aided by use of a stream function. This automatically satisfies the 2-D continuity equation. Substitute

$$u = \partial\psi / \partial y\,, \quad v = -\,\partial\psi / \partial x$$

and

$$\zeta_z = \left(\frac{\partial v}{\partial x} - \frac{\partial u}{\partial y}\right) = \left(-\frac{\partial^2 \psi}{\partial x^2} - \frac{\partial^2 \psi}{\partial y^2}\right) = -\,\nabla^2\psi$$

into (9.7) to get

$$\frac{\partial \nabla^2\psi}{\partial t} + \frac{\partial \psi}{\partial y}\frac{\partial \nabla^2\psi}{\partial x} - \frac{\partial \psi}{\partial x}\frac{\partial \nabla^2\psi}{\partial y} = \nu\nabla^4\psi \tag{9.8}$$

A numerical solution of (9.8) for flow past a sphere was obtained by Jenson (1959) and is shown in Figures 9.1a–d. The figures show streamlines for steady-state flows with Reynolds numbers $R_e = 5$ and $R_e = 40$ on the left, and the corresponding lines of constant vorticity on the right. Vorticity is made dimensionless with U/L rather than Ω^{-1}, because here, U and L, not Ω, define the basic flow phenomena.

The streamline pattern at $R_e = 5$ is only slightly different from that of inviscid flow; however, in inviscid flow, the irrotational free stream would remain irrotational, and ζ_z would be zero everywhere about the sphere. Fluid distortion (strain rate) occurs in both cases as $\dot{\varepsilon}_{xy} = (\partial v/\partial x + \partial u/\partial y)/2$, but for inviscid flow, shear stress and transverse momentum flux are zero. In viscous flow, μ diffuses momentum transverse to the flow direction. This reduces $\dot{\varepsilon}_{xy}$ and increases $\zeta_z = (\partial v/\partial x - \partial u/\partial y)$ from zero to a finite value, generating vorticity.

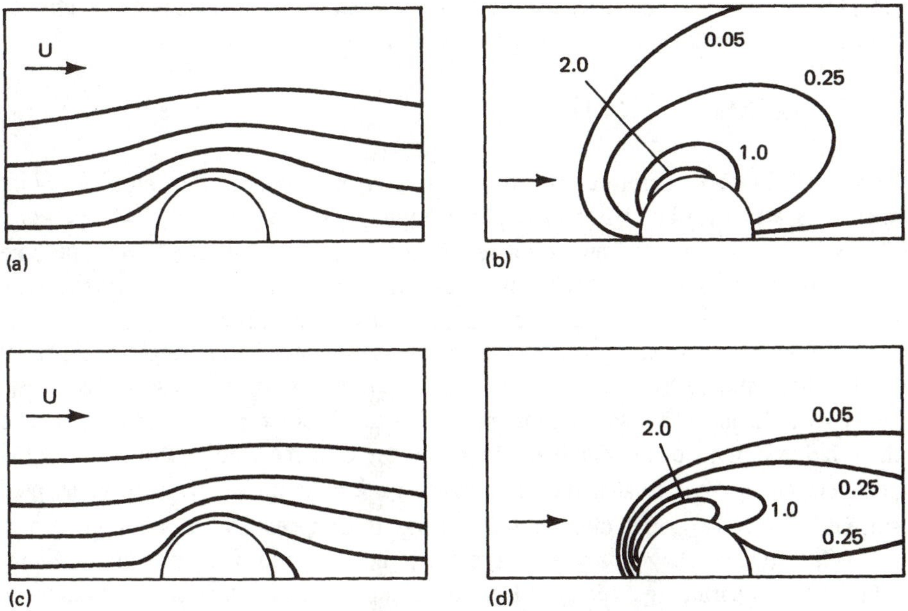

Figure 9.1: *(a) Streamlines of viscous flow past a sphere at $R_e = 5$. (b) Lines of constant vorticity in the same flow. Vorticity is normalized here as $\zeta/(U/d)$, with d for diameter. (c) and (d) illustrate streamlines and vorticity, respectively, at $R_e = 40$. At this larger value of R_e, the flow separates. Creation and convection of vorticity both increase at larger R_e. Compare with flow past a heated cylinder and distribution of temperature. Developed from equations and data in Jenson (1959).*

Vorticity generation is greatest where the product of viscosity and distortion rate is greatest. As vorticity is produced, it diffuses per the diffusion equation

$$\frac{\partial \zeta_z}{\partial t} = \nu \nabla^2 \zeta_z \tag{9.9}$$

and is then convected by the term $(\mathbf{V} \cdot \nabla)\zeta_z$ in the downstream direction. Readers familar with heat transfer will recognize the similarity with the problem of diffusion and convection of heat from a heated cylinder in a uniform flow. More vorticity is produced at higher Reynolds numbers ($R_e = 40$), and the effects of downstream convection are also more apparent. Equation (9.9) is also valid in 3-D and is known as Stokes's wave equation

$$\frac{\partial \zeta}{\partial t} = \nu \nabla^2 \zeta \tag{9.10}$$

Chapter 10 continues the discussion of vorticity, including mechanisms capable of generating and dissipating vorticity in an inviscid fluid.

9.3 Relative Vorticity

Vorticity is not invariant to a transformation to rotating coordinates, although other quantities are invariant. For example, Chapter 6 shows that ${}^{n}\nabla p_n = {}^{b}\nabla p_b$, and ${}^{n}\nabla \cdot \mathbf{V}^n = {}^{b}\nabla \cdot \mathbf{V}^b$, where frame b rotates relative to frame n. Without proof, the rate of strain tensor and the shear tensor shown in Chapter 2 are invariant to a transformation to rotating coordinates, but the rotation tensor (vorticity) is not.

Trivial examples can provide intuitive understanding. For example, if a person washes his hands, obviously there is rate of strain and shear stress in the water between his hands. The fact that that person, or an observer, rotates does not change the relative motion of his hands or the shear stress in the water. But if either the person or the observer rotates, the computation for $\boldsymbol{\zeta} = 2\boldsymbol{\omega}$ must obviously be affected since now everything, including the water, either is, or appears to be, rotating.

Geophysical flow researchers typically measure and analyze fluid vorticity relative to the earth, e.g., winds and ocean currents. Since the earth is rotating, the inertial vorticity of the fluid is the sum of the measured vorticity plus twice the angular velocity of the earth ($2\boldsymbol{\omega}$). In this text, $\boldsymbol{\zeta}$ is computed relative to the nonrotating inertial frame (n) unless specifically stated otherwise.

9.4 Application Examples

The Navier-Stokes equations are transformed so that vorticity, rather than pressure, is a primary variable. Pressure disappears completely since $\nabla \times \nabla p = 0$, but velocity remains in the definition of $\boldsymbol{\zeta} = \nabla \times \mathbf{V}$ and in the convection acceleration terms. Eight different versions of the Navier-Stokes equations are shown in Table 9.1. In later chapters fluid phenomena will be explained in one or both of $(\mathbf{V}, p)$ and $(\boldsymbol{\zeta}, \mathbf{V})$. Example 9.1 derives (9.1) in detail, Example 9.2 examines the distribution of $\mathbf{V}$ and $\boldsymbol{\zeta}$ in a classic boundary layer solution, Example 9.3 explores an important consequence of (9.5) when $R_o << 1$ and $E_k << 1$, and Example 9.4 illustrates the solution of two classic flows using vorticity as the primary variable.

Example 9.1. Derive (9.1). Show that it defines the relationship between force and rate of change of momentum in a constant density, constant viscosity fluid just like the Navier-Stokes equation also does for a fluid, or like Newton's equation, $\mathbf{F} = d(m\mathbf{V})/dt$, does for a mass particle.

Discussion. Newton's law, $\mathbf{F} = d(m\mathbf{V})/dt$, cannot be proven mathematically. It can only be verified by experiment, and then only for masses, velocities, and times that avoid relativistic or subatomic conditions. With Newton's law experimentally verified, it is straightforward, although slightly complicated, to derive the Navier-Stokes equation that says the same thing for an infinitesimal volume of fluid (Chapter 2). For a $\rho = c$, $\mu = c$ fluid, $d(m\mathbf{V})/dt$ becomes the left-hand side, and $\mathbf{F}$ becomes the right-hand side of

$$\rho\left[\frac{\partial \mathbf{V}}{\partial t} + (\mathbf{V}\cdot\nabla)\mathbf{V}\right] = \mathbf{f} - \nabla p + \mu\nabla^2\mathbf{V} \qquad (9.11)$$

If a nonzero operation is applied to each side of an equation, the result is still an equality, although the variables and units may change. The new equality continues to be based upon and to state the validity of the same physical concept. We choose now to differentiate both sides with respect to space ($\partial/\partial x$, $\partial/\partial y$, $\partial/\partial z$). This is accomplished here by taking the curl of each side and by noting that the order of differentiation (e.g., $\partial/\partial x$, $\partial/\partial t$) can be exchanged. The first term becomes $\nabla\times(\partial\mathbf{V}/\partial t) = \partial(\nabla\times\mathbf{V})/\partial t = \partial\boldsymbol{\zeta}/\partial t$.

The term $(\mathbf{V}\cdot\nabla)\mathbf{V}$ in (9.11) is a *pseudo-vector*. It behaves well in computations in Cartesian coordinates and provides a useful concept for the relevant phenomena, but it is not useful for this derivation. Instead, an alternative form is used (see Appendix A.5).

$$\rho\left[\frac{\partial \mathbf{V}}{\partial t} + \nabla\left(\frac{V^2}{2}\right) - \mathbf{V}\times(\nabla\times\mathbf{V})\right] = \mathbf{f} - \nabla p + \mu\nabla^2\mathbf{V}$$

The quantity $\nabla\times\nabla(V^2/2)$ equals zero because the curl of any gradient is zero. The next term is $-\mathbf{V}\times(\nabla\times\mathbf{V}) = -\mathbf{V}\times\boldsymbol{\zeta}$, and the curl of this is $-\nabla\times(\mathbf{V}\times\boldsymbol{\zeta})$. After use of an identity for the curl of a vector product, two pseudo-vectors can be obtained $-\nabla\times(\mathbf{V}\times\boldsymbol{\zeta}) = (\mathbf{V}\cdot\nabla)\boldsymbol{\zeta} - (\boldsymbol{\zeta}\cdot\nabla)\,\mathbf{V}$ (again, see Appendix A and remember that $\nabla\cdot\mathbf{V} = \nabla\cdot\boldsymbol{\zeta} = 0$ for this application).

Also $\nabla\times\nabla^2\mathbf{V} = \nabla^2(\nabla\times\mathbf{V})$, and of course $\nabla\times\nabla p = 0$. The term $\nabla\times\mathbf{f}$ will have a value for some applications; however, in (9.1) the arbitrary body force $\mathbf{f}$ has been assumed to be either zero or to be derivable as the gradient of some potential (e.g., conservative forces such as gravity) so that its curl is zero. Combining each term as noted and dividing by ρ finally yields

$$\frac{\partial\boldsymbol{\zeta}}{\partial t} + (\mathbf{V}\cdot\nabla)\boldsymbol{\zeta} - (\boldsymbol{\zeta}\cdot\nabla)\mathbf{V} = \nu\nabla^2\boldsymbol{\zeta}$$

$$\frac{D\boldsymbol{\zeta}}{Dt} - (\boldsymbol{\zeta}\cdot\nabla)\,\mathbf{V} = \nu\nabla^2\boldsymbol{\zeta} \qquad (9.1)$$

Example 9.2. Apply (9.7) at a point inside the boundary layer on a flat plate that oscillates in its own plane ($u = U \cos \omega t$) in a viscous liquid with ($U = 2$ cm/s, $\omega = 1$ cycle/min, $\nu = 10$ cm^2/s). What is the height of the boundary layer (δ)? Compute ζ and show that it satisfies (9.7). Use centimeter-gram-second (cgs) units.

Discussion. The basic problem in (**V**, p) was solved by Stokes (Stokes's second problem). The solution from Chapter 4 is

$$u(y,t) = U\,e^{-ky}\cos(\omega t - ky) \tag{4.4a}$$

with ($v = w = 0$) and a characteristic length $k^{-1} = (\omega/2\nu)^{-1/2}$. To obtain δ

$$\frac{u}{U} = 0.01 = e^{-ky}$$

$$k = (2\pi/60 \times 2 \times 10)^{1/2} = 0.07236\,\text{cm}^{-1}$$

$$\ell n(0.01) = -0.07236\ y, \quad y = 63.6\ \text{cm} = \delta$$

A vorticity equation, equivalent to (4.4a), is obtained as the curl of u(y, t) using

$$\zeta = \frac{\partial v}{\partial x} - \frac{\partial u}{\partial y}$$

$$\zeta(y,t) = k\,U\,e^{-ky}[\cos(\omega t - ky) - \sin(\omega t - ky)] \tag{9.12}$$

Equation (9.7)

$$\frac{\partial \zeta}{\partial t} + u\frac{\partial \zeta}{\partial x} + v\frac{\partial \zeta}{\partial y} = \nu\left(\frac{\partial^2 \zeta}{\partial x^2} + \frac{\partial^2 \zeta}{\partial y^2}\right)$$

reduces to the 1-D diffusion equation

$$\frac{\partial \zeta}{\partial t} = \nu\frac{\partial^2 \zeta}{\partial y^2} \tag{9.13}$$

Equation (9.12), obtained as the curl of (4.4a), satisfies (9.13), showing that the curl of the solution of the differential equation in **V** is equal to the solution of the curl of the differential equation. A solution can also be obtained directly from (9.13) and its boundary conditions, $\zeta(0, t) = kU = Z$ from (9.12), and $\zeta(\infty, t) = 0$.

$$\zeta(y,t) = Z\,e^{-ky}\cos(\omega t - ky) \tag{9.14}$$

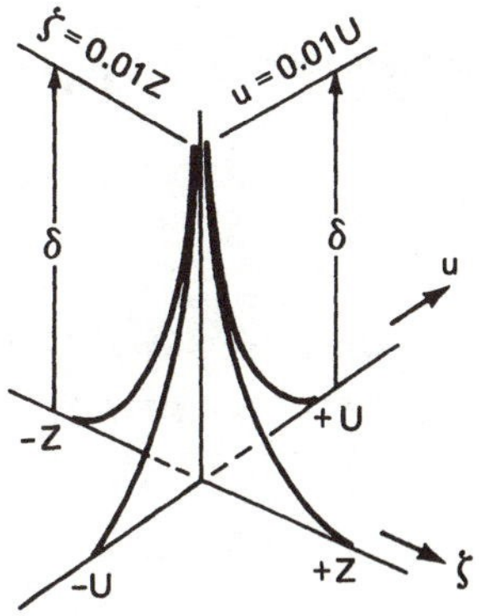

The quantity Z is a characteristic (maximum) vorticity at the plate (U/k^{-1}).

This solution is similar to (4.4a) which also was a solution of a diffusion equation. Equation (9.14) demonstrates that the vorticity boundary layer here, defined as $\delta = y$ when $\zeta / Z = 0.99$, has the same thickness as δ for the velocity field. At this depth in the liquid, 99% of the vorticity originally created by viscosity at the plate, has been dissipated by viscosity in the layer δ = 63.6 cm thick. Viscosity both produces and destroys the vorticity.

Boundary layers typically are much thinner. For example, if the liquid were water and $\omega = 10$ Hz, the boundary layer thickness would be $\delta = 0.82$ mm.

Example 9.3. Solve the classic problems of viscous, laminar, steady-state, fully developed, incompressible flow (a) between two inertially fixed parallel plates, and (b) through a circular pipe (Hagen-Poiseuille flow), using vorticity as the primary variable, i.e., use (9.1) and appropriate boundary conditions.

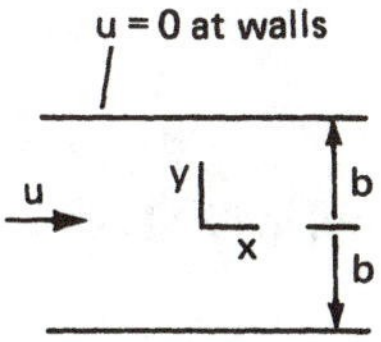

Discussion.

$$\frac{D\boldsymbol{\zeta}}{Dt} - (\boldsymbol{\zeta} \cdot \nabla)\, \mathbf{V} = \nu \nabla^2 \boldsymbol{\zeta} \tag{9.1}$$

For part (a) use Cartesian coordinates as shown, so that $\boldsymbol{\zeta} = \zeta_z \hat{\mathbf{k}}$. Then, $(\boldsymbol{\zeta} \cdot \nabla)\mathbf{V} = (\zeta_z\, \partial/\partial z)\mathbf{V} = \zeta_z\, \partial \mathbf{V}/\partial z$. But $\partial \mathbf{V}/\partial z = 0$ for this 2-D flow, so $(\boldsymbol{\zeta} \cdot \nabla)\, \mathbf{V} = 0$, and (9.1) becomes

$$\frac{\partial \zeta_z}{\partial t} + u\frac{\partial \zeta_z}{\partial x} + v\frac{\partial \zeta_z}{\partial y} + w\frac{\partial \zeta_z}{\partial z} = \nu\left(\frac{\partial^2 \zeta_z}{\partial x^2} + \frac{\partial^2 \zeta_z}{\partial y^2} + \frac{\partial^2 \zeta_z}{\partial z^2}\right)$$

In this 2-D steady-state flow, v, w, $\partial\zeta_z/\partial t$, and all changes in the z direction are zero, leaving

$$u\frac{\partial \zeta_z}{\partial x} = \nu\left(\frac{\partial^2 \zeta_z}{\partial x^2} + \frac{\partial^2 \zeta_z}{\partial y^2}\right)$$

The fully developed criteria implies no variations in the flow direction (e.g., inlet variations), so that all $\partial/\partial x$ are zero also. The final equation to be solved is

$$\frac{\partial^2 \zeta_z}{\partial y^2} = 0 \tag{9.15}$$

Integrate twice to get $\partial\zeta_z/\partial y = C_1$, and $\zeta_z = C_1 y + C_2$. Then

$$\zeta_z = \frac{\partial v}{\partial x} - \frac{\partial u}{\partial y} = -\frac{\partial u}{\partial y} = C_1 y + C_2 \tag{9.16}$$

From symmetry $\partial u/\partial y = 0$ at $y = 0$, and therefore the constant $C_2 = 0$. At the walls, $y = \pm b$, and $\tau_w = \mu \partial u/\partial y$. After substitution in (9.16)

$$\zeta_z = -\frac{\partial u}{\partial y} = -\frac{\tau_w}{\mu} = bC_1 \tag{9.17}$$

The solution then makes use of a force balance between net pressure differential across a unit section of flow of height b and viscous drag at the walls. They must be equal in fully developed steady-state flow because flow acceleration (and net force) are zero. Expressed mathematically this is

$$\tau_w = -b\frac{\partial p}{\partial x} \tag{9.18}$$

Combine (9.17) and (9.18) for τ_w

$$\mu\, b\, C_1 = -b\frac{\partial p}{\partial x}\ , \text{ or } C_1 = -\frac{1}{\mu}\frac{\partial p}{\partial x}$$

Finally from (9.16)

$$\zeta_z = -\frac{1}{\mu}\left(\frac{\partial p}{\partial x}\right) y \tag{9.19}$$

The conclusion is that vorticity has a linear distribution similar, but normal, to the shear stress distribution, and has a maximum magnitude when $y = \pm b$.

The velocity distribution is found using

$$-\frac{\partial u}{\partial y} = -\frac{1}{\mu}\left(\frac{\partial p}{\partial x}\right) y$$

and the no-slip condition at the wall to get the familiar parabolic solution

$$\mathbf{V} = \frac{1}{2\mu}\left(\frac{\partial p}{\partial x}\right)\left[y^2 - b^2\right]\hat{\mathbf{i}} \tag{9.20}$$

Note that pressure did not appear as a variable in the solution, but rather as part of a boundary condition.

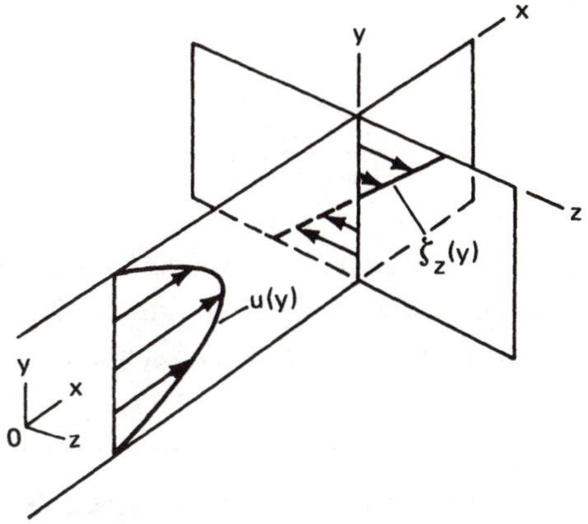

Part (b) of the example is best solved in cylindrical coordinates applied to (9.1) with flow in the z direction, and where the components of $\boldsymbol{\zeta}$ are

$$\zeta_r = \left(\frac{1}{r} \frac{\partial v_z}{\partial \theta} - \frac{\partial v_\theta}{\partial z} \right)$$

$$\zeta_\theta = \left(\frac{\partial v_r}{\partial z} - \frac{\partial v_z}{\partial r} \right)$$

$$\zeta_z = \frac{1}{r} \left[\frac{\partial}{\partial r} (r v_\theta) - \frac{\partial v_r}{\partial \theta} \right] = \left(\frac{\partial v_\theta}{\partial r} + \frac{v_\theta}{r} - \frac{1}{r} \frac{\partial v_r}{\partial \theta} \right)$$

Under conditions stated in the example, v_θ and v_r are zero everywhere, and v_z is only a function of radius, so that

$$\boldsymbol{\zeta} = -\frac{\partial v_z}{\partial r} \hat{\boldsymbol{\theta}} \tag{9.21}$$

Again, the term $(\boldsymbol{\zeta} \cdot \nabla)\mathbf{V}$ equals zero, as may be verified, and (9.1) simplifies to

$$\frac{\partial}{\partial r} \left(r \frac{\partial \zeta_\theta}{\partial r} \right) = 0 \tag{9.22}$$

Compare (9.22) with the equivalent in Cartesian coordinates (9.15).

Solution proceeds as in part (a), again using as a boundary condition that pressure differential is balanced by wall shear stress, to obtain a linear vorticity distribution

$$\boldsymbol{\zeta} = -\frac{1}{2\mu} \left(\frac{\partial p}{\partial z} \right) r \hat{\boldsymbol{\theta}} \tag{9.23}$$

and a paraboloidal velocity distribution

$$\mathbf{V} = -\frac{1}{4\mu}\left(\frac{dp}{dz}\right)(r^2 - R^2)\,\hat{\mathbf{k}} \tag{9.24}$$

where R is the cylinder radius. Compare (9.23) and (9.19), and compare (9.24) and (9.20). The lines of constant vorticity here form vortex rings symmetric about the axis and increasing in strength, linearly with increasing radius, to a maximum at the wall where $r = R$.

Example 9.4. Examine (9.5) for the case of a constant density, nearly inviscid, rotating, steady flow. Show that a necessary result for a slowly moving obstacle in the flow is that the fluid is restricted to 2-D motion; the fluid can go around the obstacle but not over it!

Discussion.

$$\frac{\partial \zeta}{\partial t} + R_o[(\mathbf{V}\cdot\nabla)\zeta - (\zeta\cdot\nabla)\mathbf{V}] - 2\frac{\partial \mathbf{V}}{\partial z} = E_k\,\nabla^2\zeta \tag{9.5}$$

The conditions state nearly steady flow so that $\partial\zeta/\partial t \sim 0$. It also implies negligible convective velocities in the flow, relative to rotational speeds, so that $R_o = U/\Omega L \sim 0$. Finally, nearly inviscid implies that $\mu \sim 0$, so that $E_k = \nu/R^2\Omega \sim 0$. These conditions lead to a result $\partial V/\partial z \sim 0$ for (9.5). The stated conditions ($\partial\zeta/\partial t \sim 0$, $R_o \sim 0$, $E_k \sim 0$) occur in the oceans and atmosphere, and can easily be produced in the laboratory.

The result ($\partial V/\partial z \sim 0$), while seemingly trivial, actually predicts some surprising phenomena. It states that rotating flows with R_o, E_k, and $\partial/\partial t \sim 0$ are restricted to 2-D motions. Note that here the restriction to 2-D motions is not an assumption but rather a physical consequence of these flow conditions. Taylor-Proudman columns are one result as shown in the drawing. The tank had been spinning at constant ω, and has just been slowed down by a small increment $\Delta\omega$, so the liquid must flow past the hemisphere glued to the bottom. A nonrotating fluid would go partly over and partly around the hemisphere, but to go over it requires $\partial V/\partial z \neq 0$. The rotating fluid will only go *around the hemisphere including the columnar space above it* as shown by the dye line half way up the cylinder. The flow behaves as if the hemisphere on the bottom surface had been replaced by a solid cylinder placed between the bottom and top surfaces. A photograph of the phenomenon is included in Greenspan (1968).

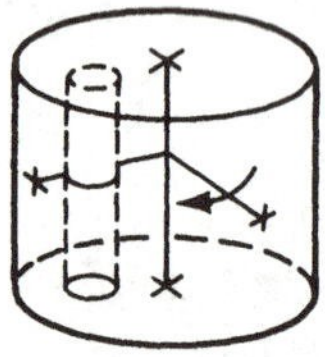

Ω liquid > Ω tank

The experiment illustrates the basic two dimensionality of all rotating flows--although this two-dimensionality characteristic can at times by suppressed by conditions violating those needed to derive $\partial V/\partial z = 0$. In Figure 1.5, the conditions are violated when the colored liquid is first dropped into the rotating water, but the slightly later photograph shows the rapidity with which a disturbed rotating fluid can return to a 2-D flow pattern.

Chapter 10

Vortex Dynamics

10.1 Vortex Interactions in Two Dimensions

This chapter introduces the topic of vortex motion. A *vortex* is a mass of fluid whose elements are moving in nearly circular pathlines about a common axis. Vortices are to be distinguished from vorticity, which is the rotation of infinitesimal fluid elements. As shown in Figure 1.1, vorticity can exist without vortex motion, and in Figure 1.2, vortex motion can exist without vorticity. A variety of simple vortex motions are introduced first; these are followed by general theorems on vortex dynamics and applications.

In two dimensions (2-D), an inviscid vortex, represented by a line with constant vortex strength κ surrounded by a circular pathline flow of constant circulation, $\Gamma = 2\pi\kappa$, produces a rotational flow field, $v_\theta = \kappa/r$, extending to infinity. It represents a flow where all fluid elements have the same angular momentum per unit mass (H/ρ), consistent with $\kappa = rv_\theta = H/\rho$. Pressures can be computed from $v_\theta(r)$ using the Bernoulli equation. Because the flow field and the pressure forces are symmetric about the vortex axis, a single line vortex in an infinite fluid cannot move itself relative to the fluid. It is, however, free to translate if perturbed by other flow fields. For example, a line vortex, in a uniform stream with velocity U, will also translate with velocity U. The vortex in turn is capable of perturbing other flow fields and their net motion.

If two *line vortices* in an infinite fluid are separated by a distance d as shown in Figure 10.1, the origin of each is in the flow field induced by the other, and both will move. Let the vortices have strength κ_1 (at 1) and κ_2 (at 2) of different magnitudes but the same direction of rotation. The velocity at 1, induced by the vortex at 2, is κ_2/d, and the velocity at 2, induced by vortex 1, is κ_1/d. Neither has a component along the line d, so d remains constant, and the net motion is a rotation of each about a stationary point on line d located as if it were the centroid of two masses of magnitude κ_1 and κ_2, respectively. Angular velocity of line d is

$$\omega_d = \frac{\kappa_1 + \kappa_2}{d^2} \tag{10.1}$$

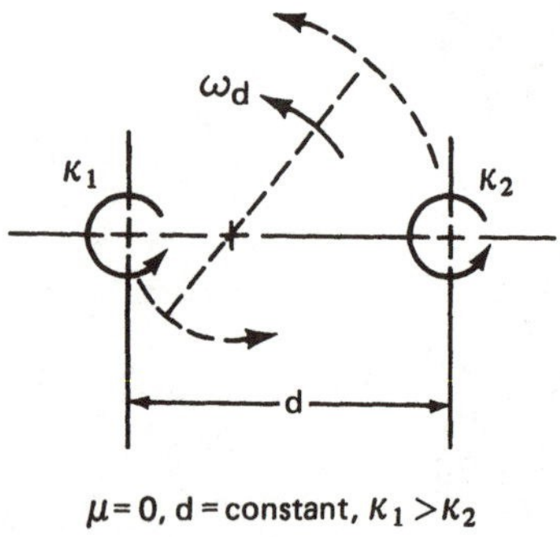

Figure 10.1: *Two vortices with the same rotation sense, in a fluid otherwise at rest, cause each other to rotate about a common center as a joined pair.*

The same concept can be applied to an arbitrary number of line vortices, but the solution is best described and quantified in the complex plane. Several applications in this text make use of the fact that the centroid of interacting vortices, all rotating in the same direction, remains at rest relative to the fluid. The centroid, representing the net effect of the vortices, can of course be perturbed by other flow fields.

Additional examples of vortex dynamics in a fluid assumed inviscid are given next, but only as descriptions and solutions. The reader interested in details should consult an advanced text on theoretical hydrodynamics. See, e.g., Milne-Thomson (1960).

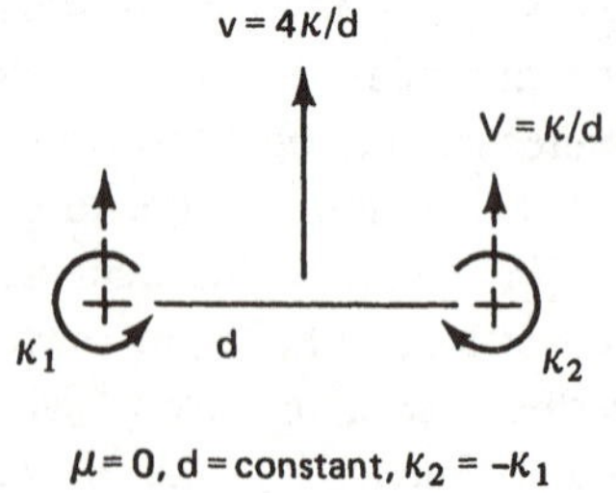

Figure 10.2: *Two vortices with opposite rotation but equal magnitude, in a fluid otherwise at rest, cause each other to translate as a joined pair.*

Two line vortices of equal strength but opposite rotations ($\kappa_2 = -\kappa_1$) form a vortex pair as shown in Figure 10.2. Each induces a velocity of the same magnitude and direction in the other, and the vortex pair as a unit moves in a fluid otherwise at rest in the direction shown with a velocity

$$v = \kappa/d \tag{10.2}$$

Velocity of fluid at the midpoint has a velocity four times as great and in the same direction

$$v = 4\kappa/d \tag{10.3}$$

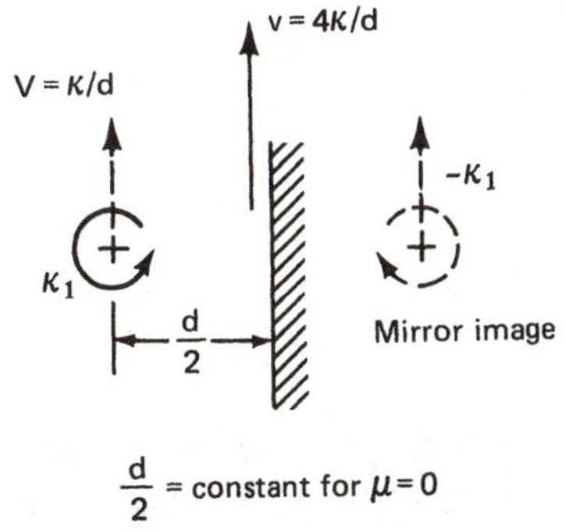

Figure 10.3: *Motion of one vortex in the presence of a wall, in a fluid otherwise at rest, is obtained by inserting a midplane in Figure 10.2.*

A more physically intuitive explanation interprets the flow as a 2-D jet with velocity ($4\kappa/d$) moving through a fluid at rest and inducing two line vortices, each with net velocity (κ/d). Because there are no induced velocities in the direction of d, the distance d again remains constant. Net pressure of each on the other therefore must be zero, as can be verified by integrating pressure along the central streamline separating the two vortices.

Since, by definition, no flow crosses a streamline, streamlines in inviscid flows can be replaced by solid walls. If a solid wall is placed at the central streamline of Figure 10.2, the previous solution continues to hold for a single vortex in the presence of a straight solid wall as shown in Figure 10.3. Net pressure on the wall is zero, and the vortex simply moves along the wall with a velocity κ/d, assuming the fluid at infinity remains at rest. The streamline at the wall has a velocity given by (10.3) relative to the fluid at rest. An equivalent explanation here assumes that the fluid near the wall has zero velocity, with the fluid at infinity moving down at velocity ($-4\kappa/d$), inducing a vortex moving down at ($-3\kappa/d$).

Solutions of flows in the presence of a solid surface are often achieved by this *method of images* where two or more flows in an unbounded fluid are arranged so that a streamline is formed coincident with the position and shape of a desired solid surface. The streamline is then replaced with the solid surface to give the desired solution. The flow structures "behind" the solid surface are referred to as images; their only purpose is to create a streamline that can then be defined as a solid surface. Chapter 3 presents a solution representing flow about a circular cylinder. A line source ($v_r = Q/2\pi r$) and a line sink ($v_r = -Q/2\pi r$) are combined to produce a *doublet*, which is then placed in a uniform stream ($u = U$). One resulting streamline forms a perfect circle ($r = R$) about the doublet. This circle can be replaced by a solid wall, and the solution for $r \geq R$ is flow about a cylinder of radius R. Flow solutions inside the cylindrical streamline ($r < R$) are not used. A line vortex ($v_\theta = \kappa/r$) located at the origin is included in Chapter 16, and the final result is flow about a cylinder with circulation ($\Gamma = 2\pi\kappa$). The result is used to prove that lift about any cylinder (circular or not) can be computed as ($L = -\rho U\Gamma$).

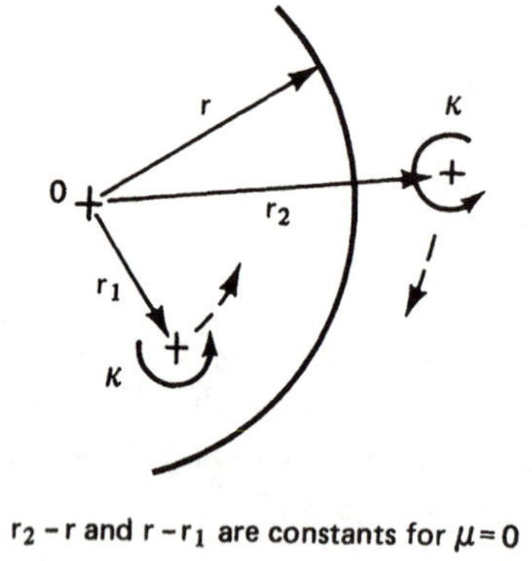

Figure 10.4: *Vortices in the presence of a circular cylinder, in a fluid otherwise at rest, translate in curved paths.*

A single vortex outside a cylindrical solid surface translates in a circular path around the cylinder, or if inside it, moves around the inside as in Figure 10.4. Two vortices of opposite sign outside a cylinder can be used to represent the vortices that form behind a moving cylinder as shown in Figure 10.5. Such vortices form at Reynolds numbers (R_e) of about 15 to 50. At higher velocities the wake region becomes completely turbulent.

A line source (or sink) combined with a line vortex is called a *spiral vortex.* See Figure 10.6. The stream function for a sink with vortex in the real plane in polar coordinates is

$$\psi = -Q\theta + \kappa \log r \tag{10.4}$$

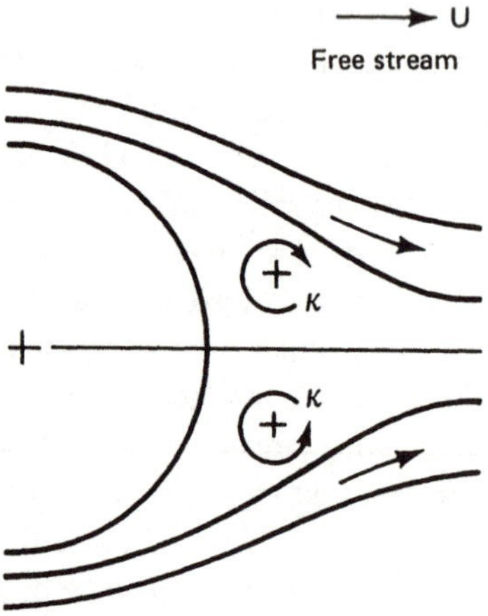

Figure 10.5: *Two vortices outside a cylinder, in the presence of a uniform stream, approximate viscous flow with R_e = 15 to 50.*

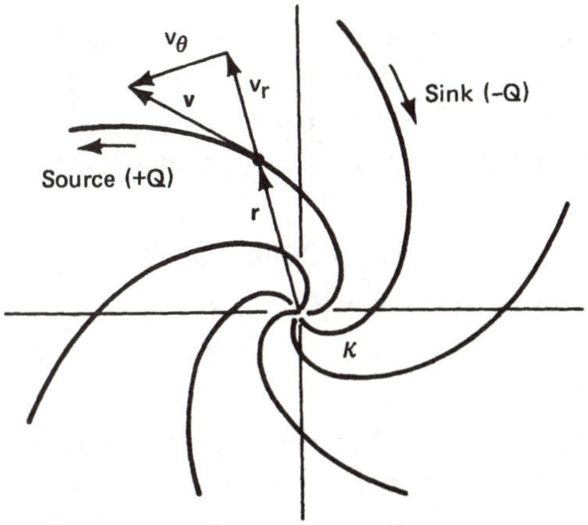

Figure 10.6: A vortex combined with a line source or sink forms a spiral vortex ($r\, v_\theta$ = constant).

The equation of the streamlines is

$$r = ce^{(Q\theta/\kappa)} \tag{10.5}$$

when ψ has a value $\kappa \log c$, with c an arbitrary constant. Again, the derivation is best accomplished in the complex plane. Stream functions and velocity potentials for important potential flows are included in Appendix B.

10.2 Conservation of Vorticity and Circulation

Additional insight into the behavior of vortices can be obtained using theorems by Helmholtz and by Kelvin. These theorems refer to vortex lines, which like streamlines, are lines connecting flow vectors (here ζ vectors) at time t. Sometimes the words *vortex filament* are used instead, and if a finite volume of fluid is represented, it is called a *vortex tube*. A surface consisting of locally parallel vortex lines is a *vortex sheet* or *vortex surface*.

Theorems attributed to Helmholtz and applicable to a $\rho = c$, $\mu = 0$ fluid are

1. A vortex filament, or vortex tube, cannot end in the fluid. It must extend to infinity, end at a solid wall, or form a closed loop within the fluid (e.g., a smoke ring).
2. The strength of a vortex filament (tube) is constant along its length. This is a consequence of theorem 1, since, if the strength varied, then some portion of the vorticity would end (or start) in the fluid.

3. Vortices remain attached to the same fluid particles as the fluid moves. Vortex strength of a flow however, like velocity, can be specified relative to a fixed Eulerian grid. For example, ζ and $\mathbf{V}$ might be constant at some fixed point (x, y, z); this implies only that all fluid as it passes that point has the same ζ and the same $\mathbf{V}$.

   The theorem by Kelvin for a $\rho = c$, $\mu = 0$ fluid states

4. The circulation around any path always enclosing the same fluid particles is independent of time. This is seen to be a consequence of Helmholtz's theorem 3 combined with Stokes's theorem.

It should be remembered that these theorems, as stated, apply only to fluids where vorticity and circulation are neither created nor destroyed, although phenomena demonstrating these theorems are often observed in nearly inviscid fluids, especially over distance and time scales too short for viscosity or other vorticity-modifying effects to dominate the result.

A more general statement on vorticity is

$$\frac{D\zeta}{Dt} = \frac{\partial\zeta}{\partial t} + (\mathbf{V}\cdot\nabla)\zeta = \nabla\frac{1}{\rho}\times\nabla p + \nabla\times\mathbf{f} + \nu\nabla^2\zeta \qquad (10.6a)$$

where variables and derivatives are relative to an inertial Eulerian grid, and $\mathbf{f}$ is force per unit mass. It is instructive to rearrange (10.6a) as

$$\frac{\partial\zeta}{\partial t} = -(\mathbf{V}\cdot\nabla)\,\zeta + \nabla\frac{1}{\rho}\times\nabla p + \nabla\times\mathbf{f} + \nu\nabla^2\zeta \qquad (10.6b)$$

which helps to visualize temporal changes in vorticity at a specific grid point in space. The terms on the right-hand side are then

$-(\mathbf{V}\cdot\nabla)\zeta$ Convection of vorticity into or out of the grid point by the velocity field. This includes stretching and tilting of vortex lines.

$\nabla\frac{1}{\rho}\times\nabla p$ This term is zero if surfaces of constant p are parallel to surfaces of constant ρ. The term obviously is zero if $\rho = c$, since then all surfaces are surfaces of constant density. Figure 10.7 illustrates the mechanism by which this term can provide a torque to a fluid element. The mass center of the control volume shown is not at its center of buoyancy, and pressure/gravity forces will rotate the control volume just as a ship rotates when its center of mass is not aligned below its center of buoyancy. A flow is called *barotropic* when the p and ρ surfaces are parallel.

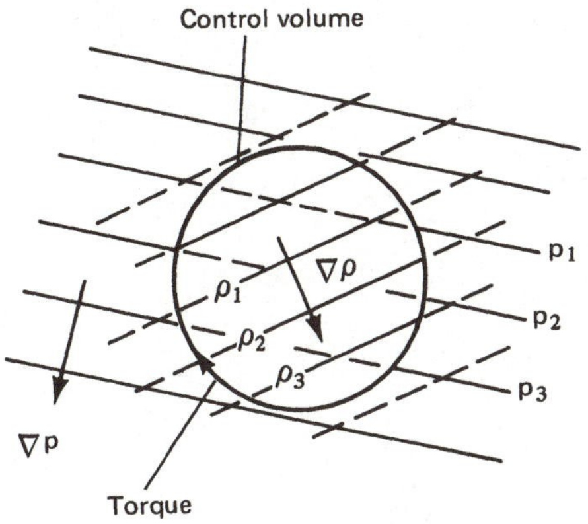

Figure 10.7: *If $\nabla(1/\rho)$ does not parallel ∇p, a torque exists on each fluid element causing it to rotate (vorticity).*

$\nabla \times \mathbf{f}$ This term is zero for conservative forces. A conservative force can always be expressed as the gradient of a scalar function, e.g., as $-\nabla\phi$, and its curl is then $\nabla \times \nabla\phi = 0$. Some magnetic and electric forces are nonconservative and can produce and/or modify vorticity in a magnetic or conducting fluid.

$\nu\nabla^2\zeta$ Production, diffusion, and/or dissipation of vorticity by viscosity is discussed in Chapter 9.

Because vorticity and circulation are defined and measured relative to a given coordinate system, their computed values depend upon the state of rotation of that coordinate system. Changes in vorticity and/or circulation caused by rotation of the coordinate system can be added or subtracted to (10.6a). This purely kinematic difference does not change $\boldsymbol{\zeta}$ or Γ relative to the nonrotating inertial frame. If $\boldsymbol{\zeta}^*$ is vorticity relative to a frame rotating at $\boldsymbol{\Omega}$, then absolute vorticity $\boldsymbol{\zeta}$ is $\boldsymbol{\zeta} = \boldsymbol{\zeta}^* + 2\boldsymbol{\Omega}$ and

$$\frac{D\boldsymbol{\zeta}}{Dt} = \frac{D\boldsymbol{\zeta}^*}{Dt} + 2\,\frac{D\boldsymbol{\Omega}}{Dt} \tag{10.7}$$

If $\boldsymbol{\Omega}$ is a constant over space and time, then $D\boldsymbol{\Omega}/Dt = 0$ and $D\boldsymbol{\zeta}/Dt = D\boldsymbol{\zeta}^*/Dt$.

In one important application $\partial\boldsymbol{\Omega}/\partial t = 0$, but $(\mathbf{V} \cdot \nabla)\,\boldsymbol{\Omega} \neq 0$. Consider the earth in Figure 10.8. Its angular velocity is assumed constant, but in atmospheric and oceanic flows it is the vertical component of the earth's angular velocity that is significant. Any north-south velocity of a parcel of air or water represents a convective change in $\boldsymbol{\zeta}$. The vertical component of angular velocity is $(\boldsymbol{\Omega}\ \sin\phi)$ as in Figure 10.8. With $\mathbf{V}$ and ∇ relative to the earth's rotating frame, then a north-south velocity (v) yields

$$\frac{v\,\partial(2\Omega\ \sin\phi)}{r\partial\phi} = \frac{2\Omega v\ (\cos\phi)}{r}$$

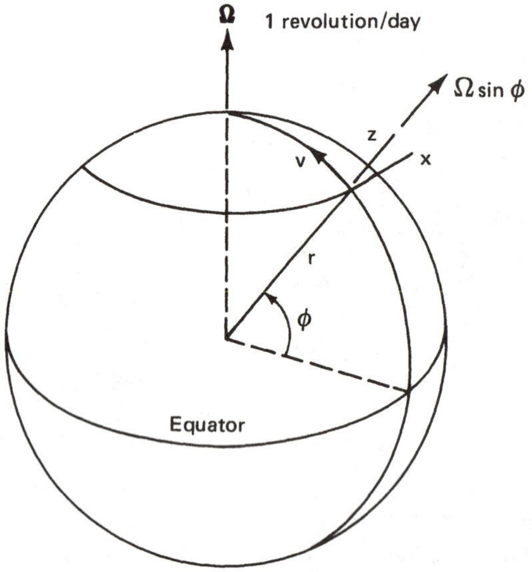

Figure 10.8: *Horizontal north-south movement of a parcel of fluid on the rotating earth represents a convective displacement of vorticity relative to the earth. Here the vorticity of interest is the local vertical component of* $2\boldsymbol{\Omega}$, *and* $(v \cdot \nabla)\zeta$ *equals* $v\partial(2\Omega \sin \phi)/\partial y$, *with* $\partial y = r\partial\phi$.

Chapters 20 and 21 examine details of atmospheric and oceanic flows on the rotating earth.

Examine next the equation ($\nabla \cdot \boldsymbol{\zeta} = 0$), valid because $\boldsymbol{\zeta} = \nabla \times \mathbf{V}$, and because the divergence of a curl is always zero. Expanded in a Cartesian coordinate system this states

$$\frac{\partial \zeta_x}{\partial x} + \frac{\partial \zeta_y}{\partial y} + \frac{\partial \zeta_z}{\partial z} = 0 \qquad (10.8)$$

Use of (10.8) is illustrated in Figure 10.9, where some of the ζ_x component of vorticity is being tilted toward the z direction, adding to ζ_z at that point in space, while removing exactly the same amount from ζ_x. This conservation of vorticity has a strong similarity to the mass conservation (continuity) equation ($\nabla \cdot \mathbf{V} = 0$). Both are statements on conservation, but their origin and implications differ. The continuity equation is derived from the mass conservation equation. When $\rho = c$, matter is not being concentrated (created) or dispersed (dissipated) within a given region, terms like $\partial\rho/\partial t$ are zero, and the equation simplifies to $\nabla \cdot \mathbf{V} = 0$. By comparison, vorticity is *defined* to be the curl of **V**, which is a spatial derivative only. Within this definition, $\nabla \cdot \boldsymbol{\zeta} = \nabla \cdot (\nabla \times \mathbf{V})$ must equal zero because it is the gradient of a curl. The *definition* of $\boldsymbol{\zeta} = \nabla \times \mathbf{V}$ is not a physical *derivation* of the temporal and spatial creation, diffusion, and/or dissipation of vorticity as in (10.6a).

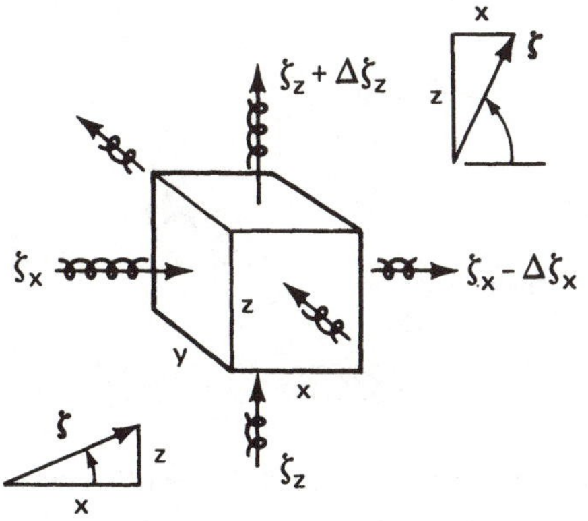

Figure 10.9: *At a region of space where vorticity is not being produced or dissipated, the equation ($\nabla \cdot \zeta = 0$) quantifies tilting or deflection of ζ. This illustrates a region where $\partial\zeta_y/\partial y = 0$, and a decrease in ζ_x appears as an increase in ζ_z.*

A more general statement on circulation, similar to (10.6a), is

$$\frac{D\Gamma}{Dt} = -\int_c \nabla\left(\frac{p}{\rho}\right) \cdot d\ell + \int_c \mathbf{f} \cdot d\ell + \int_c \frac{\mu}{\rho} \nabla^2 \mathbf{V} \cdot d\ell \qquad (10.9)$$

where c represents a closed continuous, specific material path. The path is not fixed in an Eulerian grid but is attached to specific Lagrangian fluid particles. This equation includes all mechanisms capable of producing or modifying circulation. Since circulation is the integration of vorticity over an area, and vorticity represents fluid element rotation, each mechanism shown is capable of applying a torque to an infinitesimal fluid element as in (10.6a). The last term refers to the effect of viscosity, already analyzed. It is zero only if $\nabla^2 V = 0$ or if $\mu = 0$. The middle term has zero value for conservative forces, since by definition the integral of a conservative force around a closed curve is zero. Some magnetic and electric forces are nonconservative and can produce and/or modify circulation in a magnetic or conducting fluid. The first term on the right-hand side of (10.9) is zero if surfaces of constant p are parallel to surfaces of constant ρ at all points as for (10.6a). The relationship between the inertial $D\Gamma/Dt$ and a rotating frame $D\Gamma/Dt$ follows from an analysis similar to that for (10.7).

Figure 10.10 shows an impossible situation that violates these theorems. The vortex cannot abruptly end in the fluid (even if the fluid is viscous), and the inclined junction with the solid surface is unstable. Figure 10.11 shows a real situation. Here

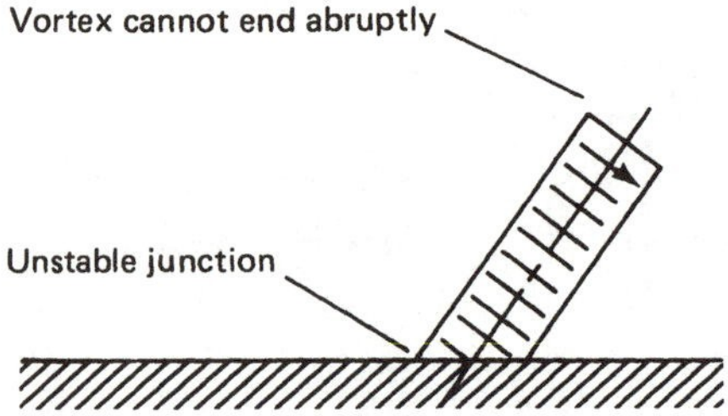

Figure 10.10: Inviscid vortex theory does not permit a vortex to end in a fluid or to be tilted at a solid surface. Such conditions can occur as transients in a real fluid, but not as stable configurations.

the vortex (a tornado) has one end at the solid surface of the earth, while the other end diffuses into a larger mass of air at a higher elevation. Rotating air masses may reach a height where the ambient pressure compares with the reduced pressure in the vortex center.

The previous discussion has analyzed behavior of line vortices that are assumed to remain straight. In general 3-D motion, the behavior of vortices becomes very complex. Later chapters show examples that illustrate the complex behavior of real vortices in low-viscosity fluids. See Figure 1.7, which illustrates aircraft wing-tip trailing vortices made visible by cooling and condensation in the low-pressure cores of the vortices. They leave the aircraft as parallel vortices rotating in opposite directions. At some distance behind the aircraft, longitudinal instabilities develop. The vortices touch, break, and recombine into large semielliptical rings that transition finally to a lower energy state as circular vortex rings, resembling smoke rings. These are best seen behind large propellor-driven aircraft. Large jet aircraft produce turbulence and condensation due to jet exhaust that also can be wrapped up into the wing-tip vortices. Jet aircraft also typically fly at higher altitudes and overall tend to produce less well-defined patterns. Wing-tip vortices are analyzed in detail in Chapter 16.

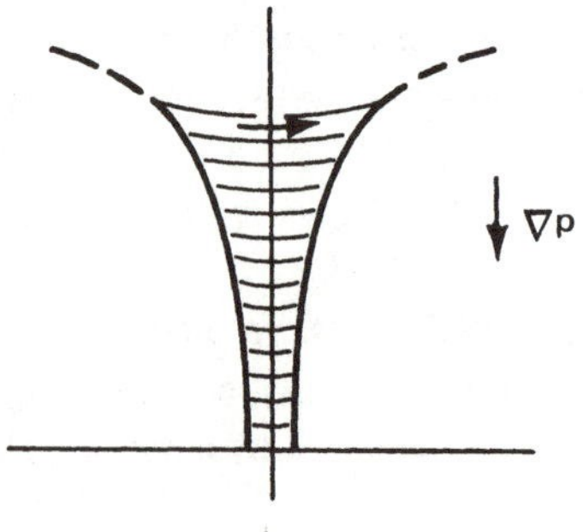

Figure 10.11: This vortex (tornado) can terminate by gradually diffusing into a low p and low ρ region of upper atmosphere or by attachment to a transient rotating mass of air (a rotator cloud).

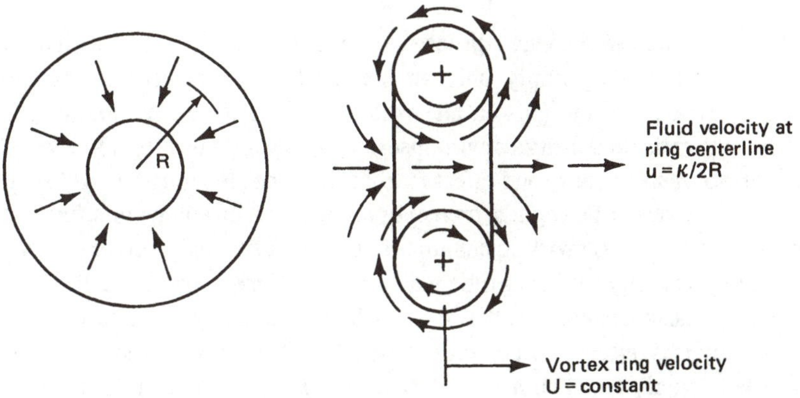

Figure 10.12: *A toroidal vortex, e.g., a smoke ring, propels itself forward. Viscosity of a real fluid causes R to increase and U to decrease.*

10.3 The Formula of Biot and Savart

A circular vortex ring, as shown in Figure 10.12, with radius R and its axis coincident with the x axis and rotating as shown, will propel itself forward along the x axis, as can be verified by integrating the net effect of each element $d\ell$ of the circular vortex line of strength κ. The induced velocity of an arbitrary point on the x axis is computed as

$$u = \frac{\kappa}{2}\frac{R^2}{(R^2 + x^2)^{3/2}} \tag{10.10}$$

At the center of the ring (x = 0) the induced velocity is $u = \kappa/2R$. The ring does not induce any component of velocity in itself that would change R, so an unperturbed ring remains constant in size in an inviscid fluid. In a viscous fluid it enlarges.

The flow field induced by a line vortex, either straight or curved, is computed using an equation similar to the formula of Biot and Savart for the magnetic field **B(R)** that is induced by a current i flowing through a conductor ℓ, with μ_o a constant

$$\mathbf{B} = \frac{\mu_o i}{4\pi}\int_\ell \frac{d\boldsymbol{\ell} \times \mathbf{R}}{R^3}$$

The formula assumes perfect conditions, equivalent in fluid theory to an unbounded constant-density inviscid fluid. Under these conditions, the formula of Biot and Savart can be used to compute the velocity field **V** induced by a line vortex ℓ of strength κ

$$\mathbf{V} = \frac{\kappa}{4\pi}\int_\ell \frac{d\boldsymbol{\ell} \times \mathbf{R}}{R^3} \tag{10.11}$$

The words *induced by*, when discussing inviscid fluids, do not imply a cause and effect relationship; they imply only an idealized flow field consistent with the mathematical model of an inviscid line vortex. The existence of finite boundaries, viscosity, and variable density compromises the application of (10.11) in real fluids. In the real world some cause and effect mechanism is needed to induce velocities, for example, conservation of angular momentum, diffusion of momentum by viscosity, secondary flows, etc. A swirling column of air behind the sharp corner of a building is initiated by vorticity in the shed boundary layer. A smoke ring (see Figure 1.3) is produced by viscous shear between a jet of blown smoke and the surrounding air. Although they may resemble line vortices, details of their flow patterns, $v_\theta(r)$, should be expected to vary from both the inviscid vortex, $v_\theta = \kappa/r$, and the rigid rotator, $v_\theta = \omega r$.

Vortex rings, both ideal and real, interact with each other and with solid surfaces. Two equal-strength ideal circular vortex rings on the same axis with opposite rotations, that induce them to approach each other, will enlarge as they approach. As they increase in size they translate more slowly; they also rotate more slowly as a consequence of conservation of energy and angular momentum. There is no flow across the plane midway between the rings, so the plane can be replaced by a solid surface. The solution then represents a vortex ring such as a smoke ring approaching a solid surface.

Two rings of similar rotation placed near each other can produce interesting results. If two rings have the same axis and interact, they induce velocities in themselves such that the forward one grows larger; the following one shrinks, and then passes through the larger, after which the process repeats again and again, alternately passing through each other. For photographs, see Van Dyke (1982). One line of current research attempts to understand turbulence as the random generation, interaction, intensification, and dissipation of vortices (*Physics Today*, March 1993).

10.4 Rankine's Combined Vortex

A closer mathematical approximation to many vortex-like motions observed in nature is a vortex consisting of a rigidly rotating core with radius R surrounded by an inviscid vortex with matching v_θ and p at R. The size of R is arbitrary in the model; in the real flow it is determined by tangential velocities, viscosity, turbulence, and/or the introduction of nonrotating fluid at the vortex center. Figure 10.13 illustrates this model. Included are graphical representations of $v_\theta(r)$ and p(r). The model is called *Rankine's combined vortex.*

A physically real model requires that both velocity and pressure be continuous at $r = R$. Within the core, velocity is given by $v_\theta = \omega r = \zeta r/2$. By Stokes's theorem, circulation is the integral of ζ over the area of the circle of radius R, so $\Gamma = \pi R^2 \zeta$. The inviscid vortex strength equivalent to this Γ is then $\kappa = \Gamma/2\pi = R^2\zeta/2$, giving

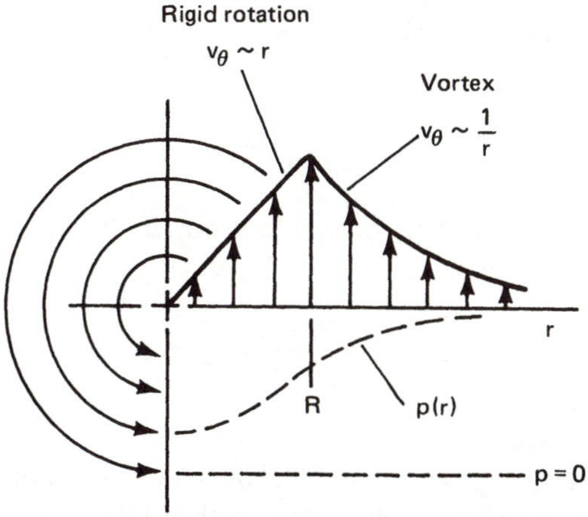

Figure 10.13: *A rigidly rotating core surrounded by an inviscid vortex is called Rankine's combined vortex and approximates many real flows.*

vorticity, $\zeta = 2\kappa/R^2$, as a function of an equivalent vortex with strength κ. Velocity inside the core as a function of this κ is $v_\theta = \kappa r/R^2$, which is then continuous with the induced velocity outside the core, $v_\theta = \kappa/r$. At $r = R$, both become $v_\theta = \kappa/R$.

Pressure is continuous over r as follows. Inside the core the radial pressure gradient at a point is equal to density times centripetal acceleration at that point $dp/dr = \rho r\omega^2$. After using the relationship of the previous paragraph, $\zeta = 2\kappa/R^2 = 2\omega$, the pressure gradient becomes

$$dp/dr = \rho r\kappa^2/R^4$$

This integrates to

$$p_1 = p_o + \frac{\rho r^2 \kappa^2}{2R^4}$$

where p_o is the pressure at the center and p_1 is p(r) inside the core. In the external irrotational flow (inviscid vortex), the Bernoulli equation applies. Equating conditions at arbitrary r to conditions at r_∞, where $q = v_\theta = 0$, gives

$$\frac{p}{\rho} + \frac{v_\theta^2}{2} = \frac{p}{\rho} + \frac{\kappa^2}{2r^2} = \frac{p_\infty}{\rho}$$

or

$$p_2 = p_\infty - \frac{\rho\kappa^2}{2r^2} \tag{10.12}$$

for the pressure field p(r) outside the core.

Equating p_1 and p_2 at $r = R$ yields

$$p_\infty - p_o = \frac{\rho \kappa^2}{R^2} \tag{10.13}$$

as a relationship between vortex strength and pressures at the center and at infinity. Therefore, inside $r = R$, pressure is

$$p_1 = p_\infty - \frac{\rho \kappa^2}{R^2}\left(1 - \frac{r^2}{2R^2}\right) \tag{10.14}$$

and outside $r = R$, pressure is given by (10.12), both now as functions of the ambient pressure at infinity. Figure 10.13 graphs p(r), and shows that at $r = R$, both (10.12) and (10.14) become

$$p = p_\infty - \rho \kappa^2 / 2R^2$$

A circular vortex can occur in a free-surface liquid, as in Figure 10.14. In this representation of Rankine's combined vortex, a pressure field $(-gz)$ in the liquid is included in the model. Pressure at the free surface is p_∞. After integration and use of the boundary conditions, the equation of the free surface becomes, with +z measured upward from the free surface at infinity,

$$r > R, \qquad z = -\frac{\kappa^2}{2gr^2} \tag{10.15a}$$

$$r < R, \qquad z = \frac{\kappa^2}{gR^2}\left(\frac{r^2}{2R^2} - 1\right) \tag{10.15b}$$

At $r = R$, both converge to $z = -\kappa^2 / 2gR^2$.

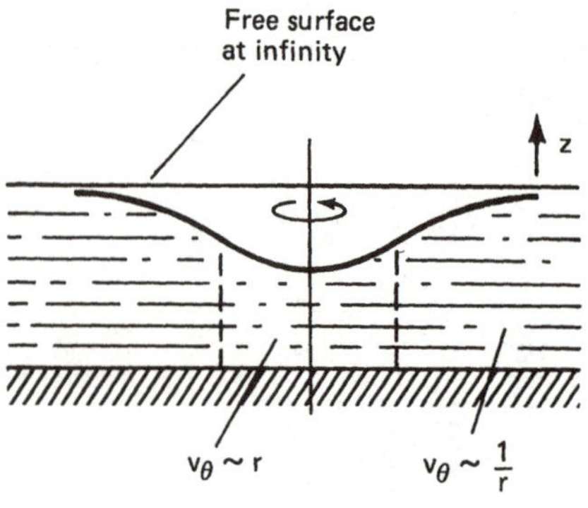

Figure 10.14: *The model of Figure 10.13 in a free-surface liquid is easily observed by depressions of the free surface and by particles floating on or suspended in the liquid.*

10.5 Vortex Intensification by Stretching

Vortex intensification by stretching does not apply to the idealized line vortex defined by $v_\theta = \kappa/r$. For example, consider a finite section, say ℓ_1 long, of an inviscid line vortex that is stretched by a length ratio of $k = \ell_2/\ell_1$. The velocity field of the vortex is given as $v_\theta = \kappa/r$ and represents a flow where each fluid element has the same angular momentum. As section ℓ_1 of the flow field is stretched, fluid moves inward, increasing its tangential velocity so that $\rho \mathbf{r} \times \mathbf{v} = \mathbf{H}$ = constant. Fluid brought in from r_2 to any smaller radius r_1 will have the same v_θ as the original fluid had at r_1. The flow pattern remains the same, and κ, Γ, and ζ also remain the same. Stretching any portion of an inviscid line vortex does not intensify the vortex because the line representing the vortex has zero diameter regardless of how much it is stretched or shrunk.

Other difficulties also exist. Because the model $v_\theta = \kappa/r$ is defined to $r \to \infty$, the included mass is not finite nor is the increase in mass due to stretching. Further, the energy required to stretch a vortex can be computed as

$$E = \int_{\ell_1}^{\ell_2} \mathbf{F} \cdot d\boldsymbol{\ell}$$

where $\mathbf{F}$ is the net force due to $p_\infty - p(r)$ on the top and bottom surfaces bounding a given length of the vortex, and $\ell_2 - \ell_1$ is the amount of stretching. Two difficulties are apparent in application to an inviscid line vortex model. One, the pressure differential $(p_\infty - p)$ at $r = 0$ must approach minus infinity; and two, the net force due to the pressure differential, even for some finite central r_o, approaches infinity as $r \to \infty$. Negative pressure is an impossibility; pressure cannot go to $-\infty$, only to zero. Obviously, r of the vortex must have a minimum (r_o), where $p = 0$, for any realizable flow. Also obviously, r never approaches infinity for any realizable flow, even on cosmological scales. The inviscid vortex is merely a mathematical convenience that sometimes approximates real physical flows and sometimes does not.

Now consider a finite sized vortex tube as in Figure 10.15. The vortex tube rotates as a rigid body, and ζ is constant over the tube. For this analysis ignore any induced velocity in the surrounding fluid; consider only fluid within the tube. The tube is stretched with tension force (F) from a length ℓ_1 to a length

$$\ell_2 = k\ell_1 \tag{10.16a}$$

The tension force does not apply a torque to the tube, so angular momentum is conserved. Next, analyze angular momentum (H) and rotational kinetic energy (E), assuming the tube continues to rotate in a rigid body mode

$$H_1 = I_1\omega_1 = \tfrac{1}{2} m r_1^2 \omega_1 \,, \qquad E_1 = \tfrac{1}{2} I_1 \omega_1^2 = \tfrac{1}{4} m r_1^2 \omega_1^2$$

where m is enclosed liquid mass, $\rho \forall$.

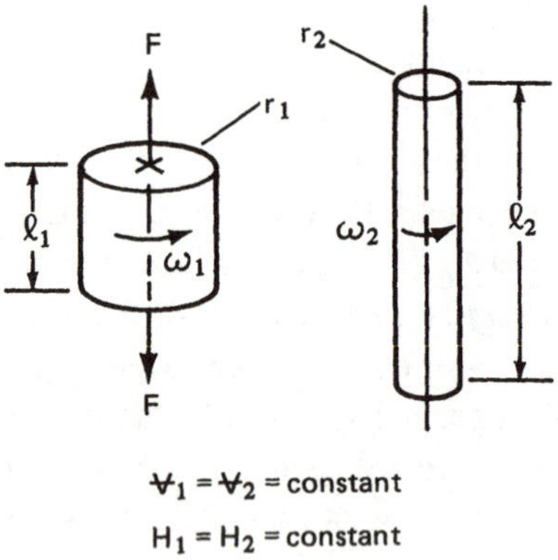

Figure 10.15: *Vorticity intensification by stretching. ζ increases but Γ remains constant for constant ρ, $\forall$, and H.*

Mass is conserved, and if $\rho = c$, volume also is conserved, so $\forall_1 = \forall_2$, and

$$\pi r_1^2 \ell_1 = \pi r_2^2 \ell_2 = \pi r_2^2 k \ell_1$$

$$r_2^2 = r_1^2 / k \tag{10.16b}$$

With angular momentum conserved, $H_2 = H_1$, and

$$\frac{1}{2} m r_1^2 \omega_1 = \frac{1}{2} m r_2^2 \omega_2 = \frac{1}{2} m \left(r_1^2/k\right) \omega_2$$

$$\omega_2 = k\omega_1 \tag{10.16c}$$

Since $\zeta = 2\omega$, then

$$\zeta_2 = k\zeta_1 \tag{10.16d}$$

with the result that vorticity is increased (intensified) proportional to increase in length (stretching). Circulation around the tube, however, remains constant because

$$\Gamma_2 = \zeta_2 \pi r_2^2 = (k\zeta_1)\, \pi \left(r_1^2/k\right) = \Gamma_1 \tag{10.16e}$$

For reference, tangential velocity at the periphery increases from $v_1 = \omega_1 r_1$ to

$$v_2 = \omega_2 r_2 = (k\omega_1) \left(r_1/\sqrt{k}\right) = \sqrt{k}\; v_1 \tag{10.16f}$$

Comparison of E_1 and E_2 yields

$$E_2 = kE_1 \tag{10.16g}$$

so that rotational kinetic energy is increased. The increase dE can be computed as $Fd\ell$, where F is the force required to stretch the vortex a distance $d\ell$

$$Fd\ell = dE = \frac{1}{4}\,\rho\pi r^4\,\omega^2\,d\ell$$

$$F = \frac{\pi}{4}\rho r^4\omega^2 = \frac{\pi}{16}\rho r^4\zeta^2 = \frac{\pi}{4}\rho\kappa^2 = \frac{\rho\Gamma^2}{16\pi} \tag{10.17}$$

using relationships of the previous paragraphs.

Vortex stretching and intensification are seen to be 3-D phenomena of finite vortex tubes. Because vorticity remains attached to material fluid particles, the vortex tube must shrink in diameter as it is stretched. With no applied moment, it must also conserve angular momentum by spinning faster. This increases ω and ζ, but not κ and Γ, which, like H, remain constant. Remember that circulation has units of angular momentum per unit mass (rv compared to ρrv), so if density is constant, $2\pi\kappa = \Gamma \propto H = \text{constant}$. The general circulation theorem (10.9) also requires this result since no mechanism has been invoked that can change Γ, i.e., $\rho = c$, $\mu = 0$, and F is conservative.

The difficulty with this model is the discontinuity in v_θ at $r = R$. Any viscosity will modify this and diffuse some angular momentum outward to round-off both the $v_\theta(r)$ curve and the p(r) curve. In this case the flow is not the induced flow of an inviscid vortex but rather a velocity field forced into existence by viscous shear and/or secondary flows. The configuration departs from any simple model, but it can be roughly approximated by Rankine's combined vortex, provided v_θ is truncated to zero within a few multiples of R.

Figure 10.16 depicts a simple laboratory demonstration of vortex intensification by stretching. A vertical vortex tube is created in a horizontal flow of water by quickly inserting and removing a plate held perpendicular to the flow. Within a short distance (i.e., time interval) the vortex shed at the plate edge becomes a vortex tube. Its diameter is apparent from the depression of the water surface, and its spin velocity is apparent from particles suspended on and/or in the water. As the vortex tube is carried over the obstruction, its length must shorten. The vortex enlarges in diameter and reduces its spin velocity. As it returns to deeper water again, it is stretched, shrinks radially, and spins faster. The effects of viscosity and secondary flows at the bottom surface prevent the vortex tube from returning to its exact original spin speed.

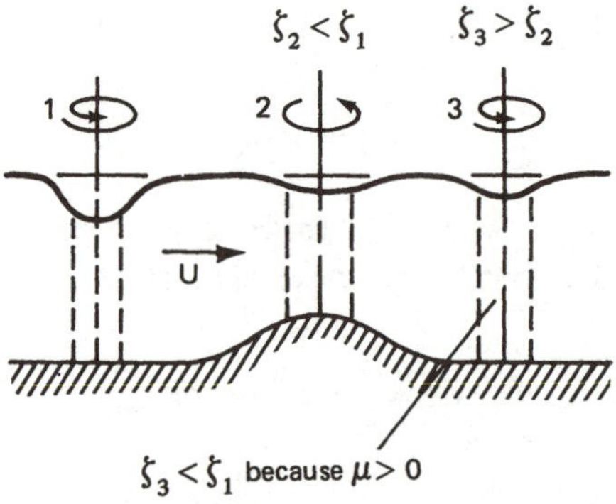

Figure 10.16: *Flow of a vortex tube over a bump illustrates vortex shortening and stretching. When its length is reduced it spins slower, and when it is stretched, it intensifies.*

10.6 Application Examples

This section introduces interactions between vortex structures, the ambient flow, and adjacent walls. Theorems by Helmholtz and Kelvin describe vorticity and circulation in an inviscid fluid. A more general theorem includes nonconservative fields and nonbarotropic pressure-density distributions as mechanisms, in addition to viscosity, capable of generating vorticity and circulation. Rankine's combined vortex and vortex stretching are analyzed. Example 10.1 develops further details of an inviscid vortex, Example 10.2 applies vortex stretching to a specific example, Example 10.3 examines a bathtub vortex, and Example 10.4 illustrates cavitation in a liquid vortex.

Example 10.1. Describe the flow field defined by a velocity potential $\phi = -\theta r v_\theta$, where $r v_\theta$ = constant. Test to verify that the flow is irrotational. Does this flow field represent flow of an incompressible fluid? Derive the equivalent stream function (ψ). Show that ϕ and ψ satisfy Laplace's equation.

Discussion. The components of velocity ($\mathbf{V} = -\nabla\phi$) in polar coordinates are

$$v_r = -\frac{\partial\phi}{\partial r} = 0\,, \quad v_\theta = -\frac{1}{r}\frac{\partial\phi}{\partial\theta}$$

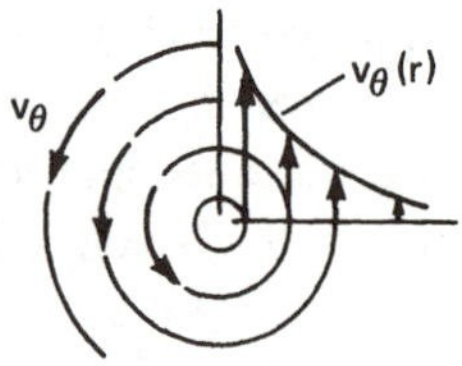

Let the symbol κ represent the constant value of $r v_\theta$ so that

$$v_\theta = -\frac{1}{r}\frac{\partial(-\theta\kappa)}{\partial\theta} = \kappa / r$$

which indicates the flow follows circular pathlines ($v_r = 0$, and $v_\theta = \kappa/r$). The constant, κ, is the vortex strength, and the flow is called a *free*, *line*, or *inviscid* vortex.

That **V** is derived from the gradient of a scalar function ($-\nabla\phi$) indicates the flow field must be irrotational. However, it is instructive to begin with given values for v_r and v_θ and ask whether the flow is irrotational. If so, all components of vorticity, here given in cylindrical coordinates, must be zero

$$\left(\frac{1}{r}\frac{\partial v_z}{\partial\theta} - \frac{\partial v_\theta}{\partial z}\right),\quad \left(\frac{\partial v_r}{\partial z} - \frac{\partial v_z}{\partial r}\right),\quad \frac{1}{r}\left(\frac{\partial(rv_\theta)}{\partial r} - \frac{\partial v_r}{\partial\theta}\right)$$

The first two terms do not apply to a 2-D flow because they involve either v_z or $\partial/\partial z$. In the last term $v_r = 0$, so $\partial v_r/\partial\theta = 0$. Next, using $rv_\theta = \kappa$ = constant indicates $\partial\kappa/\partial r = 0$. All terms therefore are zero, and the flow is irrotational. All fluid elements in this flow move in circular paths about the origin, but none of the fluid elements themselves rotate.

The continuity equation in polar coordinates is

$$\frac{\partial(r\,v_r)}{\partial r} + \frac{\partial\, v_\theta}{\partial\theta} = 0$$

Substituting in $v_r = 0$ and v_θ gives

$$\frac{\partial(\kappa / r)}{\partial\theta} = 0$$

and the flow field does represent an incompressible flow.

The stream function ψ (see Chapter 3) in polar coordinates is

$$v_r = \frac{1}{r}\left(\frac{\partial\psi}{\partial\theta}\right),\quad v_\theta = -\frac{\partial\psi}{\partial r}$$

Integrating the expression for v_θ

$$\psi = \int \frac{\partial\psi}{\partial r}\,dr + f(\theta) = \int -\frac{\kappa}{r}\,dr + f(\theta) = -\kappa\,\ell n\, r + f(\theta)$$

Use the expression for v_r to evaluate $f(\theta)$

$$v_r = \frac{1}{r}\frac{\partial\psi}{\partial\theta} = \frac{1}{r}\,f'(\theta) = 0,\ \text{therefore, } f(\theta) = \text{constant (c)}$$

$$\psi = -\kappa\,\ell n\, r + c$$

Verify the result

$$v_r = \frac{1}{r}\frac{\partial\psi}{\partial\theta} = 0,\quad v_\theta = -\frac{\partial\psi}{\partial r} = \frac{\kappa}{r}$$

Using (3.11) and (3.12) from Chapter 3, verify that both θ and ψ for inviscid incompressible flows satisfy Laplace's equation using Cartesian coordinates. The reader can easily verify the same result here in polar coordinates. Refer to Appendix A.6 for Laplace's equation in polar coordinates.

Angular momentum about the origin of a specific small mass of fluid is $\mathbf{R} \times d\mathbf{P} = \mathbf{R} \times \mathbf{V}\ dm$, or $\mathbf{R} \times \mathbf{V}\ \rho$ per unit volume. If density is uniform and constant over the flow, angular momentum per unit mass is $\mathbf{R} \times \mathbf{V}$. Further, if $\mathbf{V}$ is perpendicular to $\mathbf{R}$ (as v_θ is to r), then angular momentum per unit mass becomes (rv_θ). The problem states rv_θ = constant, which is equivalent to saying, *Each element of fluid in this flow has the same angular momentum.*

Free (inviscid) vortices are used as approximate, or at least limiting, models for many flows in nature (e.g., hurricanes) and in engineering (e.g., drain vortices, or cyclonic separators). In one important application, ϕ of a vortex is added to ϕ for flow about a cylinder to produce a model of flow with circulation past a cylinder. This is used in Chapter 16 as one method to explain and to model lift of an aircraft wing.

Example 10.2. A tornado develops consisting of a 60 m diameter column of rotating air. It extends from a 400 m high cloud to the ground, and it has a maximum tangential wind speed of 120 km/hr. During the next 10 min the parent cloud is lifted to a height of 900 m. Estimate final radius, final v_θ max, and final p_o for the tornado.

Discussion. A fairly reasonable estimate can be obtained by assuming that the initial and final columns of air are in rigid rotation ($v_\theta = r\omega$), that angular momentum is conserved during the axial stretching, and that the volume of the column remains constant, i.e., vorticity remains attached to the same particles, and density remains nearly constant.

$$\mathcal{V}_1 = \mathcal{V}_2, \quad \pi r_1^2 h_1 = \pi r_2^2 h_2, \quad r_2 = r_1\sqrt{h_1/h_2} = 20 \text{ m}$$

$$H_1 = H_2, \quad I_1\omega_1 = I_2\omega_2, \quad (1/2)\, m r_1^2 \omega_1 = (1/2)\, m r_2^2 \omega_2$$

$$\omega_2 = \omega_1(r_1^2/r_2^2) = \omega_1(h_2/h_1) = 2.5 \text{ rad/s}$$

$$v_\theta(r) = \omega_2 r = 2.5\, r \text{(m/s)}, \quad \text{final } v_\theta \max = 50 \text{ m/s} = 180 \text{ km/hr}$$

$$dp/dr = \rho r\omega^2, \quad dp = \rho r \omega^2 dr, \quad p_a - p_o = \rho\omega^2 r^2/2$$

$$p_o = p_a - \rho\omega^2 r^2/2 \quad \text{(with } p_a = \text{atmospheric pressure)}$$

$$p_o = 1.013(+5) - 1.23 \times (2.5 \times 20)^2/2$$

$$\text{Final } p_o = 0.998 \times 10^5 \text{ Pa} = 0.98 \text{ atm}$$

Chapter 22 includes further descriptions of tornado phenomena.

Example 10.3. A tank, in London, England, precisely cylindrical, 2 m diameter, and with the axis vertical, is filled 50 cm deep with water. A small circular drain pipe is precisely centered in the perfectly horizontal bottom. The water is maintained free of motion, surface air currents, and temperature gradients for several weeks. Water is then very slowly drained from the central pipe. Why does a bathtub vortex form? Estimate its magnitude when the tank is half drained.

Discussion. Relative to inertial space the earth spins at 7.27×10^{-5} rad/s (i.e., ≈ 1 rev/24hr), counter clockwise viewed from the north. The vertical component of this angular velocity vector at the latitude of London (51.5°N) is 7.27(−5) sin 51.5° = 5.69(−5) rad/s. Water, stationary in the tank, and the tank will have this vertical component of earth angular velocity. As water is drawn toward the center by the radial pressure difference due to the release of water from the pipe, its angular momentum ($\rho \mathbf{r} \times \mathbf{v}$) will be conserved as in Figure 7.3. The radial pressure gradient replaces the string in the earlier example.

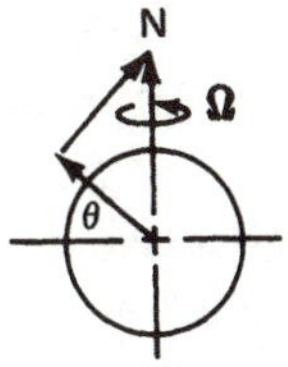

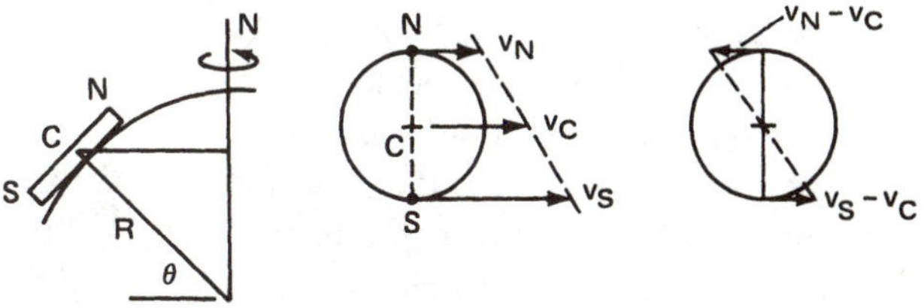

Here ($\rho \mathbf{r} \times \mathbf{v}$) must be relative to inertial space. Velocity at point C, in the tank shown, has the earth's tangential velocity at point C. The velocity at point N is a little less since it is nearer the earth's axis, and at point S is a little larger. The vertical component of the angular velocity of the tank, the water, and London relative to inertial space is more apparent by subtracting v_c from each velocity. When water is withdrawn, the remaining water can move toward the center of the tank, increasing its tangential velocity to be greater than that of the tank. The water in the tank, therefore, simply increases its counterclockwise angular velocity while the tank does not. The water will spin counterclockwise relative to the tank. A similar analysis for the Southern Hemisphere will indicate clockwise rotation relative to the tank.

The form of v(r) here differs from an inviscid vortex ($v \propto r^{-1}$), because the water in the tank, when half empty, did not all start at r_{max}. A formal solution is difficult, but some limiting cases can easily be estimated. If the bottom shaded portion drained, and the water was inviscid, the top half would simply settle down without changing its spin rate. If the central shaded portion drained, and the remainder expanded inward uniformly, the result might nearly resemble a classic inviscid vortex. If, for example,

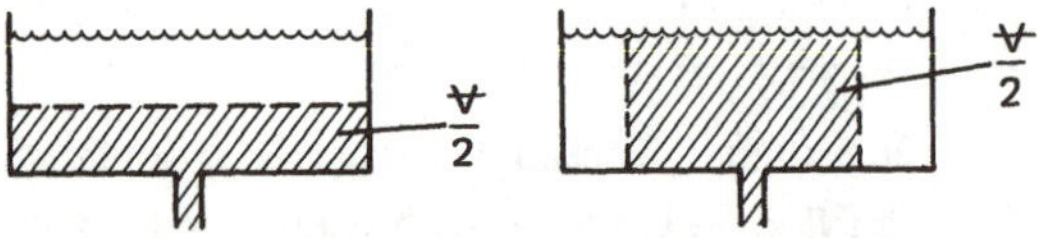

fluid moved in to half its original radius, its tangential velocity would double, and its angular velocity would quadruple (four times earth rate), with $\mathbf{r} \times \mathbf{v}$ remaining constant. Chapter 11 will show, however, that after a brief initial outflow from the central region, most of the remaining outflow will come from a thin bottom boundary layer.

Vortices caused by concentration of earth angular momentum are difficult to create on a laboratory scale because the slightest perturbation can dominate the effect of the earth's rotation, but on the scale of the large atmospheric and oceanic motions discussed in Chapters 20 and 21, earth rotation dominates all other phenomena. The vortex in the usual bathtub outflow is caused by residual motion in the water and/or by irregularities in the shape of the tub. These vortices can easily be caused to spin in either direction, i.e., with, or opposed to, the earth's rotation.

Example 10.4. The same tank as in Example 10.3 is now equipped with a pump that forces the tank to drain at a controllable rate. Assume the central outlet pipe is 8 cm diameter. The pumped water is then ejected tangentially at the outside wall through a vent 1 cm wide and fluid height (50 cm). Investigate the flow pattern as pump capacity is increased. Assume first the water is inviscid, and then examine the effect of adding viscosity. Use centimeter-gram-second (cgs) units.

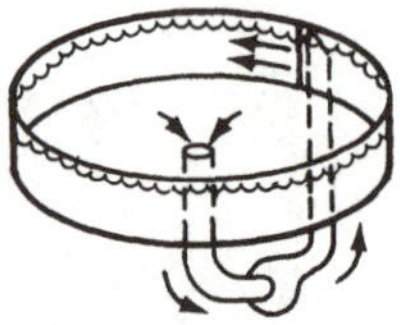

Discussion. After all the water in the tank has flowed once through the pump, it has reentered the tank with the same tangential speed, and, therefore, it all has the same angular momentum per unit mass. Angular momentum is conserved as the water, assumed inviscid, moves radially inward, and an inviscid vortex is formed with ($v_\theta = \kappa/r$). Vortex strength can be evaluated at r_{max} as (κ = 100 v_θ), with v_θ expressed as Q/A, where Q is pump capacity (cm^3/s) and A is vent area (cm^2)

$$\kappa = 100\ Q/50 = 2Q\ \text{cm}^2/\text{s}$$

Equation (10.15a) for the free-surface height of an inviscid vortex is $z = -\kappa^2/2gr^2$. Note that as $r \rightarrow 0$, $z \rightarrow -\infty$, indicating that inviscid vortices in a free-surface liquid must have a funnel-shaped depression extending to the bottom of the liquid container. Any viscosity, of course, causes the flow to resemble Rankine's model and modifies this result. In either case the pump completely cavitates when the funnel diameter reaches 8 cm, the diameter of the outlet pipe. At this vortex strength, pressures due to centrifugal force will balance pressures due to gravity, and no water can flow from the tank into the pump.

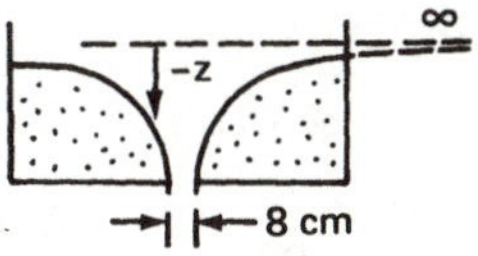

Neglecting the change in height at the periphery due to the loss of volume in the funnel, and assuming the inviscid model to be valid, the flow rate that completely cavitates the outlet is

$$-50 = -(2Q)^2/2 \times 9.81 \times 4^2$$

$$Q = 62.6\ \text{cm}^3/\text{s} = 3.76\ \text{liters/min}$$

Inlet velocity at this flow rate is

$$v_\theta = Q/A = 1.25\ \text{cm/s}$$

Both are very small quantities for most engineering operations. They illustrate the ease with which vortex cavitation occurs and the need to place vanes near an outlet to suppress vortex motions.

The viscosity of water would modify the result only very slightly for the maximum flow rate ($Q = 62.6\ \text{cm}^3/\text{s}$); however, a different phenomena would become very apparent. A thin boundary layer would form on the bottom surface. Water in the boundary layer would not be spinning, it would have no centrifugal pressure to balance gravity pressure, and it will flow radially inward. This phenomena is an example of a secondary flow and is detailed in Chapter 11.

Chapter 11

Secondary Flows

11.1 Boundary Layer Review

It is instructive to review several features of nonrotating boundary layer flows (see Chapter 4) for comparison to boundary layer flows with rotation. Laminar viscous flow past a flat plate aligned with the flow results in a boundary layer whose thickness grows according to $\delta \sim \sqrt{\nu x/U}$, while the boundary layer on an infinite flat plate, impulsively accelerated to a constant velocity (in the plane of the plate), grows according to $\delta \sim \sqrt{\nu t}$. One grows with distance, the other with time, although the difference is less impressive by realizing that $t \sim x/U$.

If the infinite flat plate is oscillated in its plane with a velocity $u = U \cos(\omega t)$, with ω the circular frequency of oscillation, an entirely different result is obtained. In this case the boundary layer thickness does not grow to infinity but rather reaches a limit known as the depth of penetration, $\delta \sim \sqrt{\nu/\omega}$. Flows, induced in opposite directions as the plate oscillates, diffuse together and cancel 99% of each other by depth δ. Boundary layers in rotating flows also reach finite depths, which also will be seen to have $\delta \sim \sqrt{\nu/\omega}$, where ω is angular velocity.

11.2 A Rotating Disk in a Stationary Fluid

The problem of an object translating at constant velocity through a stationary fluid is identical to that of a fluid moving with constant velocity past a stationary object. If the stationary frame is inertial, then any frame moving with constant velocity must also be inertial. A simple coordinate transformation transposes one problem to the other. But, because *there is one and only one nonrotating inertial frame,* the problem of a *rotating fluid* in the presence of a nonrotating object must be different than that of a *rotating object* in the presence of a nonrotating fluid. If one frame is inertial, the other cannot be.

Consider an infinite flat surface rotating at rate Ω in its plane in a nonrotating fluid ($\rho = c$, $\mu = c$) extending to infinity as shown in Figure 11.1. In this problem the

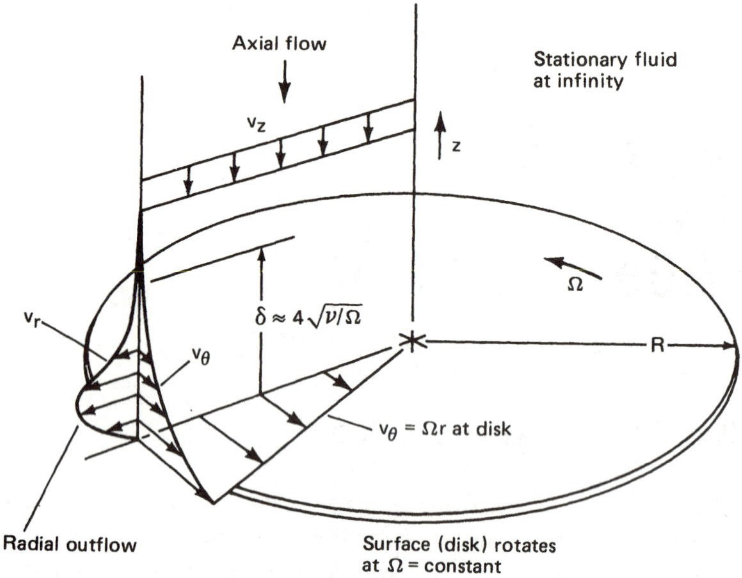

Figure 11.1: *A secondary radial flow is induced by a rotating flat surface (disk) in a nonrotating fluid, here with ρ and μ constant. The flow up to $z = \delta$ is an Ekman boundary layer.*

complete set of Navier-Stokes equations can be reduced by a transformation to an exact set of ordinary differential equations for solution by numerical or other means. Terms identically equal to zero are dropped, of course. Note that this is different than flows (e.g., the Prandtl boundary layer equations) where some terms in the basic differential equations are assumed negligible on the basis that they are smaller in magnitude than others. *Smaller in magnitude* is much different than *identically zero.*

After omitting terms identically zero, the equations of motion in physical variables and in an inertial reference system are

$$v_r \frac{\partial v_r}{\partial r} - \frac{v_\theta^2}{r} + v_z \frac{\partial v_r}{\partial z} = -\frac{1}{\rho}\frac{\partial p}{\partial r} + \nu\left[\frac{\partial^2 v_r}{\partial r^2} + \frac{\partial}{\partial r}\left(\frac{v_r}{r}\right) + \frac{\partial^2 v_r}{\partial z^2}\right] \tag{11.1a}$$

$$v_r \frac{\partial v_\theta}{\partial r} + \frac{v_r v_\theta}{r} + v_z \frac{\partial v_\theta}{\partial z} = \nu\left[\frac{\partial^2 v_\theta}{\partial r^2} + \frac{\partial}{\partial r}\left(\frac{v_\theta}{r}\right) + \frac{\partial^2 v_\theta}{\partial z^2}\right] \tag{11.1b}$$

$$v_r \frac{\partial v_z}{\partial r} + v_z \frac{\partial v_z}{\partial z} = -\frac{1}{\rho}\frac{\partial p}{\partial z} + \nu\left[\frac{\partial^2 v_z}{\partial r^2} + \frac{1}{r}\frac{\partial v_z}{\partial r} + \frac{\partial^2 v_z}{\partial z^2}\right] \tag{11.1c}$$

$$\frac{\partial v_r}{\partial r} + \frac{v_r}{r} + \frac{\partial v_z}{\partial z} = 0 \tag{11.1d}$$

Omission of the $\partial/\partial\theta$ terms implies rotational symmetry for the solution. This is valid for flows that remain laminar, but for higher angular velocities, flows that are periodic in θ and/or turbulent occur, invalidating this solution.

Boundary conditions consistent with no-slip at the rotating surface are

$$\begin{aligned} &\text{At} \quad z = 0; \quad v_r = 0,\ v_\theta = r\Omega,\ v_z = 0 \\ &\text{At} \quad z = \infty; \quad v_r = 0,\ v_\theta = 0 \end{aligned} \tag{11.2}$$

The equations are made dimensionless using

$$z^* = z / \delta = z / \sqrt{\nu / \Omega} \tag{11.3a}$$

$$v_r^*(z^*) = v_r / r\Omega \tag{11.3b}$$

$$v_\theta^*(z^*) = v_\theta / r\Omega \tag{11.3c}$$

$$v_z^*(z^*) = v_z / \delta\Omega = v_z / \sqrt{\nu\Omega} \tag{11.3d}$$

$$p^*(z^*) = p / \mu\Omega = p / \rho\nu\Omega \tag{11.3e}$$

Choice of characteristic quantities (e.g., δ, $r\Omega$, $\delta\Omega$, and $\mu\Omega$) can often be deduced correctly by a careful analysis of the physical problem. Experience of course helps, but the ultimate test is whether the transformed dimensionless equations are more accessible to solution than the original equations. In practice a researcher will attempt several reasonable choices, accepting the form that shows most promise of success.

After substituting (11.3a-e) into (11.1a-d), the following set of four coupled, dimensionless, ordinary differential equations is obtained. Asterisks are omitted for clarity. Note that the physical parameters ν and Ω do not appear in the new equations nor in the boundary conditions, and consequently do not predetermine the solution.

$$\frac{d^2 v_r}{dz^2} - v_z \frac{dv_r}{dz} - v_r^2 + v_\theta^2 = 0 \tag{11.4a}$$

$$\frac{d^2 v_\theta}{dz^2} - v_z \frac{dv_\theta}{dz} - 2 v_r v_\theta = 0 \tag{11.4b}$$

$$\frac{d^2 v_z}{dz^2} - v_z \frac{dv_z}{dz} - \frac{dp}{dz} = 0 \tag{11.4c}$$

$$\frac{dv_z}{dz} + 2 v_r = 0 \tag{11.4d}$$

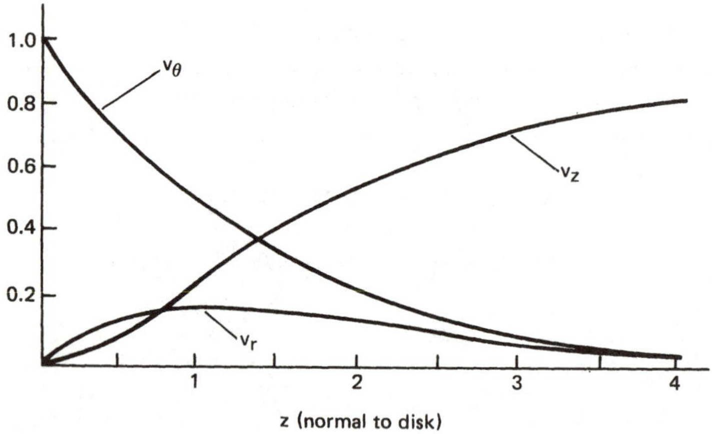

***Figure 11.2:** Dimensionless velocity components of flow (v_r, v_θ, v_z) induced by a spinning disk, in a fluid otherwise at rest, are a function of dimensionless distance (z) from the disk.*

Boundary conditions, also in dimensionless variables, are now

$$\begin{aligned} &\text{At} \quad z = 0\,; \quad v_r = 0\,,\ v_\theta = 1\,,\ v_z = 0,\ p = 0 \\ &\text{At} \quad z = \infty\,; \quad v_r = 0\,,\ v_\theta = 0 \end{aligned} \tag{11.5}$$

The equations still are not trivial, nor are they solvable in closed analytical form, but they are much more accessible to solution by numerical procedures than the original set of partial differential equations. Furthermore, the variables are now scaled to the specific problem of a surface (or disk) rotating in a fluid otherwise at rest, and the equations do not need to be solved for specific values of ν and Ω. Only after the solution has been returned to physical variables do they depend on ν and Ω.

Solutions are shown pictorially in Figure 11.1, graphically in Figure 11.2, and numerically in Table 11.1. The decision to make z dimensionless with a boundary layer thickness $\delta = \sqrt{\nu/\Omega}$ is validated by the solution. A major feature of the solution is the efficiency of the rotating surface or disk in pulling fluid axially downward and pumping it radially outward. The main flow here is tangential, and the radially outward flow is the secondary flow caused by the rotating boundary layer. The rotating boundary layer is commonly referred to as the *Ekman layer*. The tangential component of shear stress at the wall ($\tau = \mu\, \partial v_\theta/\partial z = \mu\, 0.616$), integrated over the area of the disk, is the moment (M) required to maintain the motion. Pumping capacity (Q) is computed as the integral of v_r from $z = 0$ to ∞. The data in Table 11.1 is summarized from Sparrow and Gregg (1960).

Table 11.1: Dimensionless Flow Solution of a Rotating Disk in a Fluid at Rest

z	v_r	v_θ	$-v_z$	p	$-\frac{\partial v_\theta}{\partial z}$
0	0	1.0	0	0	0.616
0.2	0.084	0.878	0.018	0.167	0.599
0.4	0.136	0.762	0.063	0.275	0.558
0.6	0.166	0.656	0.124	0.340	0.505
0.8	0.179	0.561	0.193	0.377	0.448
1.0	0.180	0.468	0.266	0.395	0.391
2.0	0.118	0.203	0.572	0.401	0.177
3.0	0.058	0.083	0.746	0.395	0.075
4.0	0.026	0.035	0.826	0.393	0.031
∞	0	0	0.886	0.393	0

A qualitative description of the flow is as follows. Because of the no-slip condition, fluid in contact with the surface rotates with the same angular velocity as the surface and experiences the same centripetal acceleration. At the start of motion, a boundary layer begins to form in the circumferential (v_θ) direction. Fluid in the boundary layer (just above the surface) begins to spin but cannot maintain the same centripetal acceleration as the surface. It acquires an outward radial component. As the radial component increases in magnitude, a secondary boundary layer develops in the radial direction with stresses centrally directed. These stresses do provide a central force and a centripetal acceleration greater than zero, but less than that of the surface. At distances greater than δ from the surface, the tangential boundary layer thickness is exceeded, that fluid has no rotation, and no mechanism is available to continue the radial flow. Continuity requires a downward flow to match the outward flow in volume. The net effect is that zero angular momentum fluid is drawn axially from infinity, given angular momentum in the boundary layer, and then pumped radially outward as high angular momentum fluid.

Figure 11.3 depicts the horizontal velocity of the fluid relative to the nonrotating inertial frame at a typical radius for various heights above the surface. The envelope of the horizontal velocity $(v_r^2 + v_\theta^2)^{1/2}$ as a function of z is called an *Ekman spiral*. Again all quantities are made dimensionless. The boundary layer is often referred to as an Ekman boundary layer with characteristic thickness $\delta \sim \sqrt{\nu/\Omega}$. Numerical results for this case yield $\delta \approx 4\sqrt{\nu/\Omega}$, with δ defined as the height where $v_\theta = 0.02\ r\Omega$ (see also Figure 11.1).

Moment required to maintain constant Ω is obtained by integrating the tangential component of fluid stress at the surface over the area to be analyzed. For a disk wetted on one side and of radius R, the required moment is

$$M = -2\pi \int_0^R \tau_{z\theta}\ r^2 dr \qquad (11.6)$$

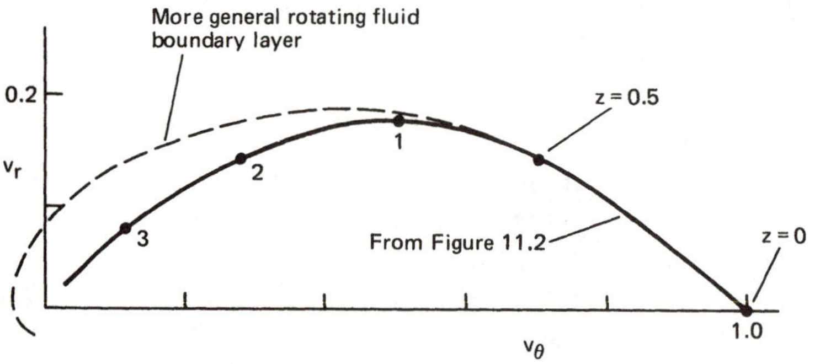

Figure 11.3: *An Ekman spiral shows the relationship of radial to tangential velocities as a function of distance from the surface in a boundary layer influenced by rotation (z becomes a polar angle).*

where $\tau_{z\theta} = \rho r \nu^{1/2}\,\Omega^{3/2}(\partial v_\theta/\partial z)_o$ with $\partial v_\theta/\partial z$ at $z = 0$ obtained from the solution as -0.616. The final value in dimensional variables is

$$M = 0.313\ \pi\rho R^4(\nu\Omega^3)^{1/2} \tag{11.7}$$

The solution assumes edge effects on the finite disk are negligible. Quantity of fluid pumped outward along one side of a disk at a radius R in dimensional variables is

$$Q = 2\pi R\int_0^\infty v_r dz = 0.886\pi R^2(\nu\Omega)^{1/2} \tag{11.8}$$

with dimensionless $v_r(z)$ obtained from the numerical solution.

11.3 A Rotating Fluid above a Stationary Surface

The alternative problem of an *infinite mass of fluid rotating at uniform rate Ω,* and in contact with a nonrotating surface, is analyzed using the same differential equations as before (11.1a-d). The equations again are set in a nonrotating coordinate system but now have changed boundary conditions of

$$\begin{aligned} &\text{At} \quad z = 0; \quad v_r = 0,\ v_\theta = 0,\ v_z = 0 \\ &\text{At} \quad z = \infty; \quad v_r = 0,\ v_\theta = r\Omega \end{aligned} \tag{11.9}$$

Height above the surface (z) and velocity components (v_r, v_θ, v_z) are made dimensionless as before, but pressure is analyzed with a different procedure.

Anticipating the existence of a boundary-layer-type solution, an assumption is made that the pressure distribution p(r) above the boundary layer is constant across the boundary layer thickness of $\delta << 1$ (refer to Chapter 4). The pressure distribution in the fluid, rotating at Ω = constant, is obtained using

$$\frac{\partial p}{\partial r} = \rho r \Omega^2$$

Anticipating the result, note that fluid in the boundary layer on the nonrotating surface has $\Omega \sim 0$, and has no ability to resist the pressure field ($\partial p/\partial r$) in the rotating mass of fluid.

After substitution, a set of three coupled, dimensionless, ordinary differential equations are obtained. Again asterisks are omitted for clarity.

$$\frac{d^2 v_r}{dz^2} - v_z \frac{dv_r}{dz} - v_r^2 + v_\theta^2 - 1 = 0 \tag{11.10a}$$

$$\frac{d^2 v_\theta}{dz^2} - v_z \frac{dv_\theta}{dz} - 2 v_r v_\theta = 0 \tag{11.10b}$$

$$\frac{dv_z}{dz} + 2 v_r = 0 \tag{11.10c}$$

The dimensionless boundary conditions are

$$\begin{aligned} &\text{At} \quad z = 0; \quad v_r = 0, \, v_\theta = 0, \, v_z = 0 \\ &\text{At} \quad z = \infty; \quad v_r = 0, \, v_\theta = 1 \end{aligned} \tag{11.11}$$

The pressure equation does not appear because pressure in the main body of the rotating fluid is known, and the boundary layer approximation assumes this same pressure is impressed into the boundary layer.

The flow field is shown in Figure 11.4. Fluid in the boundary layer near the nonrotating surface has less centripetal pressure differential than the rotating fluid. The exterior fluid leaks inwardly through the thin boundary layer toward the axis as shown. Tangential motion in the fluid distant from the surface is the main flow, and the radially inward flow is the secondary flow.

Conservation of angular momentum of the inward moving fluid causes v_θ greater than Ωr at about $z = 0.4\delta$, which leads to flow reversals as shown in Figure 11.4. Height of the boundary layer ($\delta \approx 8\sqrt{\nu/\Omega}$) is approximately twice that of the rotating

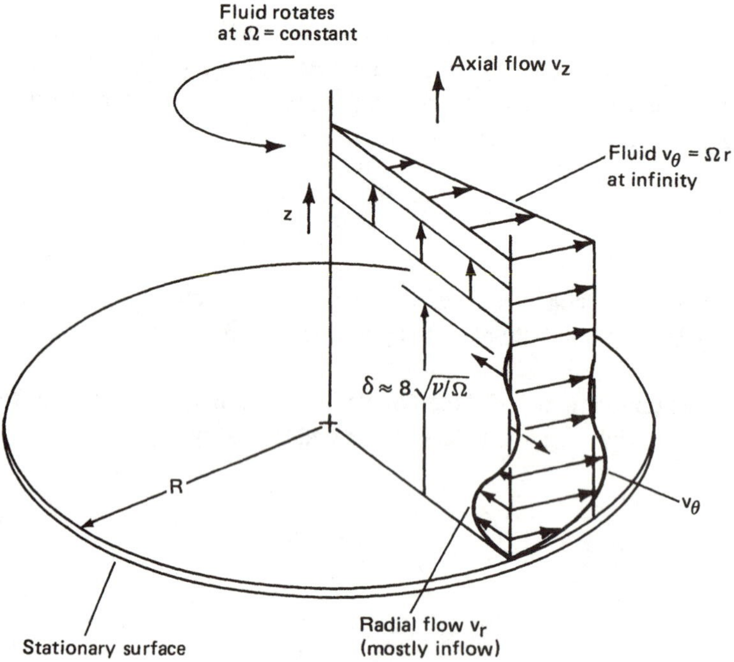

Figure 11.4: *A secondary radial flow is induced by a rotating fluid adjacent to a nonrotating flat surface. Note the flow pattern reversals between 0.4δ and δ. Compare with Figure 11.1.*

surface case, with δ defined again at a velocity differential of 0.02 rΩ. In the earlier case, axial flow toward the surface tends to suppress boundary layer height, while in the present case, axial flow away from the surface and angular momentum conservation tend to move the boundary layer upward.

The Ekman spiral showing horizontal velocity as a function of dimensionless distance above the surface is shown in Figure 11.5. Net volumetric flow through a cylinder of radius R is obtained by integrating $v_r(z)$ over the cylinder. In dimensional variables it is

$$Q = -1.387\pi R^2(\Omega v)^{1/2} \tag{11.12}$$

The rotating surface case is a basic model for many flows of engineering interest, while the rotating fluid case provides a model for understanding many atmospheric and oceanic phenomena. The analyses of this section use coordinate systems fixed relative to inertial space. Because of that, centripetal and Coriolis terms do not appear in the analyses. In these inertial reference frames, equivalent accelerations are included in the inertial DV/Dt term. It is, of course, possible to solve the same problems using a

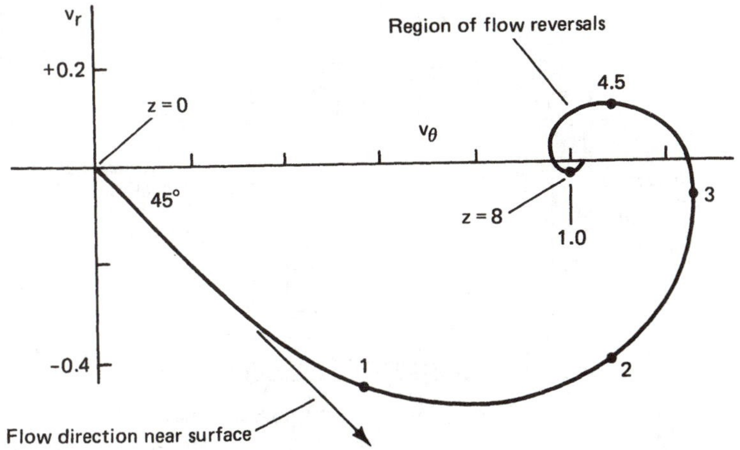

***Figure 11.5:** An Ekman spiral for flow induced by viscous fluid rotating with constant angular velocity in the vicinity of a flat surface.*

coordinate system rotating at Ω with explicit centripetal and Coriolis terms. See Example 11.4 and Greenspan (1968). Rotating coordinate system solutions are usually used in geophysical analyses, with Ω the angular velocity of the earth's rotation.

11.4 Enclosed Secondary Flows

The dramatic impact of secondary flows on rotating fluids is easily observed in the case of spin-up or spin-down of liquids in containers. Figure 11.6 shows a circular cylinder of radius R and height H completely filled with water. Initially the cylinder and the water are at rest. The cylinder is impulsively accelerated to an angular velocity Ω, and the manner and time of liquid spin-up is the topic of interest.

The top, bottom, and periphery of the cylinder have only tangential motions relative to the fluid so viscosity must be relied on for transfer of momentum from the cylinder surfaces to the water. For most laboratory-sized cylinders, for water, and for easily observed angular velocities, the Reynolds number, $R_e = RU/\nu >> 1$, and boundary layers will form. Note that the Ekman number, $E_k = \nu/R^2\Omega$, is sometimes taken as $1/R_e$ if U of the Reynolds number is taken as $\sim R\Omega$. Accordingly, $E_k << 1$ for this example.

Two mechanisms involving viscous transfer of momentum are available. One involves the boundary layer at the cylinder wall. Its effect can be approximated as a suddenly accelerated wall, similar to a suddenly accelerated plate, with boundary layers growing with time, $\delta \sim \sqrt{\nu t}$. With this mechanism, the entire mass of fluid rotates only after the boundary layer thickness has grown to the axis, $\delta \sim R$, so that

$$R \sim \sqrt{\nu t_1} \qquad \text{or} \qquad t_1 \sim R^2/\nu \tag{11.13}$$

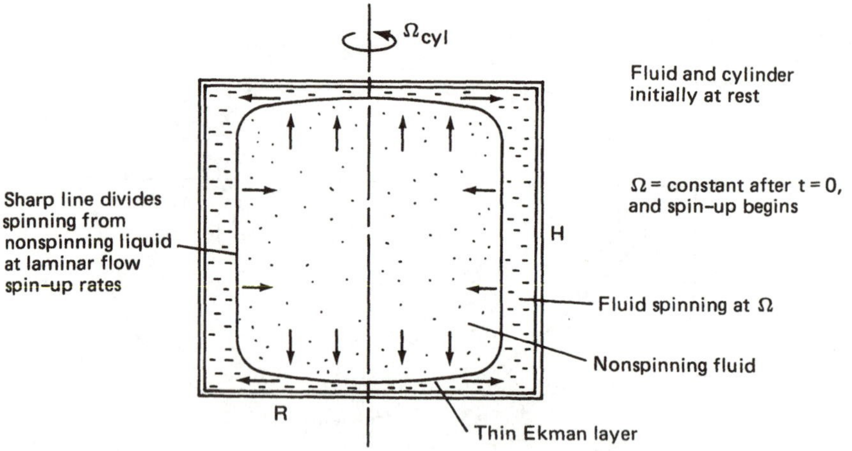

Figure 11.6: *Spin-up in a cylinder. See Figure 11.1 for an illustration of fluid motion at the top and bottom surfaces.*

The second mechanism involves secondary flow at the top and bottom end surfaces. Boundary layers form here also, initially of the form $\delta \sim \sqrt{\nu t}$. But because these spinning boundary layers of fluid produce radial pressure differentials, $dp/dr = \rho r\Omega^2$, unbalanced by pressure differentials in the central nonspinning fluid, a secondary outward flow occurs in the top and bottom boundary layers. Fluid is pumped axially from the interior, into the thin end boundary layers, and then outward to the periphery. The spinning fluid satisfies continuity requirements by forming a layer of spinning fluid along the cylinder wall. As more fluid is pumped outward, the cylindrical layer grows in thickness, finally filling the entire volume.

The question then is, *Which mechanism is most effective?* More specifically, *Will either of the mechanisms act so quickly that it spins up the fluid before the other mechanism has had time to act?* Time for the second mechanism (secondary flow) can be estimated as the time required to pump all the interior fluid through the end boundary layers. The volume of fluid is $\forall \sim R^2H$, with H = height, and the pumping rate is given by (11.8), $Q \sim R^2(\nu\Omega)^{1/2}$. Then

$$t_2 \sim \forall/Q \sim R^2H/R^2(\nu\Omega)^{1/2} = H/(\nu\Omega)^{1/2} \qquad (11.14)$$

Assuming a container with characteristic size L = 5 cm, with $L \sim R \sim H$, an angular velocity of one revolution per second, and water, yields numerical results of

$$t_1 \sim 25/0.01 = 2500 \text{ s}$$

$$t_2 \sim 5/(0.01 \times 2\pi)^{1/2} \approx 20 \text{ s}$$

In this simplified example, the secondary flow mechanism (t_2) is 125 times more effective than the direct viscous diffusion mechanism (t_1) in transferring momentum from the cylinder to the fluid. This is a typical result for most flows of interest, although note that the opposite result can occur when the product of H/R and $\sqrt{E_k}$ is >>1, as can be seen by expressing the ratio of t_2/t_1.

$$\frac{t_2}{t_1} \sim \frac{H}{(\nu\Omega)^{1/2}}\frac{\nu}{R^2} = \frac{H}{R}\left(\frac{\nu}{\Omega R^2}\right)^{1/2} = \frac{H}{R}\sqrt{E_k} \qquad (11.15)$$

This latter case might represent a long, skinny cylinder filled with a very viscous liquid that is impulsively spun-up to a low angular velocity. Spin-up of water in a circular cylinder is illustrated also in Figure 1.4.

Spin-down of a fluid-filled container is a related problem. In this case, both the fluid and container initially spin at rate Ω, and then the container is impulsively stopped. A radial pressure field now exists in the spinning fluid but not in the stationary end boundary layers. Secondary flow in the end boundary layers pump fluid radially inward and up along the axis. A central column of nonrotating fluid forms and gradually grows in diameter. Usually, however, before spin-down is complete, instabilities occur that distort and/or dominate the secondary flow mechanism. Spin-up from rest is intrinsically more stable because only nonrotating fluid is displaced.

In the more general case, both the boundary and the fluid rotate at the same angular velocity, and the boundary is caused to have an incremental change in angular velocity ($\pm\Delta\omega$). Note that this change could consist of only a change in direction of ω rather than a change in its magnitude. Precessing motion is one such example; it is applied to space vehicles and the earth in Chapters 18 and 19.

Spin-up, and spin-down to a lesser degree, are stable flow configurations if $R_o << 1$ and $E_k << 1$, but both can be very unstable otherwise. When stable, the line between the rotating and nonrotating fluids remains very sharp. "Fuzziness" of the separation surface is caused by viscous diffusion that, in the numerical example above, is only 1/125 as fast as secondary flow spin-up (spin-down). In the example, spin-up (or spin-down) terminates before the diffusion region is 0.01R thick. If motions are so fast that turbulence occurs, momentum transfer by turbulence can be more rapid than secondary flows. In addition, if the fluid is initially rotating, vortex stretching (Chapter 10) and generation of internal inertia waves due to angular momentum gradients (Chapter 13) also become effective transporters of fluid momenta and energy.

Greenspan (1968) provides information on secondary flows at a more advanced mathematical level. His derivation is set in a rotating coordinate system, and his solution is expressed in dimensionless form as a function of the Ekman number. See also Example 11.4.

11.5 Application Examples

This section outlines equations and numerical solutions of a rotating plane surface in the presence of a stationary fluid and a rotating fluid in the presence of a stationary plane surface. Radial pumping in a thin boundary layer is the primary result in both cases. The effectiveness of this secondary flow (radial pumping) is compared to viscous diffusion as a means for transferring momentum. Example 11.1 computes the radial pumping effectiveness of a spinning disk, Example 11.2 examines bottom boundary layer flow in a curved channel, Example 11.3 examines motion in a liquid-filled cylinder in a space vehicle that is impulsively spun up and then stopped, and Example 11.4 summarizes analysis of secondary flows when placed in rotating frame components.

Example 11.1. A thin 3 inch diameter steel disk on a long, thin shaft spins at 300 rpm while centered in a 5 gallon (gal) bucket (H = D) of liquid (ν = 10,000 cs, $\rho \approx 1$ g/cm^3). Its function is to keep the liquid stirred. What is its pumping rate? How often is the liquid completely recirculated? How much horsepower is required?

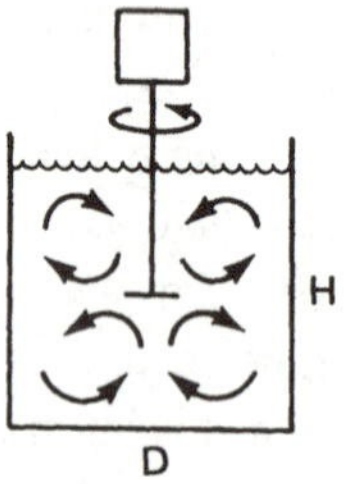

Discussion. Initially the liquid is at rest, and the spinning disk in the stationary liquid can be modelled as a spinning surface truncated at r = 1.5 inch. Equation (11.8) provides the pumping rate per side as

$$Q = 0.886\pi\, R^2(\nu\Omega)^{1/2} = 0.886\pi(1.5 \times 2.54)^2 \times (100 \times 300\pi/30)^{1/2} \text{ cm}^3/\text{s}$$
$$Q = 0.886\pi\ 14.52 \times 56.0 = 2265 \text{ cm}^3/\text{s} = 0.60 \text{ gal/s}$$

The 2.5 gal of liquid on each side of the spinning disk are circulated once each in a time

$$t = \mathcal{V}/Q = 2.5/0.6 = 4.17 \text{ s}$$

which is the recirculation time.

Power is obtained as the product of moment (M) times angular velocity. Using (11.7)

$$M = 0.313\pi\rho R^4(\nu\Omega^3)^{1/2} = 0.313\pi \times 1 \times (1.5 \times 2.54)^4\,(100 \times 1000\pi^3)^{1/2}$$
$$M = 0.313\pi\ 211 \times 10^2(10\pi^3)^{1/2} = 0.313\pi\ 211 \times 17.6 \times 10^2 = 3.65 \times 10^5 \text{ dyne-cm}$$

$$P = M\Omega = 3.65 \times 10^5 \times 10\pi = 1.147 \times 10^7 \text{ ergs/s}$$
$$P = 1.147 \times 10^7 \times 1.341 \times 10^{-10} = 1.538 \times 10^{-3} \text{ horsepower}$$

A more accurate solution requires a critical analysis of the assumptions used in this approximation. These assumptions include (1) neglecting edge effects for the finite disk and (2), after start-up, the main body of liquid is no longer at rest and may even become turbulent. Further, the top half of the liquid will behave differently than the bottom half. Interaction at the bottom surface involves the problem of a spinning fluid in the presence of a stationary surface, but the top half has no such surface, and the liquid will adopt a significant swirl about the spin axis. The free surface will deform and may even cavitate at the axis.

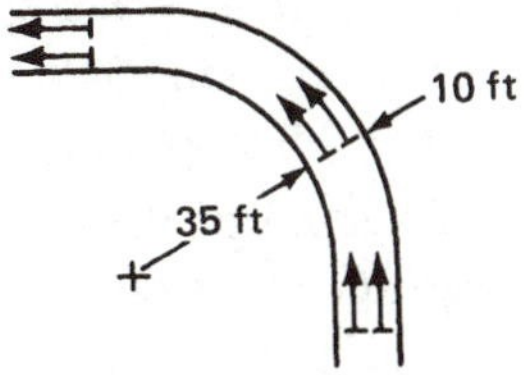

Example 11.2. Water enters a channel with a flat velocity distribution of 3 mph, except for thin boundary layers on each surface. The channel is 10 ft wide and 2 ft deep with straight side walls. The channel rotates the flow through a 90° horizontal bend with an inside radius of 35 ft. Estimate the height of the bottom boundary layer, and then estimate the secondary flow field leaving the 90° bend. Explain using (**V**, p) as the variables of interest and also using (**ζ**, **V**).

Discussion. A scale drawing helps to place the problem in perspective, and helps the formulation of useful approximations and assumptions. Some distance before and after the bend, flow velocity v is stated to have a uniform speeed of 3 mph (4.4 ft/s). A hydraulic diameter (d = 4 A/C) can be used to estimate a Reynolds number, where A is the flow cross section and C is its periphery. (For a circular pipe 4 A/C is its diameter.) The hydraulic diameter is d = 4 × 20/24 = 3.33 ft, and $R_e = vd/\nu = 4.4 \times 3.33/1.09 \times 10^{-5} = 13.4 \times 10^5$. Obviously the flow is turbulent. Since 3 mph is a fast walking speed, few rivers or channel flows will be laminar. Engineering handbooks sometimes have useful data on turbulent flow results; in other cases, turbulent theory provides accurate estimates. However, first estimates often use laminar theory. These estimates should be used with caution, of course.

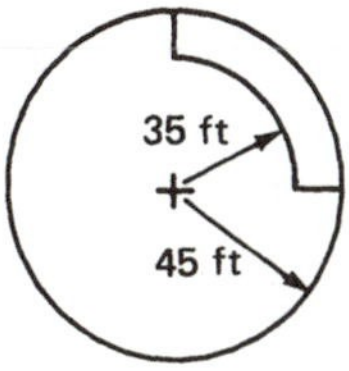

In this spirit of first approximations, flow around the bend might be considered as a rotating fluid adjacent to a stationary surface (the channel bottom). The bottom boundary layer height is $\delta \approx 8\sqrt{\nu/\Omega}$, and radially inward flux along the bottom is $Q = 1.387\pi R^2(\Omega\nu)^{1/2}$. Solution for δ requires an estimate for Ω, which is taken as $\Omega \approx 4.44/40 = 0.11$ rad/s. $\delta \approx 8\,(1.09 \times 10^{-5}/0.11)^{1/2} = 0.080$ ft. (almost 1 inch). As an estimate of Q, we take 1/4 of the difference between flow over a 45 ft diameter surface and flow over a 35 ft diameter surface. This assumes ($v_\theta = 4.4$) is equivalent to ($v_\theta = 0.11r$) across the narrow strip.

$$Q = -\frac{1}{4}\left[1.387\pi(\Omega\nu)^{1/2}\,(45^2 - 35^2)\right] = -0.954\ \text{ft}^3/\text{s}$$

The negative sign indicates a negative cross-flow (radial) within the bottom boundary layer. This secondary flow moves from the outside to the inside of the bend, and if the bottom surface were loose sand or dirt, material would be transported from the outside to the inside. The phenomenon is observed in rivers flowing down a wide, gentle valley. It causes stream *meandering* as shown in Figure 1.6.

The secondary flow velocity can be estimated (again only as a rough estimate) by assuming the flow rate Q moves through a *displacement* boundary thickness, say $\delta/3$ in thickness, and $80\pi/4$ in length.

$v_r = -0.954/(0.0267 \times 20\pi) = 0.6$ ft/s

or about 1/7 of the downstream flow value. Considering the approximations and assumptions, the estimate is fairly reasonable in magnitude.

The equations used to estimate δ and Q were derived using the Navier-Stokes equations in variables **V** and p. As a physical explanation, the water moving around the bend creates an outward centrifugal pressure field. This radial pressure field exists throughout the flow, except in the thin boundary layer on the bottom. As a result, the high pressure at the outside of the bend forces an inward flow through the nearly stationary bottom layer. The outward pressure gradient is evidenced by an increase in height of the free surface proportional to the pressure along the outer wall.

After the Navier-Stokes equations are transformed to variables ζ and **V**, the pressure terms have disappeared. Because the equations are still valid, it must be possible to explain this same phenomena in terms of ζ and **V**, and with no reference to the concept of pressure. In the (ζ, **V**) explanation, note that as the flow moves down the channel, an ordinary bottom boundary layer must form with vortex lines transverse to the flow direction as shown. In this flow, with v_θ = constant, water at the inside of the bend (both in and above the boundary layer) moves around the bend

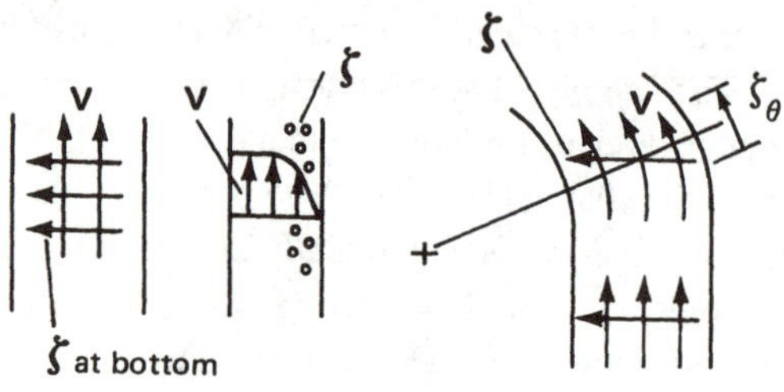

faster than equivalent water at the outside. The horizontal transverse vortex lines are fixed to these fluid particles and are *convected* to more align with the flow direction. As the vortex lines change direction, they acquire a component of vorticity aligned with the flow. This component of vorticity causes an inward flow along the bottom and an outward flow in the free stream. No reference to pressure is necessary. Chapter 14 provides further details and applications.

Example 11.3. An axisymmetric space vehicle contains a centrally located cylindrical tank filled with liquid fuel. It has reached a state of inertial rest ($\omega = 0$). Attitude control rockets then impulsively rotate the vehicle at 6 rpm for 6 s and then impulsively stop the vehicle ($\omega = 0$ again). The fuel has viscosity and density nearly those of water, and the tank is 1 m diameter by 2 m long. What does the liquid do during and after this maneuver? Does the vehicle remain at $\omega = 0$?

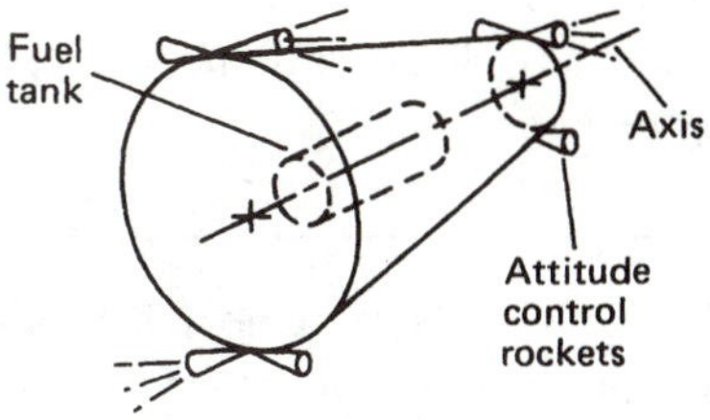

Discussion. A useful prediction of fluid-vehicle response to the forcing function requires first an assessment of the fluid-structure interaction mechanisms and their relative effectiveness. Five mechanisms are available:

1. Viscous diffusion of momentum at the tank walls
2. Secondary flows in boundary layers
3. Vortex stretching
4. Momentum transport by internal waves
5 Fluid instability and turbulence

If the tank had been nonaxisymmetric or had had internal vanes that interacted with the flow, then mechanical stirring would have been a sixth mechanism. Chapter 13 will show that in spin-up from rest the flow is basically stable and that internal waves will not be a significant factor. Vortex stretching in the stationary liquid is zero, of course, because initial vorticity is zero. A comparison of mechanisms 1 and 2 is provided by (11.15), which shows mechanism 2 to be about 100 times as effective as mechanism 1 for this problem.

$$\frac{t_2}{t_1} = \frac{H}{R}\sqrt{E_k} = \frac{2}{0.5}\sqrt{6.37}\times 10^{-3} = 1.01\times 10^{-2}$$

The pumping rate is $Q = 2 \times 0.886\pi R^2(\nu\Omega)^{1/2} = 2.01 \times 10^{-3}$ m^3/s, which will pump $2.01 \times 10^{-3} \times 6 = 1.206 \times 10^{-2}$ m^3 in 6 s. Total volume is $\maltese = \pi R^2 H = 1.571$ m^3 so only a thin layer of fluid adjacent to the outer wall, $\delta R \approx 1.206 \times 10^{-2}/(2\pi H) = 0.960$ mm thick achieves spin-up.

After the vehicle structure has been impulsively returned to $\omega = 0$, the thin layer of fluid is still spinning and must eventually transfer its angular momentum to the vehicle. This interaction is a spin-down situation with a spinning fluid adjacent to a stationary wall. The vehicle itself must start spinning again until both liquid and vehicle spin at the same angular velocity. This angular velocity could be determined, independent of the fluid-structure mechanism involved, using conservation of angular momentum if the moment of inertia of the vehicle about its axis of symmetry had been given. It obviously is difficult to control the motion of a space vehicle containing any significant quantity of liquids. Only rarely is the liquid doing exactly what the vehicle is doing. Chapter 18 deals with the general case of liquids in spacecraft.

Example 11.4. Assume a disk and a viscous fluid are both rotating at the same rate and then the rotation rate of the disk is increased slightly. Analyze the resulting flow using a coordinate system rotating at the original rate, i.e., fixed in the fluid. Refer to Greenspan (1968).

Discussion. Equation (6.9), the Navier-Stokes equation in a frame rotating at $\mathbf{\Omega} = \Omega\hat{\mathbf{k}}$ and expressed in dimensionless variables $(\mathbf{V}, p)$, is

$$\frac{\partial \mathbf{V}}{\partial t} + R_o(\mathbf{V} \cdot \nabla)\, \mathbf{V} + 2\hat{\mathbf{k}} \times \mathbf{V} = -\nabla p + E_k\, \nabla^2 \mathbf{V} \tag{6.9}$$

The equation is augmented with $\nabla \cdot \mathbf{V} = 0$, and uses $E_k = \nu/\Omega^2 L$, $R_o = U/\Omega L$, $\mathbf{\Omega} \times (\mathbf{\Omega} \times \mathbf{r})$ included in p, and all variables and derivatives relative to the rotating frame.

R_o is small if convective changes are small compared to rotational changes. On the assumption this is true, $R_o \sim 0$ and is neglected, leaving

$$\frac{\partial \mathbf{V}}{\partial t} + 2\hat{\mathbf{k}} \times \mathbf{V} = -\nabla p + E_k\, \nabla^2\, \mathbf{V} \tag{11.16}$$

E_k cannot be neglected here, but obviously if $\mu << 1$, then E_k is small and a boundary layer type solution is indicated.

At $t = 0$, the fluid and disk rotate at the same rate, so for all $\mathbf{r}$, $[\mathbf{V}(\mathbf{r}, 0) = 0]$. After the disk is impulsively accelerated to $(\Omega + \varepsilon)\, \hat{\mathbf{k}}$, the boundary conditions are $[\mathbf{V} = \varepsilon\hat{\mathbf{k}} \times \mathbf{r}]$ on the disk $(z = 0)$, and $[\mathbf{V} \to 0]$ as $(z \to \infty)$. Assume further that $\mathbf{V} \neq f(\theta)$. This assumption is valid so long as instabilities do not occur. Chapter 15 discusses instabilities that are $f(\theta)$ and that would modify the solution.

When the problem of a rotating disk in a stationary fluid is solved in inertial coordinates, a solution independent of r is obtained by a proper choice of characteristic lengths and velocities (11.3a-e). An equivalent result is obtained here using

$$\mathbf{V} = -\nabla \times \left[r\psi(z,t)\ \hat{\boldsymbol{\theta}} \right] + rV(z,t)\ \hat{\boldsymbol{\theta}}$$

where ψ is a stream function. This choice proves to be valid, but it certainly is less physically intuitive than the choices used for the inertial frame analysis. Boundary conditions for ψ and V, and a Laplace transform solution, including transient terms, are given in Greenspan (1968).

The steady-state solution with $[Z = zE^{-1/2}]$, and all quantities dimensionless, is

$$\frac{u}{r} = \sin Z\ e^{-Z}$$

$$\frac{v}{r} = \cos Z\ e^{-Z}$$

$$w = -E^{1/2}\left[1 - (\sin Z + \cos Z)\ e^{-Z}\right]$$

$$p = -2E \sin Z\ e^{-Z}$$

As before, a boundary layer is obtained

$$\delta \sim E^{-1/2} \qquad \left(\sim \sqrt{\nu / \Omega}\ \text{ in dimensional units}\right)$$

Graphs of u/r, v/r, and $-w\ E^{-1/2}$ as f(Z) and the Ekman spiral for this solution are shown in Greenspan (1968, Figures 2.1 and 2.2). Similar graphs are shown in Figures 11.2 and 11.3, respectively. Note that the Greenspan solution provides an analytical expression for the result, while the solution described in the text resulted in equations (11.4a-d), which were solved numerically. In addition, the Greenspan solution illustrates explicitly the dependence of the Ekman layer on the Ekman number.

Chapter 12

Circular Pathline Flows

12.1 Theoretical Criteria

This chapter summarizes important fluid motions consisting of circular or nearly circular paths. In each case, the axis of rotation coincides with the z axis. The flows are described by their flow patterns $v_\theta(r)$, although in some flows, secondary velocity components v_r, v_z, and/or v_ϕ may exist also. The differences in the flows are due to differences in the physical mechanisms that initiate and/or maintain the flows in the given condition. Flows are illustrated in Figures 12.1 through 12.8, and Table 12.1 summarizes important mathematical relationships.

In the fourth column of Table 12.1, $\omega_*(r)$ is the angular velocity of a fluid annulus at radius r computed as

$$\omega_* = v_\theta / r \tag{12.1}$$

It is not the angular velocity of an infinitesimal fluid element at r, which would be computed as $\omega = \zeta/2$. In column five, $\zeta(r)$ is the z component of vorticity computed using

$$\zeta = (\nabla \times \mathbf{V}) \cdot \hat{\mathbf{z}} \tag{12.2}$$

Rate of strain is computed using cylindrical coordinates and the equation

$$\dot{\varepsilon}_{r\theta} = \frac{1}{2}\left(\frac{1}{r}\frac{\partial v_r}{\partial \theta} + \frac{\partial v_\theta}{\partial r} - \frac{v_\theta}{r}\right) \tag{12.3}$$

Column seven in Table 12.1 shows circulation around an annulus of radius r. It is computed using

$$\Gamma = \int \mathbf{V} \cdot d\boldsymbol{\ell} = 2\pi r\, v_\theta \tag{12.4}$$

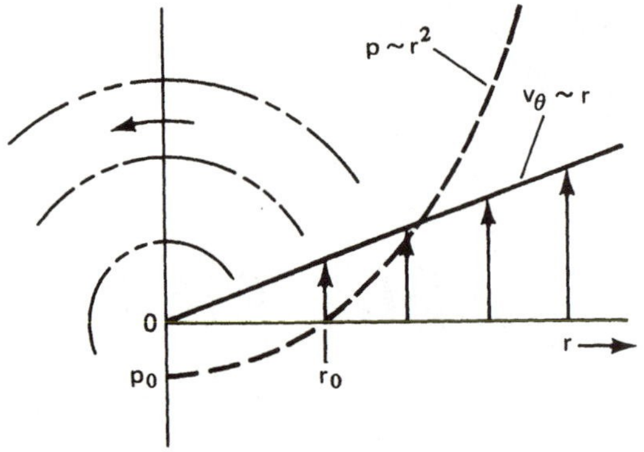

Figure 12.1: *Rigid rotation. Angular velocity may vary with time but not over space. Rate of strain equals zero.*

The next column is angular momentum per unit volume--not per annulus. The computation is

$$h = (\rho \mathbf{r} \times \mathbf{V}) \cdot \hat{\mathbf{z}} = \rho r v_\theta \tag{12.5}$$

Pressure is integrated from a reference pressure (either at r = 0 or at r = ∞), out or in as appropriate, by noting that radial changes in pressure are related to centrifugal force of a fluid element at that r.

$$\frac{\partial p}{\partial r} = \rho \omega_*^2 r = \rho v_\theta^2 / r \tag{12.6}$$

Energy in the last column represents kinetic energy per unit volume computed as

$$E = \tfrac{1}{2} \rho v_\theta^2 \tag{12.7}$$

12.2 Some Important Flows

Rigid rotation flow shown in Figure 12.1 has constant vorticity and zero rate of strain. Angular momentum and energy per unit mass, and pressure, approach infinity with increasing radius. Therefore, any real rigid rotation flow must be contained by a closed surface or an equivalent force field. There is no requirement on viscosity. An assumed inviscid rigid rotation is possible provided a mechanism is available to initiate the motion.

Table 12.1: Circular Pathline Flows

Motion	Assumed conditions	$v_\theta(r)$	$\omega_*(r)$	$\zeta(r)$	$\dot\varepsilon_{r\theta}(r)$	$\Gamma(r)$	$h(r)$	$p(r)$	$E(r)$
Rigid rotation	Ω = constant	Ωr	Ω	2Ω	0	$2\pi\Omega r^2$	$\rho\Omega r^2$	$p_o+\frac{\rho}{2}(\Omega r)^2$	$\frac{\rho}{2}(\Omega r)^2$
Inviscid vortex	$\mu = 0$, ρ = constant $\kappa = rv_\theta$ = constant	$\frac{\kappa}{r}$	$\frac{\kappa}{r^2}$	0	$-\frac{\kappa}{r^2}$	$2\pi\kappa$	$\rho\kappa$	$p_\infty-\frac{\rho\kappa^2}{2r^2}$	$\frac{\rho\kappa^2}{2r^2}$
Vortex spiral[1]	Q = constant $C=(\kappa^2+Q^2/4\pi^2)$	$\frac{\kappa}{r}$	$\frac{\kappa}{r^2}$	0	$-\frac{\kappa}{r^2}$	$2\pi k$	$\rho\kappa$	$p_\infty-\frac{\rho}{2r^2}C$	$\frac{\rho}{2r^2}C$
Circular channel	$v_\theta = V$ = constant $\mu = 0$, ρ = constant	V	V/r	V/r	$-V/2r$	$2\pi Vr$	ρVr	$p_o+\rho V^2 \ell n(r)$	$\frac{\rho V^2}{2}$
Circular Couette[2]	$\mu, \rho, \Omega_a, \Omega_b$ constant	$Ar+B/r$	$A+B/r^2$	2A	$-Br^{-2}$	$2\pi(Ar^2+B)$	$\rho(Ar^2+B)$	Note 2	$\frac{\rho}{2}(Ar+B/r)^2$
Circular orbit	$\mu = 0$ $k=\sqrt{Gm}$	$\frac{k}{\sqrt{r}}$	$\frac{k}{\sqrt{r^3}}$	$\frac{k}{2\sqrt{r^3}}$	$-\frac{3k}{4\sqrt{r^3}}$	$2\pi k\sqrt{r}$	$\rho k\sqrt{r}$	0	$\frac{\rho k^2}{2r}$
Viscous flows	$\mu > 0$ especially $\mu << 1$	In circular flows having viscosity, shear stresses can produce specific flow patterns, including core regions with rotation rates, averaged over time, approximating those of a rigid body. Secondary flows and turbulence are often apparent.							

[1] $v_r(r) = -Q/(2\pi r)$ for sink, (+) for source. Additional strain rates due to $v_r(r)$, $\dot\varepsilon_{rr} = Q/2\pi r^2$, and $\dot\varepsilon_{\theta\theta} = -Q/2\pi r^2$. Otherwise, same as inviscid vortex.

[2] $$2A=-\left(\frac{\Omega_a R_a^2}{R_b^2}\right)\frac{1-\Omega_b R_b^2/\Omega_a R_a^2}{1-(R_a/R_b)^2},\ B=\left(\Omega_a R_a^2\right)\frac{1-\Omega_b/\Omega_a}{1-(R_a/R_b)^2},\ p=p_o+\rho\left(\frac{1}{2}A^2r^2+2AB\ell n\, r-\frac{1}{2}B^2/r^2\right).$$

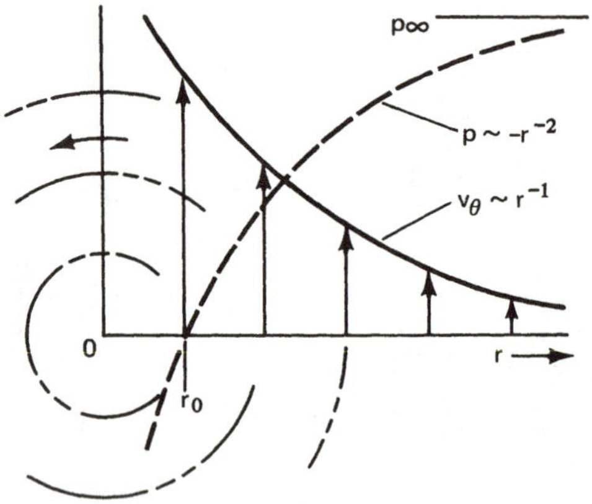

ζ = 0; K, Γ, and h are constants

Figure 12.2: *Inviscid vortex. Vorticity equals zero. Angular momentum per unit mass is constant over space.*

The two-dimensional (2-D) inviscid vortex of Figure 12.2 could not continue if $\mu \neq 0$, because each annulus rotates at a different angular velocity. Zero viscosity indicates zero vorticity for this flow, except for the singularity at $r = 0$. Circulation and angular momentum per unit mass are the same at all radii. The close relationship of circulation to angular momentum is apparent from the tabulated data for all flows illustrated. As radius approaches infinity, kinetic energy per unit mass relative to inertial space approaches zero, and pressure approaches the ambient pressure. That each is proportional to r^{-2} indicates a rapid enough decrease with radius that trivial perturbations can match an inviscid vortex to a real ambient fluid at modest values of r.

The 2-D, constant-density, inviscid vortex spiral of Figure 12.3 illustrates a mechanism responsible for many observed vortex flows, and, except for the existence of a radial flow (v_r), its features are identical to those of the inviscid vortex. A small radial flow has negligible effect on vorticity and the magnitude of the rate of strain $\dot{\varepsilon}_{r\theta}$. It does, however, introduce additional strain rates

$$\dot{\varepsilon}_{rr} = Q/(2\pi r^2) \quad \text{and} \quad \dot{\varepsilon}_{\theta\theta} = -Q/(2\pi r^2) \qquad (12.8)$$

These are associated with mass conservation for an incompressible flow. Extension in the r direction must be balanced by an equivalent contraction in the θ direction. The ratio of v_θ/v_r for the vortex spiral is Γ/Q, plus (+) if a source and minus (–) if a sink. A sink in any nearly inviscid fluid having some net circular motion (i.e., net angular momentum) will tend to produce a vortex spiral, and then if the sink is discontinued, an approximate inviscid vortex remains as suggested in Figure 1.2. This model is

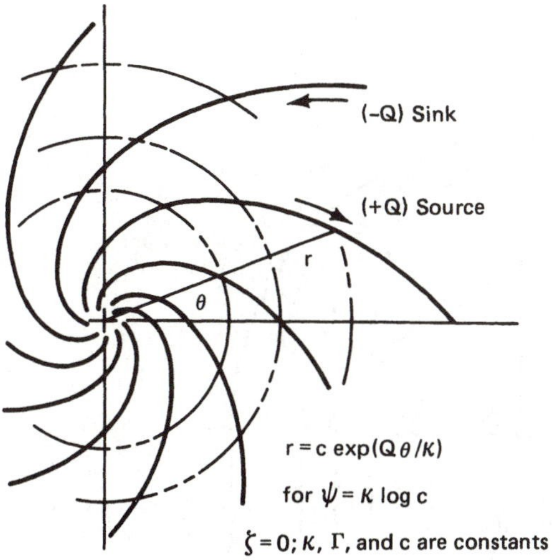

Figure 12.3: *Vortex spiral. A sink (− Q) or a source (+Q) is added to an inviscid vortex. ψ is the stream function.*

used in Chapter 22 to generate a hurricane where initial circular motion of the air is due to the vertical component of the earth's rotation. The sink at sea level is provided by thermal convection of surface-level air to the upper atmosphere.

Circular channel flow in Figure 12.4 is obtained by deflecting a flow, originally in a straight channel, around a circular bend. The straight line velocity of each fluid element is assumed here to be its tangential velocity, similar to runners assigned to lanes on a track that curves. This simplified model shows uniform velocity across the channel, $v_\theta = V =$ constant. Note that this model is simplified, and that solutions of both inviscid and viscous laminar flows, and steady-state actual flows around bends, deviate from this model as discussed further in Chapter 14. This simplified model will be used to illustrate secondary flow phenomena in later sections. Velocity and energy are constant in this model, but all other quantities vary with r.

The circular Couette flow of Figure 12.5 occurs in a fluid-filled annulus between two concentric cylinders rotating at different angular velocities. Only a few generalizations can be made because many combinations of fluids, cylinder radii, and cylinder angular velocities lead to unstable flows. An exact solution for laminar viscous stable flow is shown in Table 12.1. Simplified models are commonly used when the annular gap is very narrow relative to its radius. Unstable flows are discussed in Chapter 13, with applications and further details in Chapter 15.

Circular orbit flow, Figure 12.6, is a special case of an isolated particle of mass m (or fluid element with mass $\rho d\forall$) in orbit about a central mass M. Figure 12.6 represents the circular orbit of a small mass $\rho d\forall$ about a large central mass M. A stable circular orbit exists at that value of r where force due to centripetal acceleration

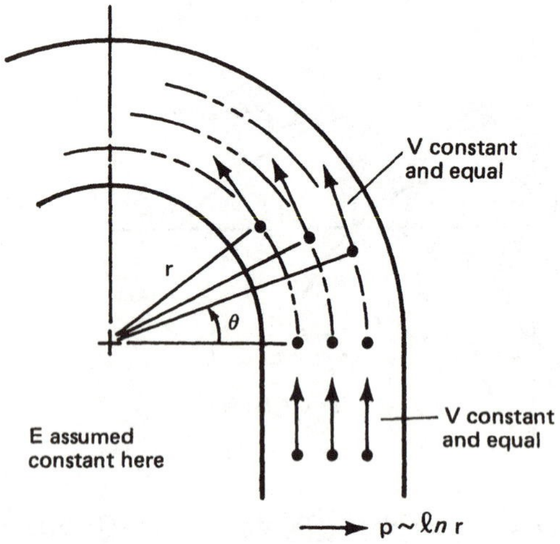

***Figure 12.4:** Circular channel. In this simplified model, V is constant over r, like equal runners on a curved race track.*

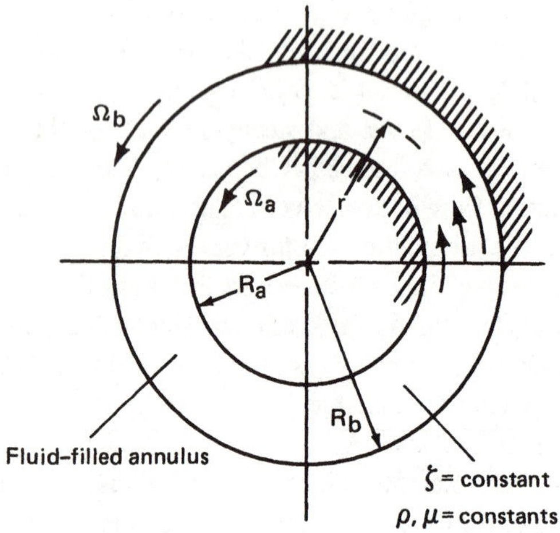

***Figure 12.5:** Circular Couette flow. Solution assumes laminar flow. Instability occurs for many real combinations of variables.*

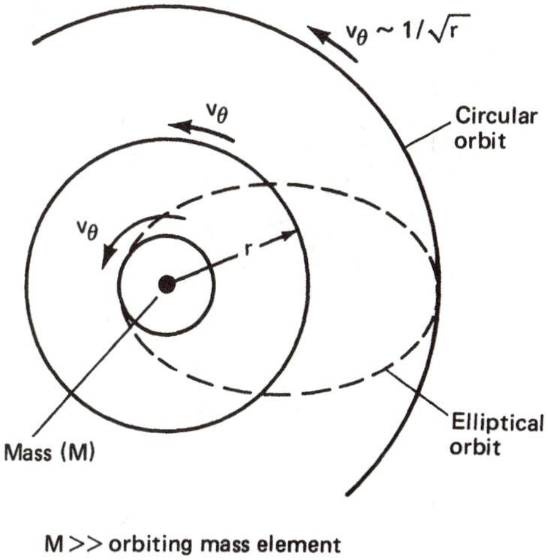

Figure 12.6: *Circular orbit. Centripetal acceleration is due to gravity, not pressure.*

is equal to the gravitational force. Equating Newton's law of gravity to force due to centripetal acceleration

$$F = G\,\frac{M\rho d\forall}{r^2} \qquad \text{and} \qquad F = \rho d\forall\,\frac{v_\theta^2}{r}$$

where G is the gravitational constant yields

$$G\,\frac{M(\rho d\forall)}{r^2} = (\rho d\forall)\,\frac{v_\theta^2}{r}$$

$$v_\theta = (GM/r)^{1/2}$$

$$v_\theta = kr^{-1/2}, \quad k = (GM)^{1/2} \tag{12.9}$$

A large quantity of small mass elements orbiting in the same plane, but at different radii, can be treated as a fluid, and provided $\mu = 0$ (i.e., no shear exists between r and $r \pm dr$), the flow can continue indefinitely. Note that fluid mass elements at different values of r have different angular momenta, so the flow is not an inviscid vortex. Generation of this flow must therefore involve a mechanism different than the central sink of the vortex spiral. In this flow the mechanism is gravity. Note also that pressure is zero in this flow because all other forces are balanced. To a first approximation this model describes motion of the rings of Saturn, and of the solar system in an early stage of evolution (see Example 12.4).

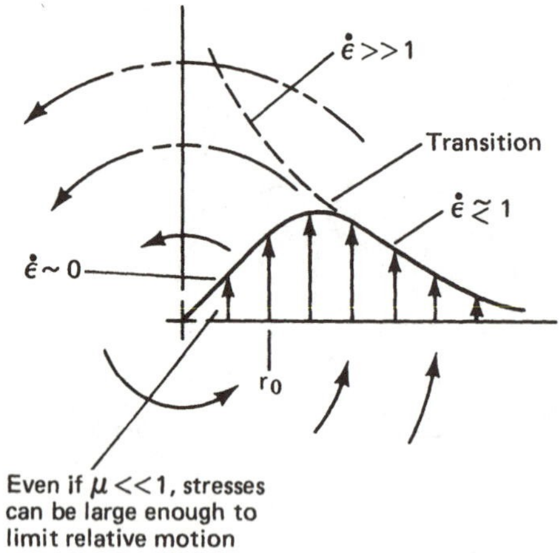

Figure 12.7: *Viscous flows. Viscosity and turbulence, in most rotating flows, reduce regions at small r to near rigid rotation.*

Figure 12.7 demonstrates phenomena introduced if the fluid has $\mu > 0$. Obviously if $\mu \rightarrow \infty$, all velocity gradients are prohibited, and all flows reduce to rigid body motion. The more interesting application is $\mu << 1$, with shear stress small enough to be negligible, except where strain rates are >> 1. In each of Figures 12.2, 12.3, 12.4, and 12.6, the rate of strain ($\dot{\epsilon}_{r\theta}$) approaches infinity as the radius approaches zero. Consequently, even if $\mu << 1$, the product $\mu\ \dot{\epsilon}_{r\theta}$ dominates other phenomena at small enough r, and a region near the origin must be constrained by viscous interaction to nearly rigid rotation.

The nonsteady-state solution for $v_\theta(r, t)$ of an inviscid vortex, assumed to have acquired viscosity at $t = 0$ is

$$v_\theta = \frac{\Gamma_o}{2\pi r}\left[1 - e^{-r^2/4vt}\right] \tag{12.10}$$

where Γ_o is vortex circulation at $t = 0$. Figure 12.8 shows dimensionless v_θ at three dimensionless times (0, 0.02, 0.10).

The rotating fluid applications presented in Part III (Chapters 14 through 22) are often modelled as one or some combination of these circular flows. The real world is finite in size rather than infinite, viscosity almost always exists, and density rarely is exactly constant. As a consequence, conditions assumed for these mathematical models are often violated in real applications. The real world, in addition, rarely exists in a stable steady-state condition. Time-dependent phenomena include the possibility of waves, oscillations, and turbulence. All of these modify basic flow models--sometimes in negligible amounts, other times in quite remarkable ways. Chapter 13 introduces a type of wave unique to rotating fluids.

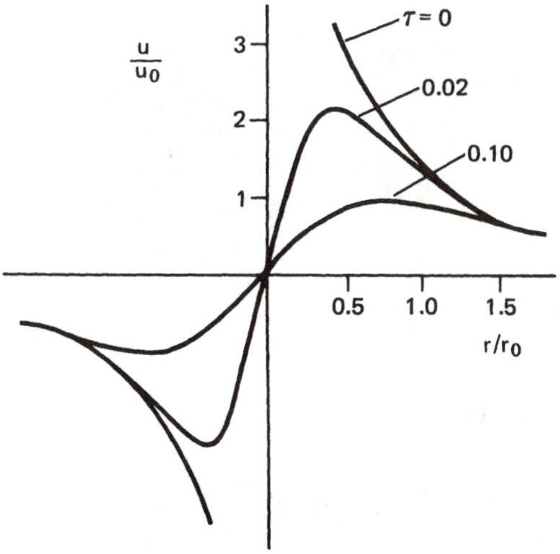

Figure 12.8: *Viscous dissipation of an inviscid vortex assumed to have acquired viscosity at* $\tau = \nu t / r_o^2 = 0$ *, with* $u_o = \Gamma_o / 2\pi r_o$.

12.3 Application Examples

Parameters that characterize circular pathline flows are explained in the text and itemized in Table 12.1 for seven important flows. These seven flows, singly and in combination, form the basic models for many of the applications in Part III. Example 12.1 examines Couette flow in narrow circular and spherical gaps; Example 12.2 assesses the feasibility of a "supersonic" tornado; Example 12.3 studies flow patterns in bends in closed rectangular ducts; and Example 12.4 briefly discusses the dynamics of a spinning disk of gas and particles constrained by angular momentum, pressure, and a central mass.

Example 12.1. The annulus between two cylinders is filled with oil (μ = 500 cp). Cylinder (a) is stationary and (b) rotates at 0.5 revolutions per second (rev/s). What is the torque on (a) assuming Couette flow? Assume laminar flow with linear $v_\theta(r)$. Next, suppose (a) and (b) are concentric spheres. What is the torque on (a) using the same radii and liquid?

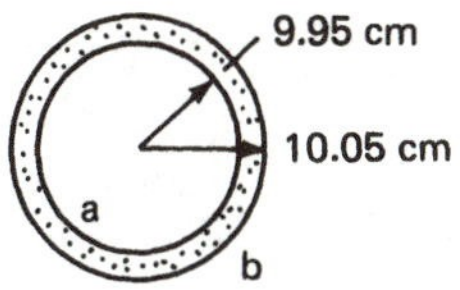

Discussion. The difference in r_a and r_b is very small, and to a close approximation with h >> r, the annulus may be assumed a flat surface ℓ = $2\pi r$ long, with a linear Couette flow distribution. Shear stress in this approximation is $\tau = \mu \frac{\partial v}{\partial r}$, with μ = 500 cp (5 poise in cgs units).

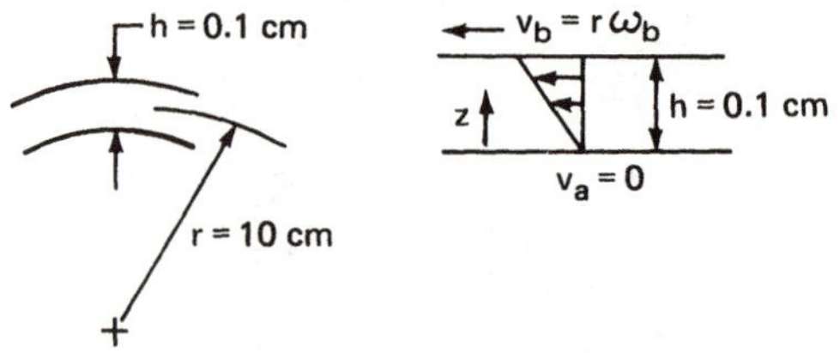

$$\tau = 5 \times \frac{10\pi}{0.1} = 500\pi \text{ dynes/cm}^2$$

Total force per 1 cm wide strip of cylinder is

$$F = \tau \ell = 500\pi \times 2\pi 10 = \pi^2 \times 10^4 \text{ dynes}$$

Total torque (moment) on (a) is then

$$M = rF = \pi^2 10^5 \text{ dyne-cm per cm length of cylinder}$$

In the same approximation using (a) and (b) as spheres, and with (a) stationary and (b) rotating about a fixed axis with arbitrary direction; at 0.5 rev/s, the relative velocity at a point is $v = y\omega_b = (r \sin\theta)\pi$. Stress at that point is $\tau = 5(100\pi \sin\theta)$ dynes/cm^2. Force on an annular strip about the rotation axis is

$$\begin{aligned} dF_s &= \tau dA = (500\pi \sin\theta)(2\pi 10 \sin\theta) r d\theta \\ &= \pi^2 \sin^2\theta \times 10^5 \, d\theta \text{ dynes} \end{aligned}$$

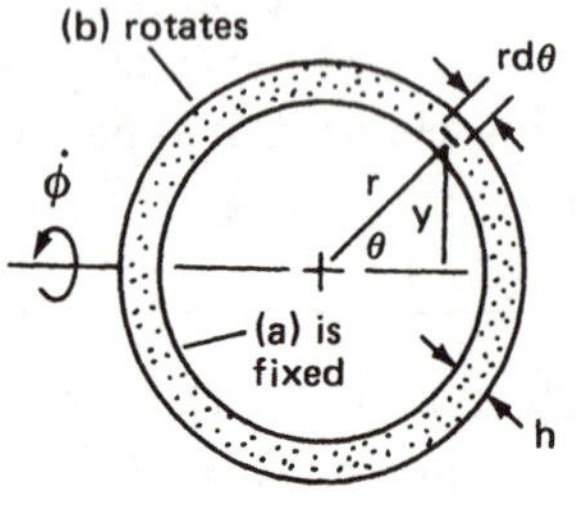

Torque (moment) on this annular strip is dM = rdF or

$$M_s = (10 \sin\theta) \times \pi^2 \sin^2\theta \times 10^5 d\theta$$

Total torque on (a) is then

$$M = \pi^2 10^6 \int_0^\pi \sin^3\theta d\theta = \pi^2 10^6 \left[-\frac{1}{3} \cos\theta (\sin^2\theta + 2) \right]_0^\pi = \frac{4}{3} \pi^2 10^6 \text{ dyne-cm}$$

Example 12.2. Einstein sometimes used "thought experiments" to help himself understand the feasibility of some model for a physical phenomenon. As a thought experiment, imagine the existence of a tornado with supersonic tangential velocities and examine its structure and feasibility.

Discussion. This thought experiment does not ask how the supersonic tornado got there or where it goes, it simply supposes its existence and that it is made available for examination and analysis. The simplest model would be a thin-walled cylindrical sheath of air, of some sheath radius (r_s) and angular velocity (ω) such that $v_\theta = \omega r_s \geq c$, where c is the speed of sound. Pressure difference ($p_a - p_o$) provides the force necessary to balance the centripetal acceleration

$$p_a - p_o = \left(\rho v_s^2/r\right)$$

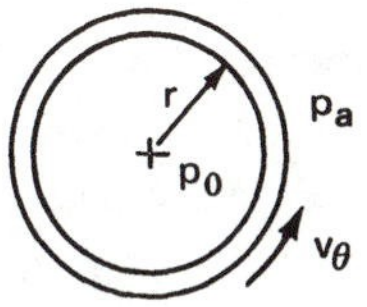

As the lower limit for supersonic v_s, we set $v_s = c = 331$ m/s, the speed of sound at sea-level standard temperature and density. Assume further that p_a is sea-level standard pressure, and then compute $p_o(r)$

$$p_o = p_a - \rho v_s^2/r$$
$$p_o = 101 \times 10^3 - 1.22 \times 331^2/r$$
$$p_o = (101 - 134/r) \times 10^3$$

Some values of tornado sheath radius and central pressure consistent with $v_s = c$ at sea level (in this simplified model) are

At	r_s	=	1.33m	13.3m	26.6m	133 m
	p_o	=	0	0.90 p_a	0.96 p_a	0.99 p_a

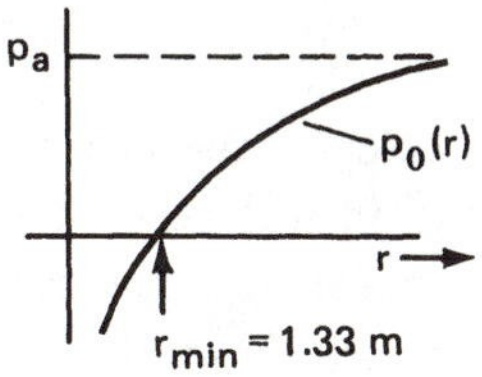

Pressure drops of 1/10 atm (p_o = 0.9 p_a) have been estimated across a tornado vortex, and vortex diameters of 60 m are common. If one assumes p_o = 0.9 p_a and r_s = 30 m, the sheath velocity is v_s = 496 m/s = 1.5 times the speed of sound.

The violence of tornadoes, their small scale, and their transient and erratic nature make them difficult to study. Estimates based on film sequences have suggested maxima of 75 to 110 m/s, or up to one-third the speed of sound. Some researchers have used the severity and types of damage observed to place maximum v_θ at or above sonic speeds. Note that there is nothing magic about a circular pathline flow exceeding the speed of sound. For example, the earth's atmosphere at the equator moves with the earth at $v_\theta \approx 460$ m/s. At the poles $v_\theta = 0$. Therefore, air at the equator has a velocity of approximately 1.4 times the speed of sound relative to air at the poles. Shock wave and other sonic phenomena require an interaction of an object and a fluid or of two fluids at or above the sonic speed. Any reasonable transition of speeds eliminates the shock phenomena. Transition of 460 m/s over a quadrant of the earth is certainly gradual enough, but so is a transition through viscous diffusion over some portion of a meter inside and outside the high-velocity sheath of a tornado. The shock-wave type damage used to estimate sonic speeds is often damage to (or caused by) small objects that interact directly with the tornado funnel (vortex sheath). Chapter 22 continues the analysis of intense atmospheric vortices.

Example 12.3. Suppose fluid in a straight, closed, rectangular channel has reached a steady-state, approximately parabolic flow distribution and then is forced to turn through a bend as shown. Examine probable flow patterns in the bend. Assume h = 4w. Next, assume h = w/4.

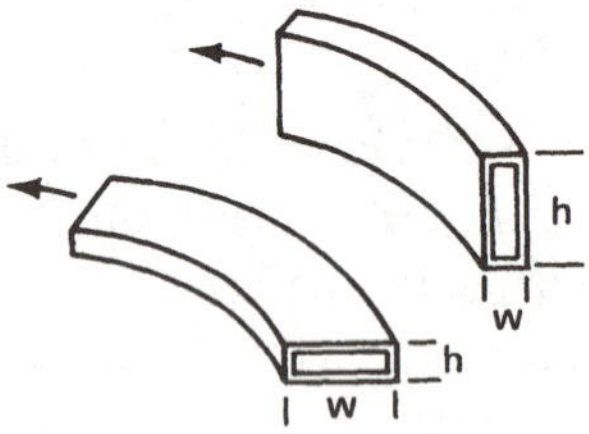

Discussion. Note that the problem *assumes* a parabolic flow distribution. Viscous laminar flow between two parallel plates is parabolic in the classic analysis, but other phenomena occur in finite-width channels, especially if the channel is curved. Side surfaces, regardless of the ratio (h/w), produce vorticity perpendicular to that produced by the top and bottom parallel plates. The interaction of these perpendicular vortex fields, and the uneven forward convection of vortex lines, cause real flows to swirl as they move down straight noncircular channels. Flow velocity in the corners is higher than might be expected as a result. If flow is turbulent, as is more often true, the mean flow has a flatter than parabolic distribution, but a similar pattern of swirling occurs.

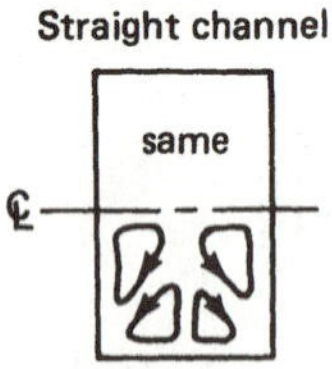

Neglect the swirl phenomena, and examine the effect of a bend in a duct with a parabolic distribution across each midplane, top to bottom and side to side. Analyzed in variables (V, p), the higher velocities in the middle (top to bottom) cause greater Δp across the width of the duct than equivalent Δp for the slower flows near the top or the bottom boundaries. This pressure differential causes longitudinal swirls much larger in magnitude than the corner swirls mentioned previously. Whether laminar or turbulent flow exists, it is apparent the phenomena will be greater when h = w/4 than when h = 4w. It is the width (w) that determines the magnitude of Δp, and therefore the magnitude of the swirling secondary flow.

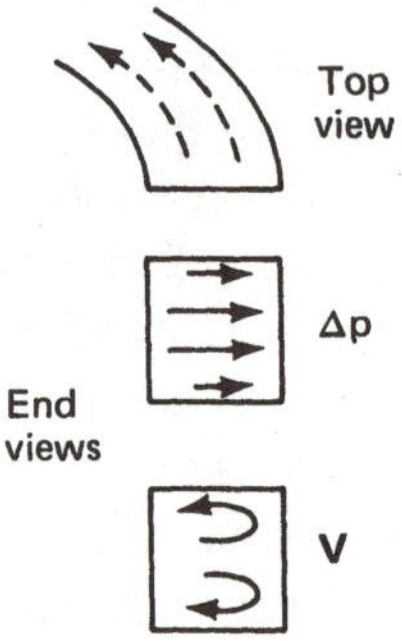

Any bends in pipes or ducts cause major energy losses. Recirculating wind tunnels, for example, are designed as shown to minimize this problem. Turning vanes are placed at each corner, with the distance between vanes (w) much less than their height (h). Longitudinal swirl, with its large energy losses and other problems, is reduced. Chapter 14 presents more rigorous analyses and experimental investigations of these phenomena.

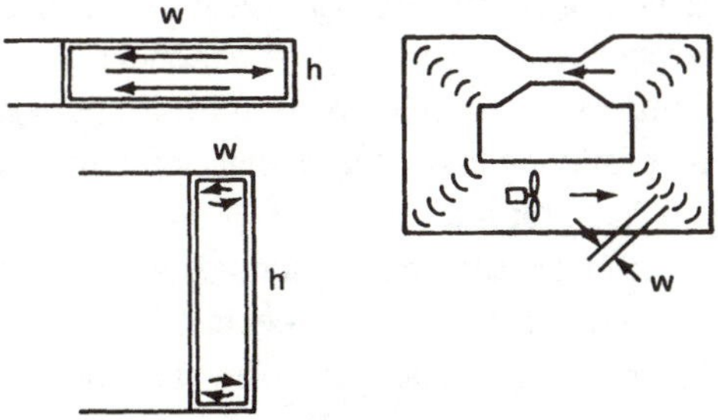

Example 12.4. Assume that a "cloud" of widely separated small particles exists in otherwise empty space, and that, because the mutual gravitational attraction of the particles exceeds their random kinetic energies, they begin to contract toward their center of mass. Assume their total mass is approximately that of the sun. Assume also that the vector sum of their moments of momentum has direction $\hat{\mathbf{z}}$ and total magnitude approximately equal to (mrv_θ) of the planet Jupiter. Briefly describe a sequence of events that culminate in our solar system. Is the same model appropriate for Saturn and its rings?

Discussion. The solar system consists of the sun, nine known planets, and many smaller objects. Viewed from the north, all planets orbit counterclockwise, and the sun and most planets rotate counterclockwise also. The sun contains over 99.8% of the total mass, and the orbital motion of Jupiter contains about 95% of the total angular momentum. All orbits and the equators of most planets and the sun lie in nearly the same plane, called the *invariant plane*. (The plane of the earth's orbit is called the *ecliptic plane* because the moon must pass through this plane for eclipses to occur.)

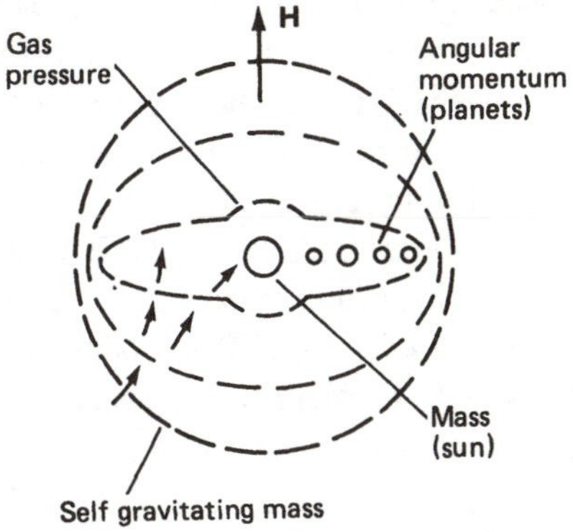

Random collisions between the particles redistribute angular momenta. Particles left with no tangential velocity move quickly to the mass center. Collisions eliminate most with orbits inclined to the (r, θ) plane. Noncircular orbits in the (r, θ) plane also cause collisions. Again, some will be left with

$v_\theta \sim 0$ and move to the center. Finally, mostly circular orbits remain, and angular momenta prevents these from moving further inward.

At some point in the collapse, the concentration of particles (or molecules) is so great that the system of particles behaves as a fluid (gas). Gravitational energy is converted to random kinetic energy (temperature), and gas pressure retards collapse rate. When the gas becomes opaque (e.g., incandescent), collapse is further retarded because energy cannot be radiated away directly and must first be conducted or convected to the outside surface. Interior temperatures reach tens of millions of degrees and nuclear (hydrogen) fusion begins, providing the energy that sometimes prevents further collapse for billions of years (e.g., for the sun).

A considerable amount of mass and angular momenta may be ejected from the system, especially as it heats up. A very small amount of mass remains with orbits highly inclined to the (r, θ) plane, e.g., some meteorites and comets. Mass left in the invariant (r, θ) plane goes through complex processes of secondary gravitational collapses, collisions, accretions, etc., until most of that mass is collected into a limited number of objects, e.g., the planets.

Evolution of the larger planet-satellite systems may have occurred in a similar but smaller scale process. Jupiter, for example, has 16 known satellites, and all but a few orbit in nearly the same plane. Saturn has millions of small "rocks" orbiting it, all nearly in the same plane giving the appearance of rings. These rings may have resulted from either a gravitational collapse or a fragmentation of one or more larger objects. Gravitational perturbations cause divisions between the rings. The rocks are closely enough spaced that noncircular orbits are rare. Although not closely enough spaced to be treated as a classic fluid, some interactions may resemble those occuring in fluids.

Random collisions tend to separate mass and angular momenta. Mass that loses angular momenta in a collision falls toward the mass center; mass that retains or gains angular momenta remains in orbit. This interaction between gravity, angular momenta, and random collisions finally concentrates most of the mass at the mass center and concentrates most of the angular momenta into a few relatively small objects orbiting in a single plane. The characteristic two-dimensionality of rotating masses (fluids) is clearly evident in these examples, and, as in all rotating fluid phenomena, it is regulated by the net angular momenta of the fluid.

The earth's moon, its mass and composition, and its orbit parameters are so unusual compared to other solar system phenomena that it must have had a different origin. Most researchers believe a Mars-sized object glanced against the early earth, and the moon formed from the resulting debris. A few researchers believe the moon was captured intact. Both theories require that a great deal of energy had to be dissipated. Dissipation of energy is the major theoretical obstacle for both of these theories of lunar formation.

Chapter 13

Rotation and Inertial Waves

13.1 Rayleigh Instability

This section introduces the topics of instability and waves in rotating fluids. These phenomena are related to angular momenta and vorticity gradients. They often dominate oceanographic and atmospheric flows. They can be produced in the laboratory, and they apply to engineering scale applications. Waves in rotating fluids and many other rotating fluid phenomena were first observed and interpreted by scientists and mathematicians as part of attempts to understand and to predict atmospheric and oceanic flows. This work is fairly recent, e.g., a wave commonly associated with rotating fluids is the atmospheric *Rossby wave*, discovered and reported in 1939 in the *Journal of Marine Research* by C.G. Rossby, a meteorologist.

There is as yet no firm consensus on the use of identifying names for all rotating flow phenomena--both among the several fields motivated to analyze such motions and even within given fields. For example, *geostrophic* appears in slightly different, but related, meanings in the literature. The same difficulty applies to *Rossby wave*, although the meaning and application in the original Rossby article is explicit.

As an example of a stability analysis based on an arbitrary displacement of a parcel of constant-density, incompressible, rotating fluid, consider the Rayleigh instability criterion. This criterion compares the balance between centrifugal force and radial pressure gradient in a circular pathline flow field. More precisely, it asks whether the force due to inward radial pressure ($-\partial p/\partial r$) is adequate to maintain inward centripetal acceleration ($-v^2/r$) for an arbitrary parcel of fluid in that flow field, as in Figure 13.1.

As with the Brunt-Vaisala atmospheric stability analysis in Chapter 5, a parcel of fluid is assumed to have been perturbed, here to a new value of r, quickly enough to preserve its original properties. In a rotating-fluid, Rayleigh stability analysis, the conserved property is angular momentum. Assume a parcel of fluid is caused to move from an initial radius (r_i) out (or in) to a final radius (r_f), and that it *retains its original angular momentum* per unit mass ($h = rv$). Its new velocity (v_*) at r_f is

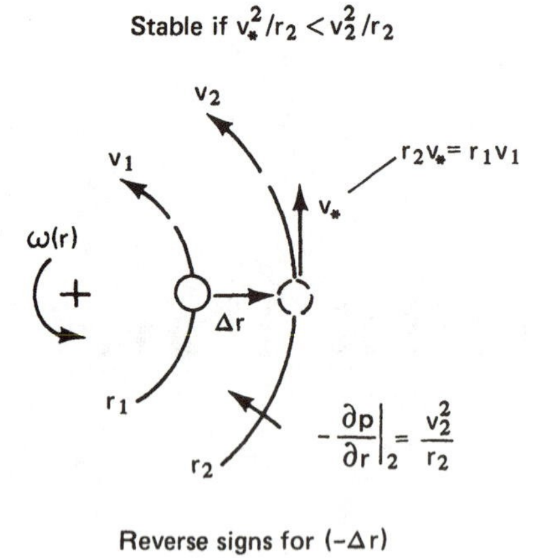

Figure 13.1: *A rotating fluid supports inertial waves if it is Rayleigh stable. Stability implies $\partial h/\partial r$ must be positive, where $h = rv$.*

$$r_f v_* = r_i v_i$$
$$v_* = r_i v_i / r_f \tag{13.1}$$

The new velocity of the perturbed parcel may be greater than or less than that of the ambient fluid (v_f) at the new radius, depending upon $\partial v/\partial r$ of the flow field. A quantitative comparison can be obtained by comparing centripetal acceleration of the perturbed parcel with the ambient pressure gradient in that flow field.

Centripetal acceleration of the displaced parcel is

$$-\frac{v_*^2}{r_f} = -\frac{(r_i v_i)^2}{r_f^3} \tag{13.2}$$

after use of (13.1). The ambient pressure gradient available to interact with this parcel of fluid is

$$\frac{1}{\rho}\frac{\partial p}{\partial r}\Bigg|_f = -\frac{v_f^2}{r_f} \tag{13.3}$$

The Rayleigh criterion then states, *If*

$$\frac{v_f^2}{r_f} > \frac{(r_i v_i)^2}{r_f^3} \quad or \quad (r_f v_f)^2 > (r_i v_i)^2 \quad or \quad |h_f| > |h_i| \tag{13.4}$$

(all of which are equivalent statements), then the flow field is stable to that perturbation. A parcel displaced outward is returned to its initial radius. A parcel

displaced inward cannot be held there, and it also returns to its initial radius. Unless the flow is heavily damped, the parcel will overshoot the undisturbed position and initiate an oscillatory wave motion, often called an *inertial* wave. Because rv is angular momentum per unit mass (h), stability and inertial waves require that h increase with r. Instability (convective motion) in an inviscid fluid results if $\partial h/\partial r < 0$. Viscosity can modify this result. If $\partial h/\partial r = 0$, as with an inviscid vortex, the flow is neutrally stable, i.e., a displaced parcel adjusts exactly to the new ambient condition.

13.2 Stability of Circular Pathline Flows

The Rayleigh stability criterion can be applied to each of the circular pathline flows of Chapter 12. The quantities v_i and v_f are replaced by the v(r) functions characteristic of each flow field. The analysis is generalized to include central forces other than a pressure gradient, e.g., gravity. It is easier to visualize the stability criterion if either an outward or an inward displacement is examined, with the knowledge that if one is stable (or unstable), then so is the other--unless of course r_i is at a flow discontinuity, in which case anything is possible. In the following discussions, the perturbation displacement will be assumed to be positive ($r_f > r_i$).

For rigid rotation (Figure 12.1), the stability criterion $(r_f v_f)^2 > (r_i v_i)^2$ with $v = \Omega r$ becomes $(\Omega r_f^2)^2 > (\Omega r_i^2)^2$. The criterion is always met with Ω constant, and rigid rotation constant-density flows are always stable to perturbations. Therefore, inertial waves and oscillations are a common phenomena in a rotating fluid having Ω = constant.

Use of the stability criterion with $v = \kappa/r$ results in neutral stability. The result might be expected because the inviscid vortex (Figure 12.2) already represents a flow with uniform angular momentum ($\partial h/\partial r = 0$). A displaced parcel preserves its angular momentum so that its new velocity (v_*) is equal to the ambient velocity for all κ, all r, and all perturbations. There are no angular momentum gradients in an inviscid vortex, and waves and/or instabilities dependent on angular momentum gradients are not possible. The vortex spiral (Figure 12.3) is similar. The radial pressure gradient differs from the inviscid vortex because of the radial flow (Q), but this is balanced at all places by the convective radial acceleration.

The circular channel flow (Figure 12.4) is assumed here to start from an initially flat velocity distribution with v_θ constant across the flow and v_r zero. Assume an inviscid fluid so that boundary layer phenomena and secondary flows can be neglected, and consider only instabilities in the interior of the flow field related to the curved pathlines of the flow. The stability criterion with $v = U$ = constant is $(Ur_f)^2 > (Ur_i)^2$, indicating stability for all U and all r. In reality, curved channel and pipe flows tend to be unstable, with the instabilities related to boundary layers and interior shear stresses. Chapter 14 analyzes similar flows in curved pipes, channels, and rivers.

Circular Couette flow (Figure 12.5), existing in the annulus between two concentric cylinders, tends to be stable with the outer cylinder rotating and the inner stationary, and unstable for the reverse case, i.e., outer stationary and inner rotating. In practice there are many exceptions to both rules, and of course there is the possibility that both rotate--even in opposite directions. Stability is dependent upon specific values of μ, ρ, Ω_a, Ω_b, r_a, and r_b, and their ratios. Details are discussed in Chapter 15, including equivalent flows between parallel rotating disks.

Circular orbit "flow" (Figure 12.6) is stable for an inverse square law (Newtonian) gravitational field, and, even though $\partial p/\partial r = 0$, the Rayleigh stability criterion is meaningful. The flow can be analyzed in terms of the stability of a small mass orbiting about a much larger central mass. A radial perturbation of the small mass will deform a circular orbit into an elliptical orbit as follows. If an object in circular orbit is perturbed radially outward to apocentron (largest distance from the central mass, r_{max}), while preserving its angular momentum, its tangential velocity will be smaller than that required for a circular orbit at r_{max}, and gravitational force (rather than $\partial p/\partial r$) will accelerate it back toward the initial circular orbit. It overshoots this orbit and reaches pericentron (r_{min}) with a tangential velocity greater than that of a circular orbit at r_{min}. Gravitational force cannot keep the object at this r, and the object returns back toward r_{max}. Each orbit represents one cycle of oscillation.

The introduction of viscosity into the flows of Figures 12.2, 12.3, 12.4, and 12.6 results in shear stress and transfer of angular momentum in the radial direction. Unless some mechanism maintains $\omega_*(r)$ for these flows, they must eventually reach rigid-body rotation with the regions having the highest strain rates the most quickly affected. Shear stress can affect v(r) in local regions and excite instabilities.

An analysis by Chandrasekhar, equivalent to that given here, concludes that any flow with $v \propto 1/r^n$ with $n > 1$ is unstable per the Rayleigh criterion because $(r_f v_f)^2 \gtrless (r_i v_i)^2$ becomes

$$\text{For } n < 1, \quad \left(r_f^{1-n}\right)^2 > \left(r_i^{1-n}\right)^2 \qquad \text{(Stable)} \qquad (13.5a)$$

$$\text{For } n = 1, \quad \left(r_f^{0}\right)^2 = \left(r_i^{0}\right)^2 \qquad \text{(Neutrally stable)} \qquad (13.5b)$$

$$\text{For } n > 1, \quad \left(r_f^{1-n}\right)^2 < \left(r_i^{1-n}\right)^2 \qquad \text{(Unstable)} \qquad (13.5c)$$

As seen in Table 12.1, each of the stable circular flows have $n < 1$, and each flow with $n = 1$ is neutrally stable. Obviously flows such as (13.5c) do not exist in nature except under very transient conditions; convective motions will quickly modify these flows either to some alternate steady-state pattern or to turbulence.

In a contained incompressible flow, the creation of either a wave or an unstable convective motion requires successive displacements of adjacent fluid because of mass conservation, $\nabla \cdot \mathbf{V} = 0$. Taylor vortices forming in the annulus between rotating

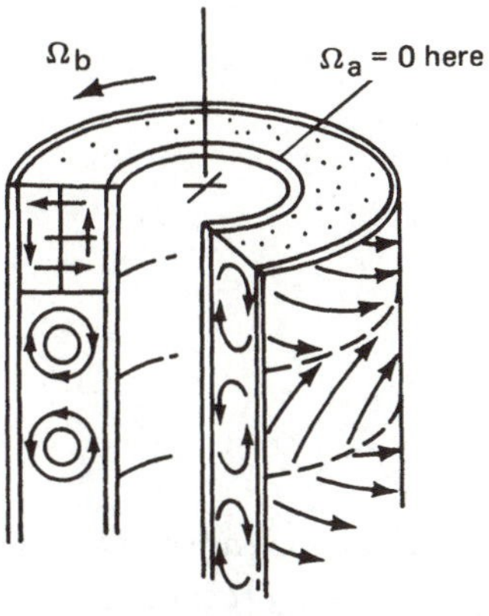

***Figure 13.2:** Taylor vortices resulting from Rayleigh instability in circular Couette flow.*

concentric circular cylinders (circular Couette flow) are a classic example. Figure 13.2 shows a successive displacement of fluid rings consistent with mass conservation. Circular Couette flow has become unstable by the Rayleigh criterion, and a new stable flow pattern is created. The outward and inward convective flows alternate axially, while being transported tangentially by the spinning cylindrical surfaces.

Simple explanations of complex flows sometimes are helpful, even though they neglect important details. For example the previous explanation neglects viscosity, yet the Taylor vortices are driven by viscous shear flow between the cylinders, and they occur only after shear rates reach fairly high values. In terms of momentum flux, the vortices represent secondary flows that transport momentum between the inner and outer cylinders. Using vorticity as the variable of interest, the vortices represent deformation and stretching of vortex lines from their original axial direction (laminar circular Couette flow) to a circumferential direction. In the process the parallel vertical vortex lines coalesce into large, organized toroidal vortices that fill the available annular clearance.

13.3 Rossby Waves and Inertial Waves

Fluid motions consistent with mass conservation in a contained fluid consist of two types, usually referred to as inertial modes and a geostrophic mode, with the latter degenerating under some conditions into alternate wave forms. The inertial modes are those corresponding to the natural resonant modes for that geometry. In a sphere, for example, spin of the interior fluid about a single axis as if it were a rigid body is the first inertial mode. Higher modes are obtained by seeking smaller subdivisions of the sphere with corresponding motions that satisfy both continuity and momentum considerations. Eigenfunctions of the equations of motion describe the mode shapes (geometry), and eigenvalues describe the mode frequencies.

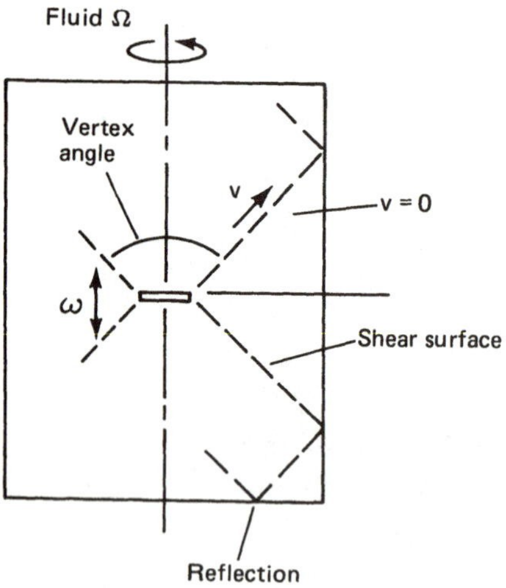

Figure 13.3: *Progressive inertial waves produce characteristic surfaces (cones) when the disturbance frequency* $\omega \leq 2\Omega$.

Energy must be introduced into the flow to excite the modes. The forcing function for a sphere can be the viscous boundary layer. For a nonspherical surface the forcing function also includes normal velocities of the boundaries relative to the fluid mass. Solutions are typically expressed as Legendre polynomials and Bessel functions, or other functions characteristic of the container geometry. Inertial waves based on the inertial modes occur easily in a fluid in rigid rotation because of its inherent stability.

Inertial waves can be produced and observed in the laboratory in the experiment illustrated in Figure 13.3. Aluminum powder is suspended in water in the form of microscopic disks or flakes. Orientation of the flakes is random in the water, except at shear surfaces where they align and are made visible by appropriate illumination. The cylinder and water rotate at a constant angular velocity Ω. A small disk is suspended centrally in the cylinder and caused to oscillate vertically with frequency ω. The vertical oscillations produce radial disturbances as the water flows around the edge of the disk. If the water were not rotating, internal waves would not be possible, but because it is rotating, inertial waves form per the Rayleigh criterion. The waves move both axially and radially, producing conical shear surfaces. A value $\omega = 0$ corresponds to a Taylor-Proudman column, and as ω increases to $\omega = 2\Omega$, the vertex angles of the cones increase to 180°.

In this application, the local (temporal) acceleration $(\partial V/\partial t)$ is balanced by the Coriolis acceleration $(2\mathbf{\Omega} \times \mathbf{V})$. The *frequency ratio* (ω/Ω) measures their ratio as discussed in Section 6.3.

Geostrophic flow represents a balance between the pressure field and Coriolis acceleration; in dimensionless form this is

$$\nabla p = 2\hat{\mathbf{k}} \times \mathbf{V} \tag{13.6}$$

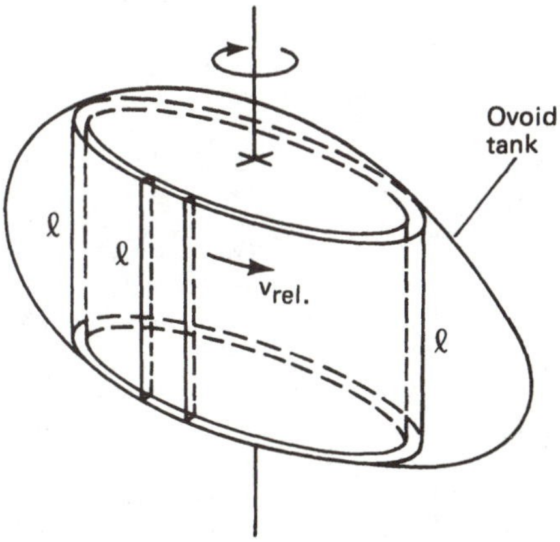

Figure 13.4: *Flow along a geostrophic contour with* ℓ *= constant. Compare with Taylor-Proudman columns.*

With R_o and E_k << 1, an inherently two-dimensional (2-D) flow is indicated because $\partial V/\partial z \approx 0$, as shown in Example 9.4. The result was used there to deduce the existence of Taylor-Proudman columns. Figure 13.4 shows an equivalent result in an egg-shaped container whose motion differs slightly from that of the enclosed rotating liquid. The streamlines are horizontal closed contours of constant-height liquid. See Greenspan (1968, Figure 2.6) for a photograph.

Geostrophic flows dominate large-scale motions in the atmosphere and oceans, but in some cases the geostrophic mode is not available due to a lack of constant-height closed contours. H. P. Greenspan and J. Pedlosky have shown that Rossby waves are a type of inertia wave that can be created in the laboratory by rotating a nearly inviscid fluid in a closed container that has no available geostrophic mode due to container geometry.

Figure 13.5 shows a circular cylinder whose bottom is tilted at an angle α from the horizontal. The cylinder and enclosed liquid rotate uniformly about the central axis after which the angular rate of the cylinder is impulsively changed. No constant-height geostrophic flow is now possible, but an infinite number of special low frequency oscillations are generated that participate in the resulting interior flow field. Top views of the cylinder sketch the formation and propagation of a Rossby wave in this application. Again see Greenspan (1968, Figure 2.16) for a photograph.

When both the cylinder (c) and liquid (ℓ) rotate in unison, vorticity is uniform throughout the liquid, $\zeta^{\ell} = 2\omega^{\ell}$. After a change in angular velocity of the cylinder, $\omega^{c} \neq \omega^{\ell}$, and the mass of liquid is rotating relative to the tilted bottom. As a vertical vortex tube travels from the shallow end around to the deep end, it is stretched and

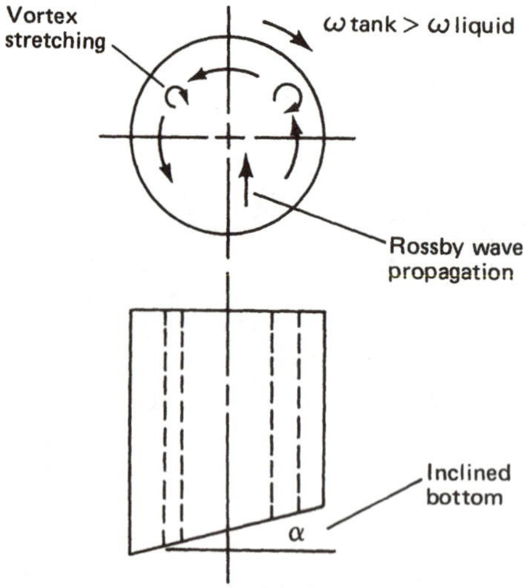

Figure 13.5: *Laboratory demonstration of Rossby wave propagation. See Pedlosky and Greenspan (1967).*

intensified, and on completing its circuit back to the shallow end, it is shortened with a return to its original vorticity. This creates a transverse vorticity gradient. In addition, volume in the shallow half is less than volume in the deep half. As the liquid mass rotates relative to the container, there is a continuous perturbation of liquid velocity relative to the cylinder to satisfy mass conservation. These transverse flow perturbations relative to the transverse vorticity gradient create Rossby waves as detailed in the following section.

13.4 Atmospheric Rossby Waves I

A concise mathematical analysis and physical description of the nature of Rossby waves are found in Rossby's original presentation. He was concerned with large (continental-sized) masses of air rotating as stationary patterns relative to the earth. The following summarizes his presentation. The model is illustrated in Figure 13.6.

Vorticity relative to inertial space of a parcel of air can be expressed as a sum of vorticity of the air relative to the earth (ζ) plus the vorticity of the earth relative to inertial space (2Ω). The atmosphere is usually stably stratified. This suppresses vertical velocities, so that atmospheric flows are mostly horizontal. Therefore, only the vertical component of atmospheric vorticity is necessary to describe phenomena relevant to this discussion. More details are included in Section 20.5, Atmospheric Rossby Waves II.

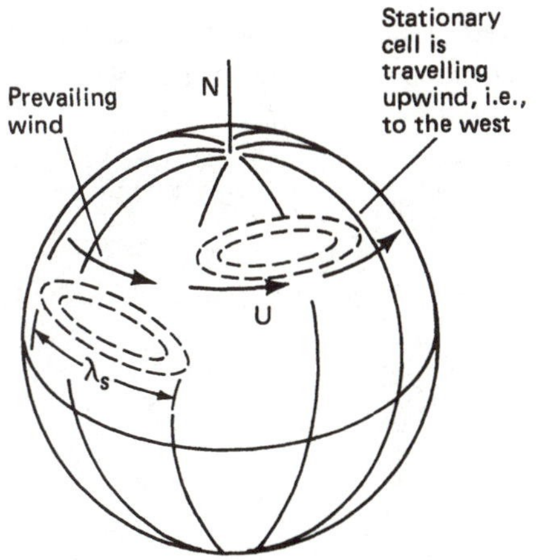

Figure 13.6: *Geographically stationary perturbations in an eastward wind indicate a wave moving upwind in Rossby's original analysis.*

Writing ζ_z as ζ for simplicity, and the vertical component of earth vorticity at latitude θ as

$$f = 2\Omega \sin\theta \tag{13.7}$$

the vertical component of total vorticity becomes $f + \zeta$. In atmospheric and oceanic analyses, f is known as the Coriolis parameter. During an arbitrary displacement of a parcel of air with no applied torques $(f + \zeta)$ must remain constant so that $d(f + \zeta)/dt = 0$. Accordingly

$$\frac{d\zeta}{dt} = -\frac{df}{dt} = -\frac{dy}{dt}\frac{\partial f}{\partial y} = -v\frac{\partial f}{\partial y} = -\beta v \tag{13.8}$$

with y and v northward direction and velocity, respectively. Here β has been written for $\partial f/\partial y$, the rate of change of earth vorticity relative to latitude, computed as

$$\beta = \frac{2\Omega\cos\theta}{R} \tag{13.9}$$

where R is the radius of the earth.

The earth's east-west trade winds result from the poleward flow of solar energy modified by Coriolis phenomena. Over the midlatitudes (30° to 60°, both hemispheres) the surface trade winds blow from the west to the east (+x direction), although there are temporary dislocations, and the patterns shift north or south about 10° with the seasons. The flow can be represented by

$$u = U + u', \qquad v = v' \tag{13.10}$$

where u is velocity to the east, v is velocity to the north, U is the trade wind, and u' and v' are perturbations. In this model relative vorticity is

$$\zeta = \left(\frac{\partial v}{\partial x} - \frac{\partial u}{\partial y}\right) = \left(\frac{\partial v'}{\partial x} - \frac{\partial u'}{\partial y}\right) \tag{13.11}$$

Equation (13.8) is placed in Eulerian form and combined with (13.10) to get

$$\frac{\partial \zeta}{\partial t} + U\frac{\partial \zeta}{\partial x} + u'\frac{\partial \zeta}{\partial x} + v'\frac{\partial \zeta}{\partial y} = -\beta v' \tag{13.12}$$

which reduces to

$$\frac{\partial \zeta}{\partial t} + U\frac{\partial \zeta}{\partial x} = -\beta v' \tag{13.13}$$

after neglecting products of small quantities. Next, if a wave exists, it is representable as a function of $(x \pm ct)$. Representing a vorticity wave as $\zeta(x - ct)$, where c is the eastward velocity of the wave, leads to

$$\frac{\partial \zeta}{\partial t} = -c\frac{\partial \zeta}{\partial x} \tag{13.14}$$

as may be verified by substitution. Use of (13.14) in (13.13) yields

$$(U - c)\frac{\partial \zeta}{\partial x} = -\beta v' \tag{13.15}$$

Assuming the perturbation is only a function of x requires modifying (13.11) to be $(\zeta = \partial v'/\partial x)$ so that

$$(U - c)\frac{\partial^2 v'}{\partial x^2} = -\beta v' \tag{13.16}$$

Finally, substitute into (13.16) a sinusoidal perturbation in v

$$v' = \sin\frac{2\pi}{\lambda}(x - ct) \tag{13.17}$$

where λ is wavelength to get

$$(U - c)\frac{4\pi^2}{\lambda^2}\sin\frac{2\pi}{\lambda}(x - ct) = \beta \sin\frac{2\pi}{\lambda}(x - ct)$$

This simplifies to

$$c = U - \frac{\beta\lambda^2}{4\pi^2} \tag{13.18}$$

Wave speed is zero when ($U = \beta\lambda^2/4\pi^2$) or

$$\lambda_s = 2\pi\sqrt{U/\beta} \tag{13.19}$$

and (13.18) can be rewritten as

$$c = U(1 - \lambda^2/\lambda_s^2) \tag{13.20}$$

When $\lambda = \lambda_s$, the wave moves westward at a velocity U and is stationary over the earth. If $\lambda > \lambda_s$, c is negative and the wave has a net westward motion relative to the earth. This is a classic Rossby wave. The wavelength is very large, and it travels up-current (here westward).

A similar analysis can be applied to the rotating cylinder with tilted bottom. There is a vorticity gradient across the cylinder due to vortex stretching (equivalent to β), and a sinusoidal perturbation in v caused by mass conservation (tilted bottom). The vortex wave moves perpendicular to the vorticity gradient similar to the westward movement of atmospheric Rossby waves.

13.5 Instability and Turbulence

Turbulence is and has been a persistent and pervasive problem in the study of fluid motion. Simple questions such as, *What is turbulence?*, and, *When is a flow turbulent?*, continue to be unanswerable except on an ad hoc basis. There have been some impressive successes but as many impressive failures. As Tritton (1985) notes, concerning turbulence in both nonrotating and rotating flows, ". . . there are many interesting observations and ideas but no real unification . . . ."

The difficulty may be a need for a more accurate or useful physical model appropriate to the flow regime. Much current research, analytical and experimental, is directed to an understanding of the successive flow regimes as fluid motion degenerates from smooth laminar flows through a succession of intermediate flows (e.g., waves or vortex rolls) to a state of apparently random chaotic motion, typified as turbulence. This *route to chaos* model is now accessible to computer analysis. It has had some success in predicting, for example, transitions from laminar circular Couette flow through a succession of Taylor vortex patterns to a final turbulent state. See, e.g., Donnelly (1991).

A particular difficulty has been a unique phenomena, in flows where rotation of the main flow is significant, that turbulence exhibits the 2-D patterns common to rotating flows in general. Figure 1.5 illustrates the two dimensionality of rotating fluids in a simple yet dramatic experiment. Rotation in this application can be considered significant when the rotational Rossby number defined as

$$R_o = \frac{1}{2\Omega}\frac{\partial u}{\partial y} \tag{13.21}$$

is small.

At one time it had been common to restrict a definition of turbulence to only those chaotic flows that exhibited 3-D random motion. Also it was generally believed, consistent with the observation of nonrotating fluids, that energy of the main flow travelled downward through a succession of smaller and smaller eddies. This maintained the turbulence and finally dissipated in molecular viscosity. Recent experiments, supported by some analysis, appear to have demonstrated that, in rotating flows, the opposite can also occur, i.e., small-scale eddies can feed energy from the turbulent regime into the mean flow.

13.6 Application Examples

Several wave classifications are summarized, and wave mechanisms available to both nonrotating fluids and rotating fluids are reviewed. The Rayleigh criterion for stability/instability of rotating fluids and the production of inertial waves from angular momentum gradients are presented and applied to the circular pathline flows of Table 12.1. Geostrophic flow and Rossby waves are described. Example 13.1 analyzes frequencies and speeds for several waves available to both rotating and nonrotating flows. Example 13.2 illustrates a computation of surface wave mode shapes in a constrained space, and Example 13.3 investigates frequencies, speeds, and mode shapes of inertia waves in a stable rotating fluid.

Example 13.1. Examine an internal gravity wave. Assume that a 20 m thick layer of water with density $\rho = 1.01$ g/cm^3 very slowly flows over a 1 km deep stationary layer of saline water with density $\rho = 1.06$ g/cm^3. They are covered with a rigid layer of ice. What is the wave speed of a perturbation with $\lambda = 80$ m at the density discontinuity? Next, examine a buoyancy oscillation caused by vertical perturbations in a static atmosphere with a nonadiabatic vertical temperature gradient. Is a standard atmosphere at 1000 m stable to vertical perturbations, and if so, what is the Brunt-Vaisala frequency?

Outside sources useful here are Milne-Thomson (1960, pp. 404–405) for the internal gravity wave, and Holton (1979, pp. 49–51, 159–165) and Pedlosky (1987, pp. 351–354) for the buoyancy oscillation. Greenspan (1968, Section 1.4) compares motions available to a stratified liquid to those available to a rotating liquid.

Discussion. Milne-Thomson (1960) outlines the solution for a general case of one fluid in motion over another denser fluid also in motion, with both held between rigid parallel planes. Here c is the wave velocity, and a coordinate system moving with velocity c is chosen so that the wave appears stationary. Assuming $\eta = \eta_{max} \sin(\alpha x - \beta t)$, where the wave number $\alpha = 1/\lambda$, the frequency $\beta = 1/\tau$, and the amplitude of η is small, the solution for c is given by

$$(\rho/\lambda)(V - c)^2 \coth(h/\lambda) + (\rho_*/\lambda)(V_* - c)^2 \coth(h_*/\lambda) = g(\rho - \rho_*)$$

If $\rho_* = 0$ and $V = 0$, the equation reduces to the surface wave studied in Chapter 5 approximated as

$$c^2 = g\lambda \tanh(h/\lambda)$$

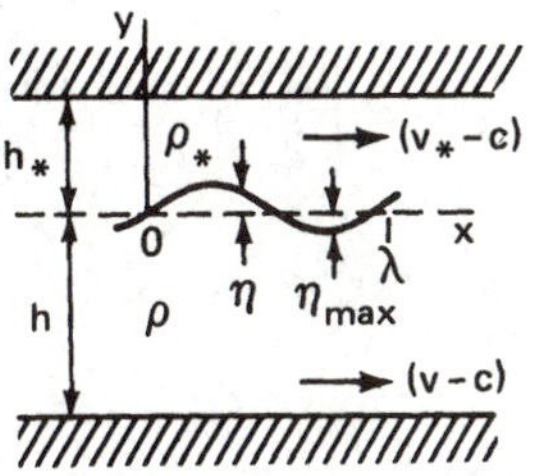

If h and $h_* \rightarrow \infty$

$$(\rho/\lambda)(V\text{-}c)^2 + (\rho_*/\lambda)(V_* - c)^2 = g(\rho - \rho_*)$$

and also if $V = V_* = 0$, then

$$c^2 = (g\lambda)\frac{(\rho - \rho_*)}{(\rho + \rho_*)}$$

In this problem, $V = 0$ and $V_* \rightarrow 0$. Remember that an ordinary surface wave of length λ causes negligible disturbances below depths of about $\lambda/2$. By this criteria, $h \sim \infty$ if $\lambda \lesssim 2$ km, and $h_* \sim \infty$ if $\lambda \lesssim 40$ m. Since waves with $\lambda = 80$ m are of interest here, we choose

$$(\rho/\lambda)c^2 + (\rho_*/\lambda)c^2 \coth(h_*/\lambda) = g(\rho - \rho_*)$$

$$c^2 = \frac{g(\rho - \rho_*)}{(\rho/\lambda) + (\rho_*/\lambda)\coth(h_*/\lambda)} = \frac{981(1.06 - 1.01)}{(1.06/8000) + (1.01/8000)\coth(2000/8000)}$$

$$c = 2.75 \times 10^2 \text{ cm/s} = 2.75 \text{ m/s}$$

Note that because the top layer is less than about $\lambda/2$ thick, a surface wave would appear in the absence of the rigid ice cover, and the problem would be beyond the theory presented here.

In the second part of this example, a static atmosphere is stable if (5.4a) is met.

If the atmosphere is stable and static, the Brunt-Vaisala buoyancy frequency is given by (5.4b). The atmosphere is stable if

$$\left(1-\frac{1}{\gamma}\right)\frac{T}{p}\rho g > \left(-\frac{\partial T}{\partial z}\right)$$

where ρg has been used for $(-\partial p/\partial z)$. Using approximate data for the standard atmosphere, the criteria is whether

$$\left(1-\frac{1}{1.40}\right)\frac{282}{89{,}900}1.112\times 9.81 > \frac{6.5}{1000}$$

The left-hand side equals 9.78 (–3) and the right-hand side equals 6.5 (–3), so the stability criteria is met. The buoyancy frequency is then

$$n^2 = \frac{g}{T}\left(\frac{\partial T}{\partial z}+\frac{g}{c_p}\right) = \frac{9.81}{282}\left(\frac{6.5}{1000}+\frac{9.81}{1005}\right) = 5.66\,(-4)$$

and n = 0.0238 Hz

The corresponding period for one oscillation is about 42 s which is quite short compared to many other atmospheric phenomena.

Example 13.2. What are the first and second slosh mode frequencies for 6 m deep water in a 5 m by 12 m rectangular tank? See Chapter 5 for a review of surface waves.

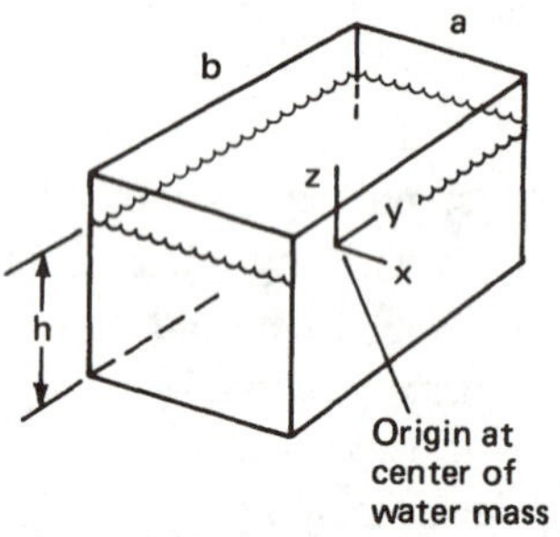

Discussion. The classic solution assumes an incompressible, uniform-density, inviscid liquid in a gravity field (g). The liquid has a free surface at the top, the tank walls are rigid and inertially fixed, and liquid displacement is $\eta \ll h/2$. Under these conditions the liquid is irrotational, it is a potential flow problem, and Laplace's equation is satisfied for the velocity potential ($\nabla^2\phi = 0$, with $\mathbf{V} = -\nabla\phi$). Solution of Laplace's equation with boundary conditions as shown provides a solution

$$\eta = \sum_{i=0}^{\infty}\sum_{j=0}^{\infty} A_{ij}\cos\left[\frac{i\pi}{a}\left(x+\frac{a}{2}\right)\right]\cos\left[\frac{j\pi}{b}\left(y+\frac{b}{2}\right)\right]$$

$$\omega_{ij}^2 = gk\tanh(kh), \quad k^2 = \pi^2\left(\frac{i^2}{a^2}+\frac{j^2}{b^2}\right)$$

where i and j are indices appropriate for each mode shape. The solution can also be derived in terms of waves reflecting from wall to wall in phase with the tank size.

The solution shows that an arbitrary disturbance of the fluid level (η) above or below the equilibrium height (h/2) can be represented as the sum of two orthogonal sets of waves, one moving in the x direction and one in the y direction. Each set of waves is composed of one or more (up to ∞) principal mode shapes, each with a separate amplitude (A_{ij}). The first mode shape in the x direction has i = 1, and the second mode shape has i = 2. In the y direction, the first mode shape has j = 1 and the second mode shape has j = 2. Any combination is possible, so that the first and second modes include all of (1, 0), (0, 1), (1, 1), (0, 2), (2, 0), (1, 2), (2, 1), and (2, 2). Consider as specific examples the principal (i.e., pure) mode shapes (i, j) = (0, 1) and then (1, 2).

With, a = 5 m, b = 12 m, h = 6 m, and with (i, j) = (0, 1)

$$\eta_{01} = A_{01} \cos\left(\frac{\pi}{12} y + \frac{\pi}{2}\right) \quad \text{as shown}$$

$$k = \pi/12$$

$$\omega_{01} = \sqrt{(9.81\pi/12)\tanh(6\pi/12)}$$

$$\omega_{01} = 1.53 \text{ rad/s}, \qquad f = 0.24 \text{ Hz}$$

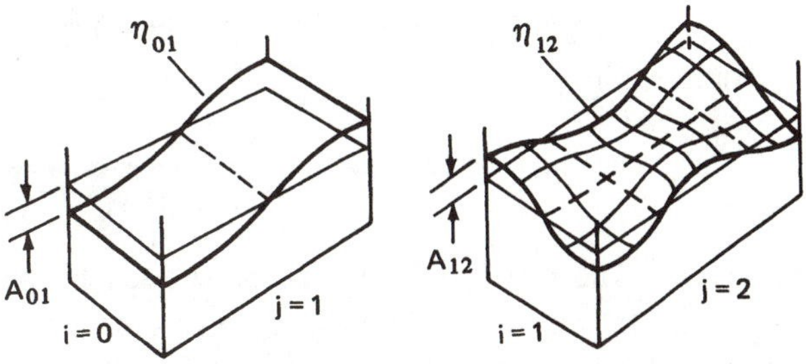

With (i, j) = (1, 2)

$$\eta_{12} = A_{12} \cos\left(\frac{\pi}{5} x + \frac{\pi}{2}\right) \cos\left(\frac{\pi}{6} y + \pi\right)$$

$$k = \pi\sqrt{(1/25) + (1/36)} = 0.818$$

$$\omega_{12} = \sqrt{9.81 \times 0.818 \tanh(6 \times 0.818)}$$

$\omega_{12} = 2.83$ rad/s, $\quad f = 0.45$ Hz

The dashed lines locate nodes where η remains zero during sloshing at these principle modes.

The lowest frequencies have the lowest energy and are the easiest to excite. In this application (0, 1) has the lowest frequency. Also in this application (1, 2), as shown, would be easier to excite than (2, 1) as may be verified. Note that when (kh) $\lesssim 0.3$, tanh (kh) $\approx$ kh, and $\omega_{ij} \approx k\sqrt{gh}$. When (kh) $\gtrsim 3$, tanh (kh) ≈ 1, and $\omega_{ij} \approx \sqrt{gk}$.

For further details on sloshing, see Lamb (1932) and Abramson (1966). The latter also illustrates space vehicle applications.

Example 13.3. A closed cylinder (10 cm radius, 20 cm high) is filled with water and rotated about its axis at 90 rpm. Describe an experimental technique that could be used to excite and observe, (1) the (γ_1, η_2) inertial mode and, (2) progressive wave characteristic cones with angles of $\pm 30°$. Include quantitative estimates of the apparatus and fluid motions. Fultz (1959) and Greenspan (1968, pp. 3, 10, 81–85) provide information useful for solving this problem.

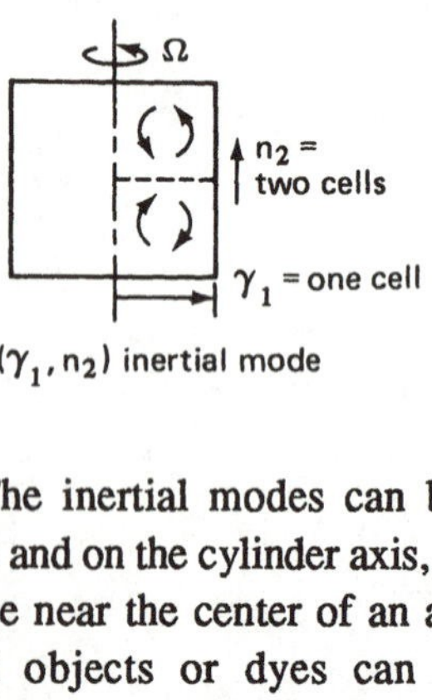

(γ_1, n_2) inertial mode

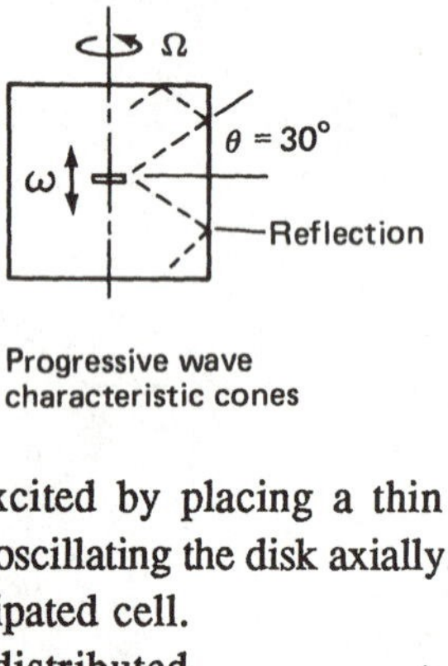

Progressive wave characteristic cones

Discussion. The inertial modes can be excited by placing a thin circular disk perpendicular to and on the cylinder axis, and oscillating the disk axially. The nominal position must be near the center of an anticipated cell. Small buoyant objects or dyes can be distributed throughout the fluid to observe motion. Fultz (1959) showed that the resonant inertial modes occur sharply at precise frequencies (period = τ) of the disk. The experimentally determined eigen-periods (τ_e) agreed very well with theoretical values (τ_t) for an inviscid fluid given by

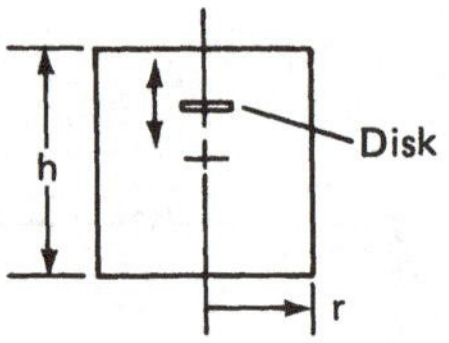

$$\tau_t = \frac{1}{2}\left[(\gamma_i/n\pi)^2\,\alpha^2 + 1\right]^{1/2} \tag{13.22}$$

where $\alpha = h/r$, γ_i is the ith root of the Bessel function $J_1(\gamma)$, and i and n are indices representing the number of integer cells in the radial and axial directions, respectively. The eigen-period (τ) has been normalized by the rotation period $\tau_0 = 2\pi/\Omega$ to get $\tau_t = \tau/\tau_0$. For reference, the roots (zeros) of $J_1(\gamma)$ are $\gamma_1 = 3.832$, $\gamma_2 = 7.016$, $\gamma_3 = 10.174$, and $\gamma_4 = 13.331$. Typical of the results were

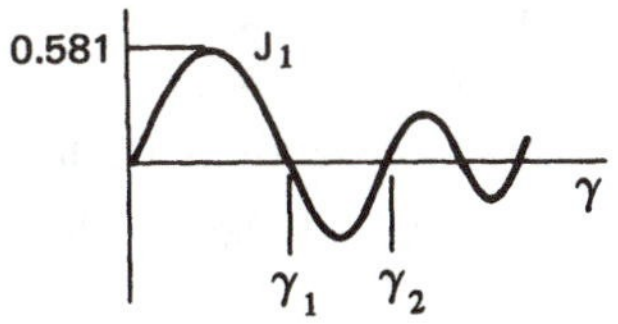

Mode	α	τ_t	τ_e
$(\gamma_1, 1)$	1.50	1.042	1.040
$(\gamma_2, 1)$	1.50	1.748	1.725
$(\gamma_2, 2)$	1.50	0.975	0.985

Fultz used an open-top cylinder with radius 4.23 cm and filled with water to obtain the specified α, and a 1.9 cm diameter disk with vibration amplitude of about 1 cm. Methylene blue dye was used for observation. Presumably the same theory should give equally good results for the problem as given, $\alpha = 2.0$, $\gamma_1 = 3.832$, and $\eta_2 = 2$.

$$\tau_t = \frac{1}{2}\left[\left(\frac{8.832}{2\pi}\right)^2 2^2 + 1\right]^{1/2} = 0.789$$

The eigen-period $\tau = 0.789 \times 2\pi/\Omega$. With $\Omega = 90$ rpm $= 3\pi$ rad/s, $\tau = 0.526$ s. A reasonable experimental plan might use a 4 cm diameter disk placed at 0.75 h and oscillated vertically at 0.53 Hz with about a 1 cm amplitude.

Note that the size of the tank is not relevant, only the ratio $\alpha = h/r$. A low viscosity fluid (water) should be used since μ was assumed zero, but ρ has cancelled out since mass inertia provides both forcing function and momentum storage. The forcing function, of course, is displacement of low-angular momentum fluid to a higher angular momentum region and vice versa. This occurs only in the radial flows. Both the axial and radial flows store linear momentum to carry the oscillation over to the next half cycle. After an oscillation is started, the central disk provides only enough energy to compensate for viscous losses.

Greenspan (1968, p. 10) shows the linear, inviscid, time-dependent flow equations for a rotating fluid to have wave-like solutions. The governing differential equation is hyperbolic with $\omega \le 2\Omega$, where ω is the frequency of an assumed disturbance. In this connection note that (13.22) has a lower limit when $h << r$ of $\tau_t = 1/2$, and that $\tau_t = \tau/\tau_0 \ge 1/2$ is equivalent to $\omega/\Omega \le 2$. When a disk, similar to that used here to excite the inertial eigen-modes, experiences nearly infinitesimal vertical oscillations at frequency ω, with $\omega/\Omega \leq 2$, progressive waves produce conical surfaces that define flow discontinuities. These thin shear layers can be observed by suspending microscopic flakes in the water. The flakes align at the shear layers, but not otherwise, so that the shear layers can be observed by preferential reflection of light.

The cones originate at the disk edges as shown at angles

$$\cos\theta = \frac{1}{2\tau_t} = \frac{\omega}{2\Omega} \tag{13.23}$$

Experimental values should agree well with theory. For the problem given, $\pm\theta = 30°$ is desired and

$$\tau_t = 1/(2\cos 30°) = 0.577 \text{ s}$$

$$\tau = \tau_t \tau_0 = 0.577 \times 2\pi/\Omega = 0.385 \text{ s}$$

Therefore, the disk must be oscillated at a frequency of

$$1/\tau = 2.60 \text{ Hz}$$

It is, of course, τ_t that cannot be less than 1/2, not τ. When $\tau_t < 1/2$, i.e., $\omega/\Omega > 2$, the defining equations are elliptic, and the previous phenomena do not exist. Also of course, if the tank is not rotating, potential flow solutions with fluid agitation in the vicinity of the oscillating disk are the only result.

This phenomenon and other rotating fluid phenomena are illustrated in experiments filmed by D. Fultz titled, *Rotating Flows*, and available as film or as video tape from Encyclopaedia Britannica Educational Corp., Chicago, Il.

PART III

Rotating Fluid Applications

Chapter 14

Pipes, Channels, and Rivers

14.1 Swirl in Straight Sections

It is well documented that turbulent flow in straight rectangular and triangular ducts is accompanied by longitudinal swirl that tends to move fluid away from wall midpoints and into corners. Although less well documented, longitudinal swirl also exists in laminar flow in straight noncircular channels. These secondary flows can be explained and computed in terms of pressure-induced transverse velocities by using the Navier-Stokes equations in variables **V** and p. This explanation requires no reference to or knowledge of vorticity. Alternatively, the secondary flows can be explained and computed in terms of convected vorticity by using the Navier-Stokes equations in variables $\boldsymbol{\zeta}$ and **V**. In this case, no reference to, or knowledge of, the pressure field is necessary. See Table 9.1 for a listing of the constant density and constant viscosity Navier-Stokes equations set first in the variables (**V**, p) and then in the variables ($\boldsymbol{\zeta}$, **V**). The variables (**V**, p) are usually chosen for analysis, partly because p can easily be measured, while $\boldsymbol{\zeta}$ cannot, and partly because the initial and boundary conditions are usually given as f(**V**, p).

Chapter 11 introduces the fundamental mechanism responsible for the existence of secondary flows. See especially Example 11.2 that illustrates creation of secondary flows caused by radial pressure differentials in the vicinity of boundary layer flows, and then alternatively, as caused by the differential convection of transverse vortex lines. Chapter 13 illustrates a further example of a secondary flow created by the amplification of a perturbation in an unstable flow, i.e., the development of Taylor vortices in an unstable, circular Couette flow. See Figure 13.2.

In the case of a straight circular pipe, flow velocities and vorticity generation are symmetric about the flow axis. Because there are no azimuthal pressure gradients and no azimuthal vorticity gradients, secondary flows are not generated. If the flow is turbulent, the mean flow also is symmetric about the flow axis, and again there are no net secondary flows. There will, of course, be small azimuthal fluctuations in velocity and vorticity when the flow is turbulent, but these occur on shorter time scales than typical secondary flow phenomena and are omitted from this discussion.

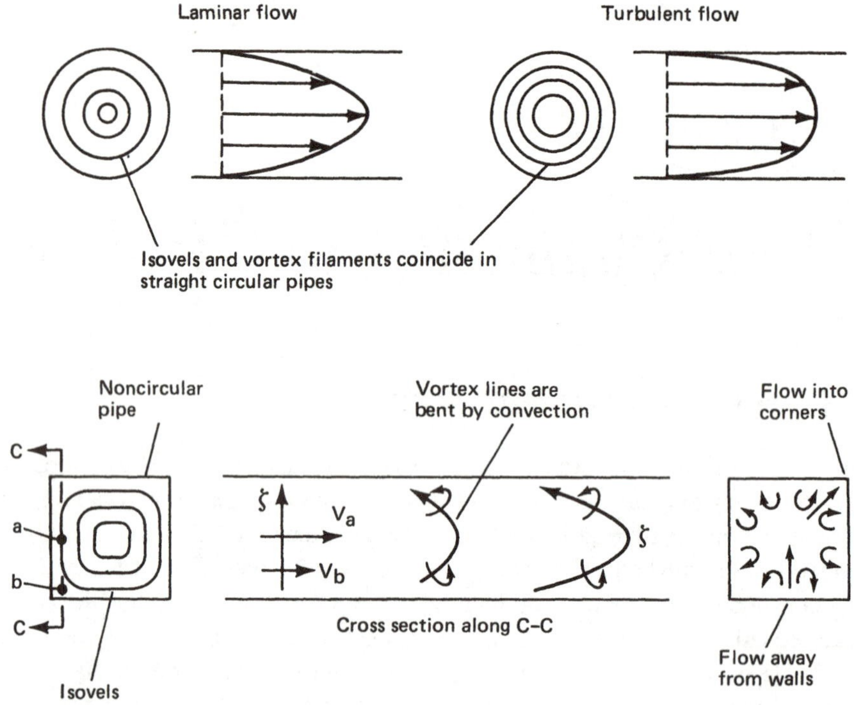

Figure 14.1: *Convective distortion of vortex lines causes secondary flows (longitudinal swirls) in straight noncircular pipes.*

Figure 14.1 illustrates the critical difference between flow in straight circular and straight noncircular pipes. In a circular pipe (or circular annulus), lines of constant velocity coincide with lines of constant vorticity. The terms $(\mathbf{V} \cdot \nabla)\ \boldsymbol{\zeta} - (\boldsymbol{\zeta} \cdot \nabla)\mathbf{V}$ of the $(\boldsymbol{\zeta}, \mathbf{V})$ Navier-Stokes equation define uniform convection of $\boldsymbol{\zeta}$ by $\mathbf{V}$ and $\mathbf{V}$ by $\boldsymbol{\zeta}$. In a noncircular pipe, here a square pipe, lines of constant $\mathbf{V}$ do not align with lines of constant $\boldsymbol{\zeta}$, and differential convection of each by the other must occur.

The cross section along surface C-C in Figure 14.1 shows the differential convection of $\boldsymbol{\zeta}$ by $\mathbf{V}$. Vorticity is created transverse to the flow direction by viscous shear. Velocity and vorticity are greatest near wall midpoints (V_a) as shown, while flow near the ends is retarded by the end walls (here top and bottom) and moves slower (V_b). The transverse vortex tubes ($\boldsymbol{\zeta}$) are bent and differentially convected by longitudinal velocity ($\mathbf{V}$), as shown in section C-C, leading to longitudinal components of vorticity, i.e., longitudinal swirl.

The net effect of the four pairs of swirls shown in Fig. 14.1 (lower right) is to convect low-momentum fluid near the wall midpoints into the central region of higher velocities and to convect high-momentum fluid into the more stagnant corners. This secondary flow represents the convection of $\mathbf{V}$ by $\boldsymbol{\zeta}$. It tends to make the constant

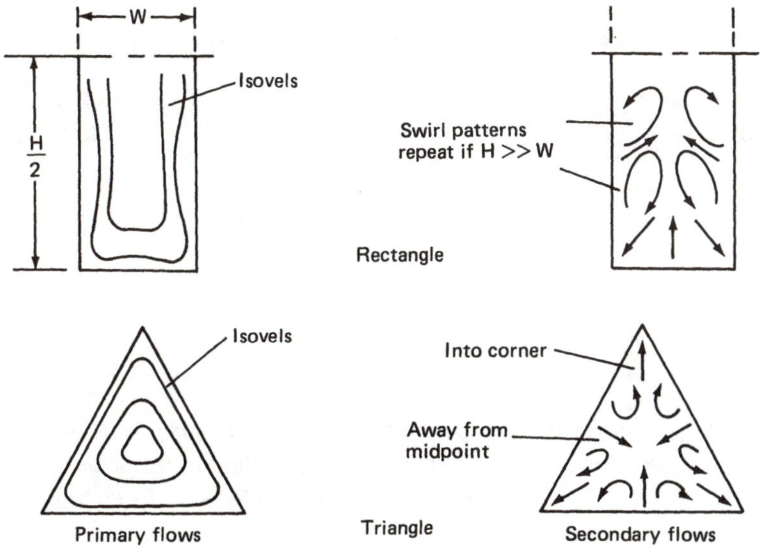

***Figure 14.2:** Turbulent mean flow in straight rectangular and triangular ducts showing lines of constant velocity (isovels) and secondary flow (swirl) patterns. (After Nikuradse 1926, 1930.)*

velocity lines, in the steady state, align more closely with the wall geometry than might appear obvious from a simple solution based only on a momentum balance in the flow direction. The swirls increase momentum transfer, indicating also increased pressure drop along the pipe and increased power required to maintain the flow. Figure 14.2 shows velocity contours and secondary swirl in straight noncircular ducts as experimentally observed by Nikuradse (1926, 1930).

Most data available on noncircular pipes and ducts is based on a mix of calculations, experiments, and empirical approximations. Usually the data is tabulated or graphed as friction or head-loss coefficients in terms of some equivalent hydraulic diameter or hydraulic radius. The hydraulic diameter is usually defined as

$$D_h = 4A/P \tag{14.1}$$

where A is the cross-section flow area and P is the wetted perimeter. For a circular cylinder, $A = \pi R^2$ and $P = 2\pi R$, so that $D_h = 4\pi R^2/2\pi R = 2R = D$. For a square pipe of size $H \times H$, $D_h = H$, implying that the addition of corners to a circular pipe is relatively ineffective in increasing flow rate for a given pressure drop. For an arbitrary rectangle with size $H \times W$, the hydraulic diameter is

$$D_h = 2H/(1 + \alpha) \tag{14.2}$$

where $\alpha = H/W$. The concept fails for extremely narrow sections, for example, in the case of a rectangle, it is useful only for $0.33 < \alpha < 3$, approximately.

Table 14.1: Values of (fR_e) with $R_e = \rho D_h \bar{V} / \mu$

Rectangular Duct		Concentric Annulus	
α = H/W	fR_e	r_1/r_2	fR_e
0.05	87	0	64
0.125	82	0.0001	72
0.25	73	0.01	80
0.5	62	0.1	89
1.0	57	0.4	95

Energy loss along a pipe due to momentum loss to the pipe walls or to fittings, bends, or other factors is expressed as a dimensionless fluid friction coefficient. The coefficient represents pressure loss per length normalized by the velocity head

$$f = -\frac{\Delta p \frac{D_h}{L}}{\frac{1}{2}\rho \bar{V}^2} \tag{14.3}$$

with $\bar{V}$ the averaged flow velocity (= Q/A) and L the total length.

Instead of pressure loss, an equivalent head loss is sometimes given as

$$h_f = f \frac{L}{D_h} \frac{\bar{V}^2}{2g} \tag{14.4}$$

with head loss defined now in terms of the fluid elevation (head) equivalent of the pressure loss. There are slight variations of these expressions in practice. Because they are invariably obtained from experiments, they include the effects of all flow phenomena whether understood analytically or not. In addition, they may be compromised by unspecified special conditions and/or experimental inaccuracies.

Table 14.1 shows fluid friction coefficients (f) normalized by R_e^{-1} as a function of α for rectangular ducts and as a function of r_1/r_2 for a concentric annulus, with r_1 as inside diameter and r_2 as outside diameter. A circular pipe with laminar flow has fR_e = 64.

Similar results hold for channels or pipe flows with a free surface. Figure 14.3 shows measured contours of constant velocity (isovels) in typical channels. Some longitudinal swirl exists in each case. These secondary flows (swirls) can be factors in eroding a stream or river bed; however, they are much less severe than the secondary flows (also a longitudinal swirl) caused by bends in the flow direction. Note that because of longitudinal swirl, the highest velocities typically are not at the stress-free

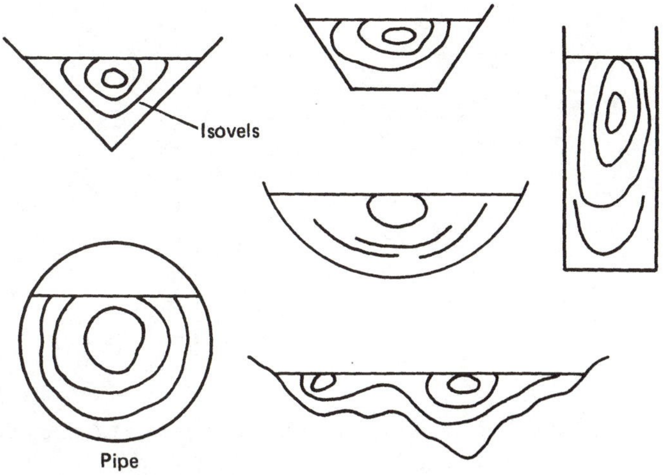

Figure 14.3: Isovels in typical manufactured and natural open-channel sections. Vorticity can be estimated approximately as proportional to fluid strain rates, i.e., relative spacing of isovels. (After Chow 1959).

top surface but are some distance down into the flow. Projects 14.2 and 14.3 in Section 14.5 include additional references to the professional literature, and suggest analytical and experimental studies on longitudinal swirl. Project 14.1 offers suggestions on the importance of literature searches.

14.2 Secondary Flow in Curved Sections

In a pipe, duct, channel, or river that is also curved or bent in the flow direction, three other secondary flow/vorticity phenomena cause major flow energy losses. These three phenomena are

1. Separation on the downflow side of a bump or sharp bend partly related to concentrations of vorticity in the upstream boundary layer. Three examples are shown in Figure 14.4. Separation can be reduced or eliminated by reducing the flow Reynolds number and by minimizing sharp bends and obstructions.

2. Secondary flows caused by Ekman layer phenomena on surfaces approximately parallel to the plane of the bend. This vorticity-related phenomena occurs at all Reynolds numbers and is a common and serious source of energy loss in bends. Figure 14.5 illustrates the flow pattern. Theory is presented in Chapter 11. See especially Example 11.2.

3. Görtler vortices caused by Rayleigh instability in the curved flow field. These occur at the outside of a bend in the boundary layer. See Figure 14.6, and refer to Chapter 13 for a discussion of Rayleigh instability. Note that in the outside-wall

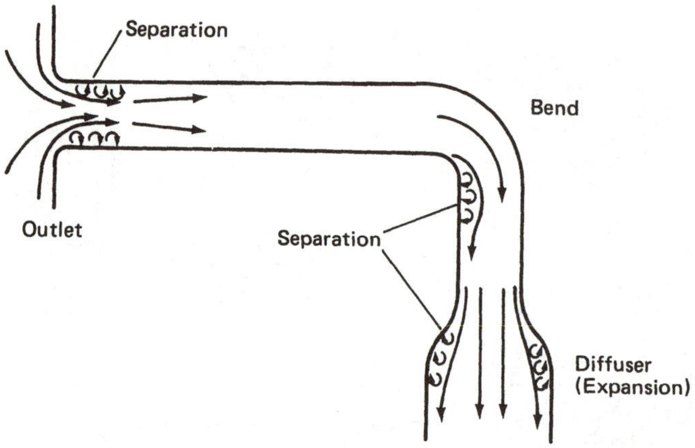

Figure 14.4: *Separation can occur at any change in pipe section or direction if* $R_e >> 1$.

boundary layer, angular momentum per unit mass will decrease with r, i.e., velocity drops off faster than r^{-1}. This is the criteria given in Chapter 13 for flow instability and generation of vortices, there called *Taylor vortices* and here called *Görtler vortices.*

The Rayleigh instability criteria presupposes an inviscid fluid. Viscosity tends to stabilize the flow. In a conventional flow, the Reynolds number is a ratio of inertial to viscous forces and provides a criteria for flow instability. The Görtler number also is a ratio of inertia (centrifugal) forces to viscous forces and provides a criteria for transition from a concave laminar boundary layer to Görtler vortices. The Görtler number is computed as

$$G_o = \frac{U\delta}{\nu}\left(\frac{\delta}{R}\right)^{1/2} \tag{14.5}$$

where δ is the momentum thickness. When $G_o \gtrsim 0.3$, Görtler vortices appear. At even higher values the vortices progress through modified forms. It is interesting here that viscosity produces the boundary layer form of v(r) that leads to Rayleigh instability, but at the same time viscosity stabilizes the flow and delays the onset of Görtler vortices until after $G_o > 0.3$. See Project 14.10 in Section 14.5 for additional details and references.

The fluid friction loss coefficients given in Figure 14.7 for various bends and flow obstructions include effects of all phenomena described here. Note in particular the reduction in loss coefficient caused by placing vanes in the 90° miter (square corner) bend. The loss is reduced by over a factor of 5 compared to the unvaned miter bend and compares favorably with the best ratios of r/d for a 90° smooth bend. The additional drag of the vanes is more than compensated for by reductions in separation losses, reductions in turbulence, and reductions in the large swirls that would be

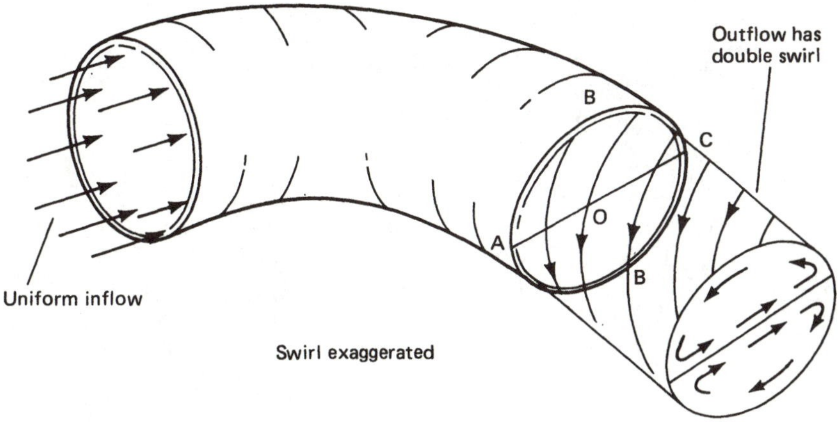

Figure 14.5: *Centripetal Δp along A-O-C exceeds Δp in boundary layers A-B-C and induces secondary flows (swirl).*

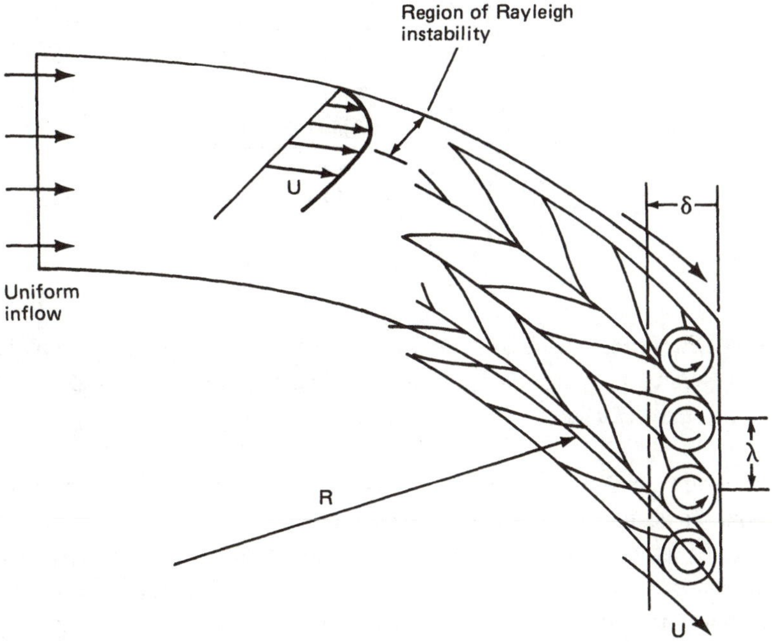

Figure 14.6: *Formation of Görtler vortices on exterior (concave) wall due to Rayleigh instability in boundary layer.*

(a)

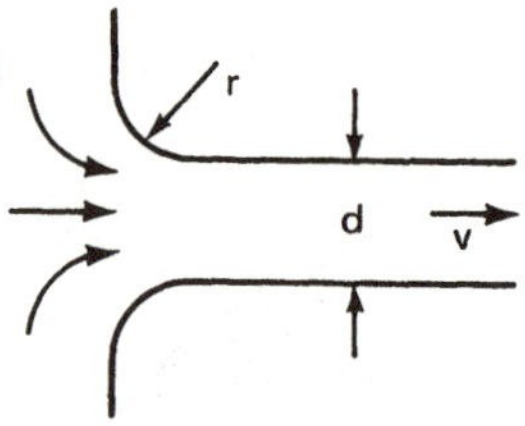

r/d	f
0.0	0.50
0.1	0.12
>0.2	0.03

(b)

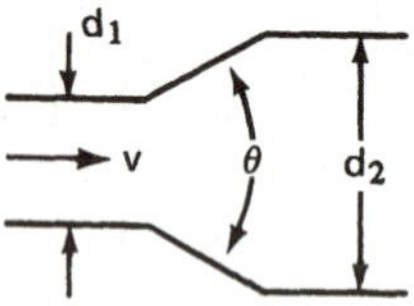

	f	
d_1/d_2	$\theta = 10°$	$\theta = 180°$
0.2	0.13	0.92
0.4	0.11	0.72
0.6	0.06	0.42
0.8	0.03	0.16

(c)

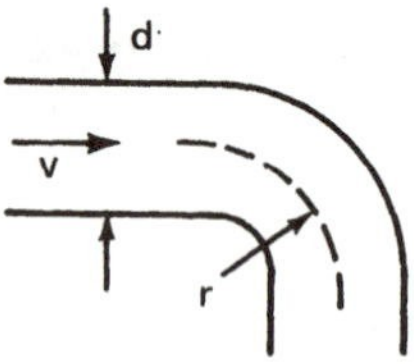

r/d	f
1	0.35
4	0.18
10	0.32

(d)

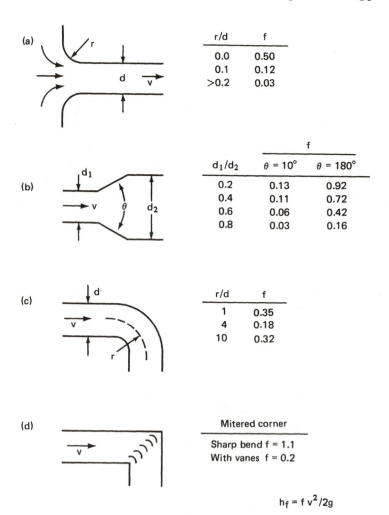

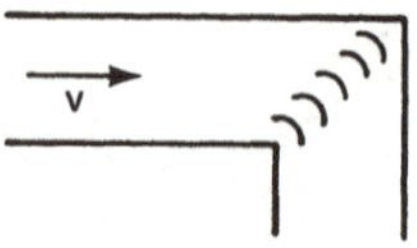

Mitered corner
Sharp bend f = 1.1
With vanes f = 0.2

$$h_f = f\,v^2/2g$$

Figure 14.7: *Experimental fluid friction loss coefficients for turbulent flow (ρ = c, μ = c) due to changes in pipe diameter and direction; (a) and (b) (ASHRAE 1977), (c) (Beij 1938; Idel'chik 1966), (d) (Streeter 1961).*

induced by the unvaned sidewall boundary layers as in Figure 14.5. For additional tabulated pipe and channel flow data, consult the references.

Figure 14.8 summarizes measurements by Shukry (1950) of near-surface forward water velocities in a 180° bend in a flume 30 cm wide and 30 cm deep with 30 cm centerline radius curvature. The entrance velocity of 77.8 cm/s, nearly uniform across the channel at the entrance, indicates $R_e \approx 73{,}500$ and turbulent flow. As flow enters the bend ($0° < \theta < 90°$), flow approximating an inviscid vortex is produced (i.e., $v \propto r^{-1}$). Mean flow in this region, although nominally turbulent, can be approximated

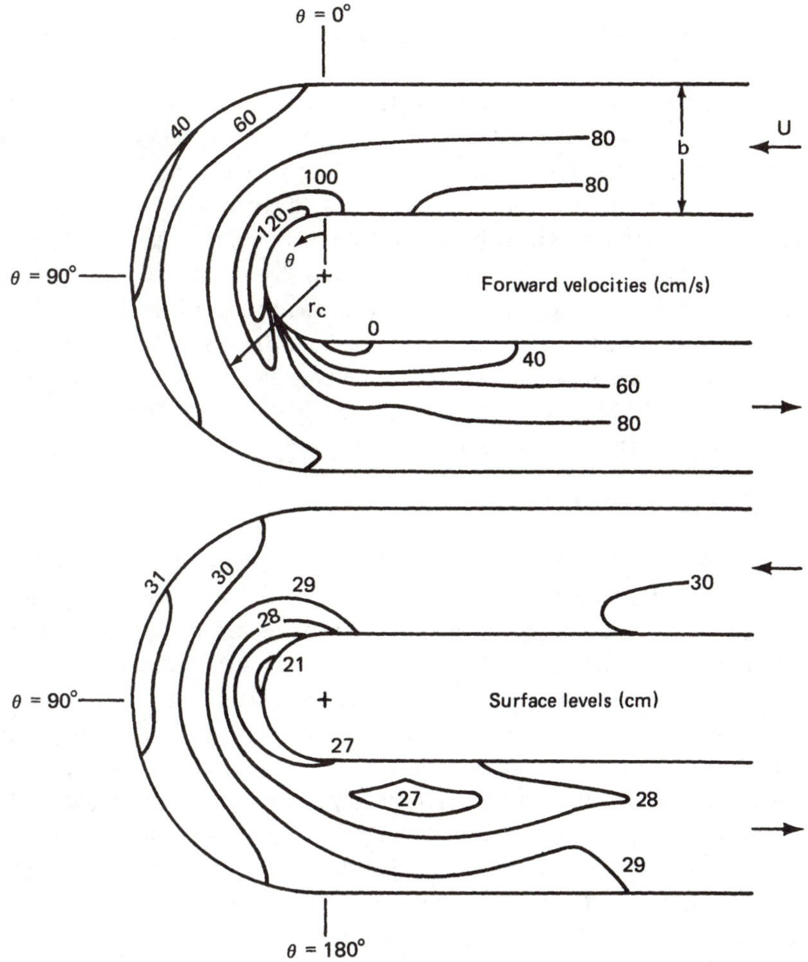

Figure 14.8: *Measured surface levels and forward velocities for turbulent flow around a 180° bend, $b = h = r_C = 30$ cm, $U = 77.8$ cm/s, $R_e = 73{,}500$. (After Shukry 1950).*

using the Bernoulli equation with pressure provided by the centripetal acceleration.

During this first quadrant ($0 \le \theta \le 90°$), an inward flow in the slower moving bottom boundary layer is initiated with a corresponding outward flow in the upper levels, i.e., a secondary flow produced by Ekman layer pumping. The outward secondary flow at the top convects the higher velocity fluid toward the outer wall between $\theta = 90°$ and 150°. Between 150° and 180° the major downstream flow is nearly parabolic across the channel. Soon after leaving the bend the secondary flow (swirl) transports the highest flow velocities almost completely to the outside wall. Note the separation region that has formed at the inside wall just past the bend.

Obviously any solution for flow in a curved channel that neglects secondary flows is inadequate.

Figure 14.8 also shows a plot of changes in free-surface elevation (often called *superelevation*). Elevations are higher over the slower moving flows and lower over the faster flows, and are systematically higher nearer the outer wall due to the centrifugally induced radial pressure differentials. Fluid elevation, of course, is only an indication of fluid pressure caused by internal flow conditions; elevation is not the cause of the pressure, only its indicator. The cause is that the fluid is being forced to follow a curved path. Secondary flows are then driven by the difference between the flow-induced radial pressure differentials in the main flow and the reduced radial pressure differentials in the slower moving bottom boundary layer. See Example 11.2 and Projects 14.4 and 14.5 that suggest experimental and numerical studies, including references to the professional literature on secondary flow in pipes and channels.

14.3 River Meandering

Flow in natural (not artificially channeled) rivers is complicated by irregular boundaries and the movement of sediment. Longitudinal swirl due to a noncircular cross section plus the secondary flows due to bends or curves in the downstream direction will tend to redirect transported sediment to one side or the other of the river bed. The net effect is to increase stream curvature so that small departures from straight downslope flow grow into approximately sinusoidally shaped patterns. Conditions are often such that the curves or meanders become more and more accentuated, developing into *oxbows* that eventually break through to the next curve. The breakthrough is likely to occur during a period of high flow rate that completely realigns the river, sometimes to a nearly straight flow, after which the process repeats. Figure 1.6, showing the Chena River just east of Fairbanks, AK (August 1986), illustrates oxbows, a breakthrough, and the traces of old oxbows from an earlier time.

Meandering is a phenomena common to all streams and rivers, and is of major importance in maintaining navigability, flow (flood) capacity, and utility of adjacent land sites. The ease with which meandering occurs can be demonstrated in a simple laboratory experiment. A trough about 1 m wide and 2 to 3 m long with 10 to 20 cm side walls is filled with loose dirt and sand of varying grain sizes. The trough is given a slight downslope, and water is permitted to overflow gently from a reservoir at the high side near the centerline. Within minutes a major channel will form and quickly begin to meander. The experiment can be modified by manually depressing a gentle sinusoidal path of arbitrary wavelength. The meander pattern will adjust until the wavelength is appropriate to the dynamics of that flow rate, slope, and soil characteristics. At appropriate conditions the pattern will move downslope as a fixed pattern. Under other conditions, oxbow development can be accentuated. Shallow braided flow or a single, deep, nearly straight channel can be produced by continuing to vary the conditions.

Repeated bends, reversing each time in direction, as in meandering, produce internal flows similar to, but more complicated than, that shown in Figure 14.8. The secondary flow in each new bend must first eliminate the swirl of the previous bend (of opposite direction) before it can generate a new swirl. In a deep channel, two swirls of opposite direction sometimes occur for a portion of the bend, with the residual swirl on top and the newly generated swirl on the bottom. In a shallower channel, the swirls sometimes exist side by side for a portion of the bend. Constant recontouring of the bottom also adds to the complexity of the problem. Some general models appear to have limited success in predicting river meandering, although most research and prediction is done on an ad hoc basis. Projects 14.6 and 14.7 provide references and suggestions for further studies on river meandering.

14.4 Abutment Undercutting

Abutment undercutting can be demonstrated in the experiment described in the previous section and is illustrated in Figure 14.9. A structure with a flat or smooth front face placed in a stream with an erodible bottom causes the bottom surface to erode immediately in front of the structure and to pile up sediment immediately behind the structure. The structure, if not sunk deeply into the bottom, will be excavated from the front and will eventually fall forward into the oncoming stream.

The flow mechanism can be analyzed in terms of variables (**V**, p) by noting that stagnation pressure just above the bottom boundary layer must be higher than stagnation pressures in the slower-moving boundary layer fluid. The pressure differential causes a downward flow along the front of the structure that moves down and then curls forward and up within a slightly enlarged boundary layer at that point. The phenomena continues around the sides, providing at the rear a mechanism for dumping sediment in the separated wake region. However, if the structure has a sharp leading edge, stagnation pressures do not occur, and therefore there is no induced vertical flow. The danger in using long streamlined piers in a natural river is that the river may reorient its flow direction so that the streamlined pier is no longer aligned with the flow.

Abutment undercutting is understood in terms of variables ($\boldsymbol{\zeta}$, **V**) by noting that the boundary layer consists of vortex filaments transverse to the flow direction and that these vortex filaments are fixed in and travel with the fluid. When they encounter a round or flat vertical surface, they are stopped along with the fluid that transports them and are then bent around the structure along with the flow field. This stationary vortex tube wrapped around the front and sides of the structure then has the potential for eroding bottom material away from the structure, digging a hole in front of it and eventually under it. The vortex can easily be observed in a laboratory experiment by injecting dye (or hydrogen bubbles) immediately in front of the structure at about boundary-layer height. A sharp-edged structure cuts the vortex tube so that it passes

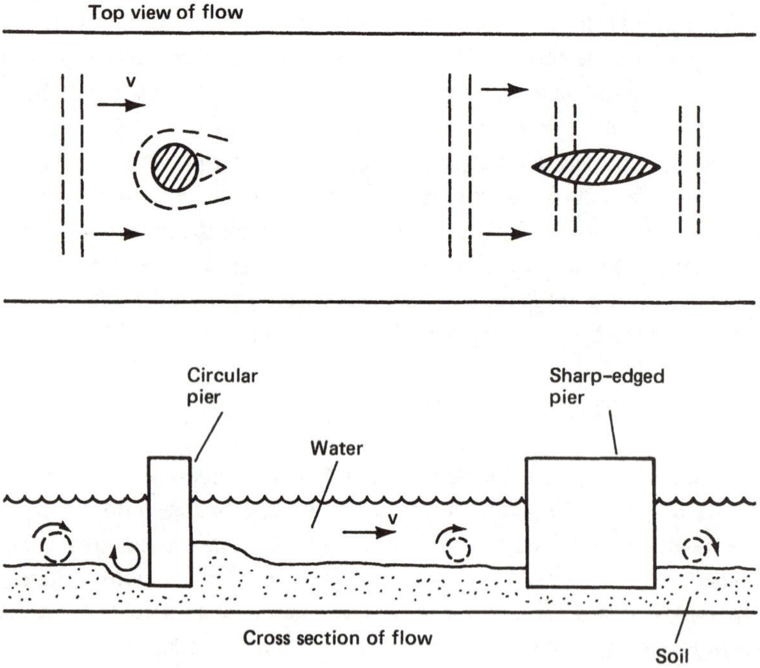

Figure 14.9: Flat-faced or round-faced piers hold stationary vortex rolls at their front surface which tend to undercut the front and deposit sediment at the rear. Sharp edges cut the vortex rolls.

harmlessly to either side. Projects 14.8 and 14.9 provide suggestions for further studies of abutment undercutting and the associated vortex structures.

14.5 Study and Research Projects

The projects suggested here are intended for individuals or groups. Some can be completed in a week or two, others are term or longer projects. Only a few will have exact answers. References to other texts and to professional journals are included.

Project 14.1: Before starting any serious study or research, the researcher needs to survey the professional literature. Maybe that research has already been completed and published, or if not, then almost certainly similar research or supporting studies have been published. Discuss your idea with a reference librarian first, then use a card catalog or computer search, as available, and finally locate the professional journals most likely to publish applicable results. One good research technique is to physically examine each volume from the prior decade for articles of interest. For this project, select any area of interest to you, search the professional literature, and then write a brief report on your findings. Only rarely does one find too little material--usually you are deluged with information and need to carefully separate out the most useful and relevant. A literature search is an important part of each suggested project in this text, even when some references are included.

Project 14.2: Nikuradse (1926, 1930) has shown that longitudinal swirl is created in a filled, straight, noncircular pipe (duct) when flow is turbulent. Equivalent explicit proof for existence of swirl when the same flow is laminar appears to be lacking, as is explicit proof that swirl does not exist in a filled, straight, circular pipe for either laminar or turbulent flow. Rigorous proofs include a mathematical derivation or numerical solution, starting with the Navier-Stokes equations, or a very careful set of experiments. Project 14.3 discusses a set of experiments. This project suggests analytical and numerical solutions. Any preconceived assumption about a flow field tends to compromise the validity of any solution or proof. For example, the typical derivation for fully developed laminar flow in a straight circular pipe places the Navier-Stokes equations in cylindrical coordinates, then assumes the equations for radial and tangential flows are not needed, and finally solves only for axial flow as a function of radius. The typical solution for flow in a rectangular duct uses rectangular coordinates and solves only for longitudinal flow (u) as a function of width and height (y, z). In each case, these solutions will indicate that swirl does not occur--that possibility is eliminated by omitting two of the three Navier-Stokes momentum equations. Absence of swirl in a straight circular pipe in laminar flow is difficult to question, but not so for a rectangular duct as indicated in Figure 14.1 and discussion. A full solution here requires solving for u(x, y, z), v(x, y, z), and w(x, y, z), perhaps impossible analytically but not impossible numerically. Consider also solving for an elliptical cross section using an elliptical coordinate system (see Hughes and Gaylord 1964). Turbulent flow solutions are more important for real applications but are far more difficult to analyze in practice. The evidence for swirl in rectangular-duct turbulent flow is experimental--not analytical. Consider the possibility of ordered or repeating structures in turbulent flow that might indicate a net or averaged swirl in even a circular cross section.

Project 14.3: Set up an experiment to test for swirl in straight pipes or ducts. Investigate four conditions: circular vs. rectangular cross section with each flow, first laminar and then turbulent. Select a test procedure to visualize the flow, e.g., dye injection, laser doppler, high-speed photography of neutral buoyancy particles, etc. Be very careful to start with a smooth, uniform flow and eliminate entrance transient conditions. As a first step, design the experiment by selecting fluids and channel dimensions. Include sketches and calculations. As a second step, build an inexpensive trial setup to verify your design. Given funds and time, complete the formal experiments. Some references are Dryden et al. (1956), Berker (1963), Schlichting (1979), Olson (1980), and White (1974, 1986). Refer also, of course, to Nikuradse (1926, 1930) and Shukry (1950).

Project 14.4: Experimentally investigate secondary flow in a curved cylindrical pipe. The pipe should bend about 90° and have a bend radius two to three times the pipe diameter. It must have a smooth interior surface and preferably should be transparent. Smooth glass elbows may be available from a laboratory supplier. Achieve fully developed laminar flow prior to the curved portion as in Figure 14.5. The swirl can be visualized by dye injection or by inclusion of nearly buoyant tracer particles. Observe changes as flow velocity is increased. Would there be flow advantages or disadvantages in suppressing these secondary flows with internal vanes? Will the secondary flows of two oppositely curved sections placed one after the other cancel? See Schlichting (1979), Herring et al. (1979), Van Dyke (1982), White (1974, 1986), Granger (1988), and Roberson and Crowe (1990).

Project 14.5: Predict analytically and/or numerically the secondary flow swirl generated by the Project 14.4 experiment. Use the theory and equations of Chapters 10 and 11. Repeat using curved rectangular ducts with various width-to-height ratios. How much flow energy is lost? Would it be possible to reclaim some of the energy associated with the swirling secondary flow?

Project 14.6: Perform the river meander experiment described in Section 14.3 of this chapter. Experiment with various bed inclinations, flow depths, soil types, and soil compactions. After performing several tests, develop an empirical equation and then use the equation to predict performance under some new condition. Check the validity of your prediction by then completing a test at that new condition. See ASCE (1983), Smith and McLean (1984), and Nelson and Smith (1989a, 1989b).

Project 14.7: River meandering and the preferential depositing of various minerals is a subject of interest for mineralogists and paleontologists (paleo = ancient, relative to geological units of time). See, for example, Flores et al. (1985), Rubin (1987), and Ikeda and Parker (1989). Locate a topological map (e.g., from the U.S. Geological Survey, USGS) or a large-scale photograph of a meandering river in your area. Study the present topology of the land and try to estimate where heavier minerals (or aggregates of various sizes) would have been deposited during the life of that river. After you feel confident in your assessments, try it in Alaska in gold country--and good luck!

Project 14.8: Use the experimental setup of Project 14.6 to investigate abutment undercutting. Undercutting occurs in both shallow and deep flows if flow near the bottom is fast enough. Revise the Project 14.6 setup to create deeper channel flows with an erodable bottom and observe the undercutting. Duplicate the observational technique illustrated in Van Dyke (1982, p. 55) and measure velocities in and near the vortex.

Project 14.9: Investigate quantitative approximations for the size and strength of the *horseshoe* vortex formed at the base of a circular abutment sitting on a solid, flat bottom surface in a horizontal flow of water. Compute the total vorticity in the bottom boundary layer just prior to the abutment. What would be the strength of a vortex with the diameter of the boundary layer thickness and the total vorticity of the boundary layer? Can this be approximated best by a rotating cylinder of water ($v_\theta = \omega r$), by an inviscid vortex ($v_\theta = \kappa/r$), or by neither? Examine how and why the original plane flow is transformed to a stationary vortex tube. Can the tube be stretched and/or broken? Examine Chapters 8 and 10 for theorems and equations. Some references are Van Dyke (1982), Lugt (1983), and Schlichting (1979).

Project 14.10: The formation and structure of Görtler vortices, as flow instabilities on concave walls, have been analyzed analytically by Görtler (1941) and Schultz-Grunow and Behbahani (1975), and experimentally by Clauser and Clauser (1937) and Liepmann (1943, 1945). See Schlichting (1979, pp. 532–535) for a summary. Assume flow past a concave wall and estimate the boundary layer thickness. Then apply the Rayleigh instability criteria to flow inside and outside the boundary layer. Compute vorticity in a boundary layer assumed laminar. What is the vorticity when transition to Görtler vortices occur? Estimate total vorticity after the formation of Görtler vortices. Do the Görtler vortices aid or hinder the transfer of momentum from the free stream to the wall? Can Görtler vortices occur on a flat or convex surface?

Chapter 15

Rotors and Centrifuges

15.1 Thin Disk in a Housing

This section examines spinning disks and cylinders, and their interaction with adjacent and enclosed fluids. Later sections include devices manufactured to separate fluid mixtures into their separate components through use of extremely high centripetal acceleration fields. Secondary flows are usually an important factor. Theory needed to analyze fluids adjacent to rotors and/or disks spinning in concentric housings is introduced in Chapters 11, 12, and 13. Equations are given in Chapter 11 for laminar flow adjacent to a rotating flat surface in an otherwise stationary fluid and for a rotating fluid adjacent to a nonrotating surface. Chapters 12 and 13 introduce the basic theory of flow between concentric cylinders rotating at different angular velocities. This chapter continues the discussion and includes details and specific application examples. It concludes in Section 15.5 with a set of 10 suggested projects that lead to a study of the professional literature.

Figure 15.1 shows a thin disk of radius R in a close-fitting concentric housing. The disk is assumed thin enough that flows induced at the periphery can be neglected. Figure 15.2 shows a long cylinder of radius r_1 in a close-fitting concentric housing of radius r_2. For this analysis the cylinder is assumed long enough relative to its radius that end effects can be neglected.

Examine first Figure 15.1 and the case $h < \delta$ (gap at the bottom of the disk). The gap is too narrow for secondary flows to exist, and the fluid adopts a Couette flow pattern with a nearly straight-line, tangential velocity profile between the fixed housing and the disk at each point, $v_\theta(r, z)$. The flow may be laminar, with $h < \delta$, depending on the local Reynolds number ($R_e = v_\theta h/\nu$). This type of flow occurs typically in close-fitting rotating machines where it can be a major cause of inefficiencies.

Shear stress is $\tau(r) = \mu r\omega/h$, and moment on one side of the disk is

$$M = 2\pi \int_0^R \tau r^2 dr = \frac{\pi\mu\omega R^4}{2h} \tag{15.1}$$

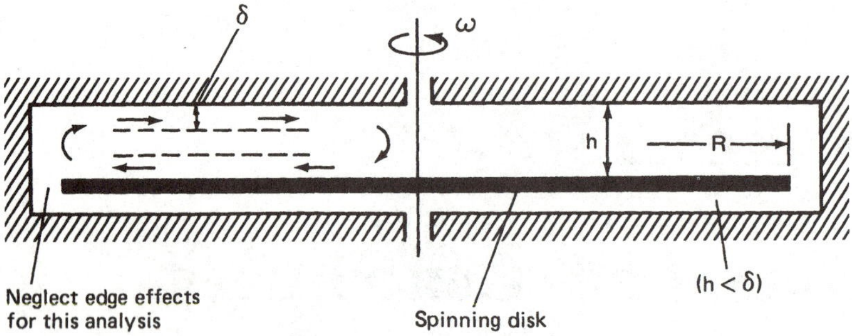

Figure 15.1: A spinning disk in a stationary housing. The top side illustrates a wide gap and the bottom a narrow gap.

It is conventional to define a dimensionless moment coefficient for a disk wetted on both sides as

$$C_M = \frac{2M}{\frac{1}{2}\rho\omega^2 R^5} \tag{15.2}$$

In terms of an Ekman number $E_k = \nu/\omega R^2$, often identified in engineering practice as a type of inverse Reynolds number ($R_e = E_k^{-1}$), C_M becomes

$$C_M = 2\pi\left(\frac{R}{h}\right) R_e^{-1} \tag{15.3}$$

For h large enough to approximate a disk in an infinite fluid, the value of C_M for laminar flow is

$$C_M = 3.87\, R_e^{-1/2} \tag{15.4}$$

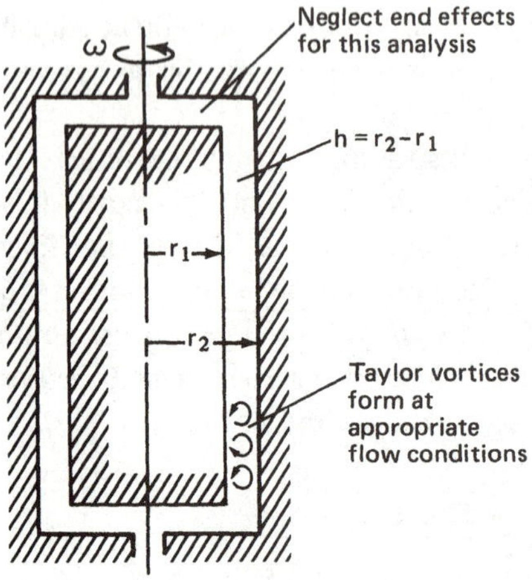

Figure 15.2: A spinning cylinder in a stationary housing. Couette flow occurs when h << r; for larger h, Taylor vortices (see Figure 13.2) provide the major mechanism for momentum transfer and energy loss.

These estimates agree well with experimental results when the clearance at the periphery of the disk is large enough for edge effects to be negligible. When edge clearance is small, C_M becomes slightly larger than these estimates.

If the gap thickness is several times as large as δ, say three to five times δ, the flow pattern is more complex. Secondary flows, radially outward near the disk and inward at the housing, dominate the flow. They are connected by axial flows near the periphery and near the axle. When h is even larger, a significant quantity of fluid in the interior region is essentially isolated from the main flow. It rotates approximately as a rigid body at one-half the angular velocity of the disk.

The thickness of the boundary layers on the disk and on the housing can be estimated using the equations of Chapter 11, modified in each case by the fact that the fluid between the boundary layers rotates at approximately ω/2. Results are obtained experimentally and theoretically. When flow is laminar, the moment coefficient is approximately

$$C_M = 2.67\, R_e^{-1/2} \tag{15.5}$$

and when flow has become turbulent, it is

$$C_M = 0.0622\, R_e^{-1/5} \tag{15.6}$$

Equation (15.3) is valid for ($R_e \lesssim 1.5 \times 10^4$), (15.5) can be used for ($1.5 \times 10^4 \lesssim R_e \lesssim 3 \times 10^5$), and (15.6) for R_e above 3×10^5. Equation (15.6) may understate C_M slightly in some applications, depending on (h/R) and surface roughness. See Schlichting (1979, pp. 647–652) for further details.

Spiral vortices are formed during transition from laminar Ekman layer flow to turbulence. At least two types of vortex rolls, designated class A and class B, appear as radial spirals in the Ekman boundary layers. Spacing of the rolls is approximately δ. Class A rolls appear as instabilities above a Reynolds number of

$$R_e = 56.3 + 58.4\, R_o \tag{15.7}$$

with class B rolls appearing above a Reynolds number of

$$R_e = 124.5 + 3.66\, R_o \tag{15.8}$$

In (15.7) and (15.8), R_o is a Rossby number (V/Ωr), with V the relative azimuthal component of velocity of the interior fluid, and R_e is computed using (Vδ/ν), with δ estimated as 4z/π, where z is height to maximum radial velocity. Alternatively, if δ is defined in rotational parameters $\delta \sim (\nu/\Omega)^{1/2}$ then roll instabilities first appear at

$$R_o E_k^{-1/2} \approx 56 \tag{15.9}$$

See Greenspan (1968, pp. 275–281) for photographs and further details.

15.2 Flow in a Cylindrical Annulus

Flow in a cylindrical annulus, caused by the differential rotation of inner and outer cylinders as in Figure 15.2, is first introduced in Chapters 12 and 13. In the general case, the angular velocities of the inner and outer cylinders can have any magnitude and either direction. When $\Omega_1 = \Omega_2$, the steady-state solution is rotation as a rigid body at that same angular velocity, and the analysis of Chapter 13 indicates flow stability against arbitrary perturbations for any Ω and r. Secondary flow during spin-up and spin-down is discussed in Chapter 11.

Chapter 12 summarized circular pathline flows, including circular Couette flow between two concentric cylindrical surfaces rotating at different angular velocities. Because many ratios of ν, r_1, r_2, Ω_1, and Ω_2 lead to instability, only a few analytical solutions for moment of one cylinder on the other are available. When the inner cylinder is at rest and flow is laminar, transmitted moment is

$$M = 4\pi\mu\omega L \frac{(r_1 r_2)^2}{r_2^2 - r_1^2} \tag{15.10}$$

where L is length of the cylinders. For a cylinder in an infinite fluid, drag moment on the rotating cylinder is

$$M = 4\pi\mu\omega L r_1^2 \tag{15.11}$$

Equation (15.11) assumes laminar flow, and the solution approximates a steady-state vortex flow ($v_\theta = \kappa/r$) from the cylinder to infinity.

Lubrication is an important application for the situation where the inner cylinder (shaft) rotates and the outer cylinder (journal bearing) is stationary. In this case the clearance is typically small enough, the lubricant viscous enough, and the speeds slow enough that the flow is laminar. The flow is complicated by the fact that loading on the shaft (e.g., rotor weight) attempts to reduce the bearing clearance on one side to zero. However, in a properly designed bearing, the shaft, while turning, will not contact the bearing, because viscous drag between the shaft and lubricant carries lubricant into this space. (See Figure 15.3.) Lubricant leaks out the ends, of course, so the bearing must have a minimum length. If the shaft is never stopped, it never contacts the bearing, and wear is zero. Ball and/or roller bearings have replaced journal bearings for very high shaft speeds, very high shaft loads, and/or where shaft/housing concentricity must be exact.

Other flow conditions for circular Couette flow are best resolved empirically. They are often expressed in terms of the Taylor number, with h the gap thickness $(r_2 - r_1)$

$$T_a = \frac{r_1 \Omega_1 h}{\nu} \sqrt{\frac{h}{r_1}} \tag{15.12}$$

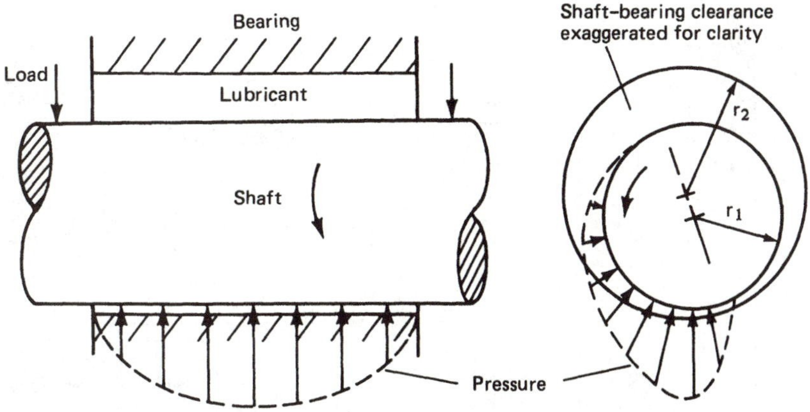

Figure 15.3: Lubricant in a journal bearing is forced beneath the rotating shaft by viscous drag. High viscosity and a narrow gap maintain laminar flow for practical speeds.

Using this criteria, laminar flow exists for $T_a < 41.3$, laminar flow with Taylor vortices (see Chapter 13) exists for $41.3 < T_a < 400$, and turbulent flow occurs for $T_a > 400$. Details and references to the professional literature are included as suggested projects in Section 15.5.

Fluids are often enclosed within a rotor for various purposes, such as cooling, viscous damping of vibration, separation of fluid mixtures by centrifuging, enclosed fuels in a fuel tank, etc. Chapter 7 introduces Coriolis effects, applicable, for example, to a fluid moving through radial coolant holes in a rotor; Chapter 11 details the mechanism of fluid spin-up and spin-down; and Chapter 18 will discuss enclosed liquids in a container that spins and wobbles, for example, a spinning satellite containing liquid fuels. The construction and analysis of centrifuges are discussed next.

15.3 Centrifuges

Centrifuges can be classified, for ease of discussion, into two types.

1. Discrete sample volume in a spinning rotor
2. Continuous flow through a spinning rotor

All centrifuges rely on sedimentation of the heavier parts of a mixture of two or more substances. A settling tank in the earth's gravitational field achieves a similar, but much less effective, result. A discrete sample volume (batch process) settling tank and a continuous flow tank are shown in Figures 15.4a and 15.4b. Several basic phenomena compete in this process. In Figure 15.4a the denser components are more affected by gravity (ρg) and tend to sink, with lighter components rising. Molecular

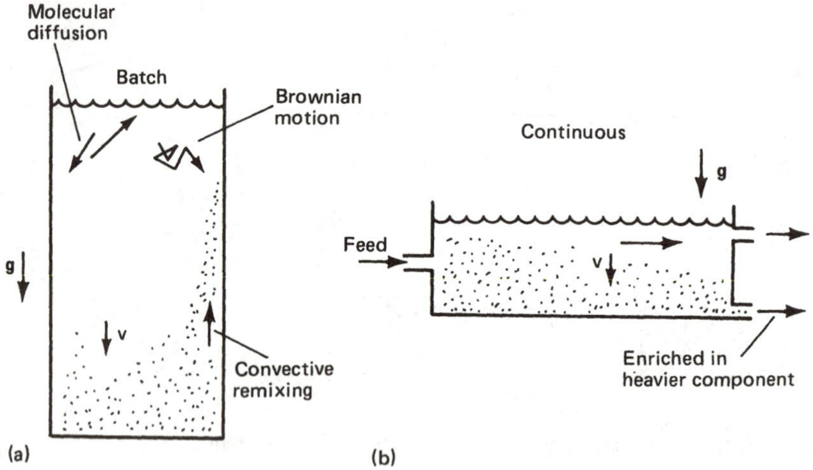

Figure 15.4: *Batch (a) and continuous flow (b) settling tanks. Settling velocity (v) in a static fluid is determined by $\Delta\rho$ and drag coefficient of particles in that fluid.*

and/or small particle motion due to the mixture's internal energy (e.g., Brownian motion) oppose the separation process through diffusion. Gross convection of the fluid mixture (e.g., by thermal gradients) is a third mechanism, and one that can easily dominate other phenomena, especially when the density variation of the components is small.

In the continuous flow settling tank (b), a mixture is introduced at one end and heavier and lighter components partially separate over the length of the flow. Fluid enriched in the desired component is withdrawn, with the remainder shunted to a different output. The enriched fluid can be reprocessed in successive settling tanks arranged in series to achieve greater enrichment of the desired component. Parallel arrangements are used to increase capacity.

The centrifuge does exactly what the settling tank does, except it replaces earth gravity (g) with the centripetal acceleration ($\omega^2 r$) in a rotating fluid. It is very easy to obtain centripetal accelerations much greater than earth gravity, and it becomes practical then, for both analysis and operation of the centrifuge, to ignore earth gravity.

Centripetal accelerations greater than 10^6 g can be achieved, making it possible to separate mixtures with very small density differences or very small, settling drag coefficient differences in a reasonable time. For example, a virus can be separated from biological fluids; a salt (NaCl) solution can be separated into more concentrated and less concentrated portions; and different isotopes of the same element can be selectively concentrated. In one important industrial application, uranium is converted to a gaseous compound, and its two isotopic components, with a difference in molecular weight of less than 1%, are separated in continuous flow centrifuges for use

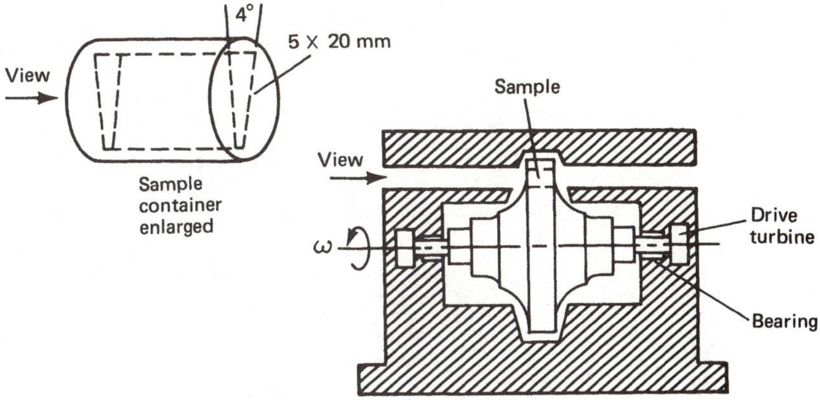

***Figure 15.5:** The Svedberg centrifuge first successfully separated small $\Delta\rho$ with large drag coefficients, e.g., virus in a biological fluid ($\omega^2 r \sim 50{,}000$ g).*

as a fuel in nuclear energy reactors.

The first theoretical and experimental studies of the use of centrifuges to separate, or at least concentrate, molecular-sized gaseous particles appeared in the late 1800s. These centrifuges were hand-operated. It was not until the mid 1920s that T. Svedberg and his associates developed the first practical high-speed centrifuges that could isolate large molecules and cell-sized structures in liquids. (See Figure 15.5.)

It had earlier become apparent that the slightest convective currents could disrupt the separation process. Thermal gradients, from bearing heat, for example, produce convective flow patterns. These can be eliminated by careful design, but another serious cause of convection is more troublesome. These are the secondary flows established during spin-up and spin-down. A sample is placed in a finite-volumed test cell, and then centrifuged at an adequate speed and for an adequate time to ensure effective separation. If the machine is stopped to retrieve the sample, the fluid must spin down with the rotor; secondary flows are induced; and by the time the rotor has stopped, the sample has been remixed.

Svedberg and his associates developed a batch-process, liquid centrifuge that used small sector-shaped test cells placed in a high-strength steel alloy rotor. The sector shape allows free radial movement of the suspended particles. Secondary flow problems are minimized or eliminated by observation of the results while the rotor is spinning or by selective extraction of the desired components. An article by Aston (1978) reviews the history and some details of the many variations of the ultracentrifuge. These devices are used extensively for biological and chemical research and production.

The need for large volume separation of the two isotopes of uranium metal (U^{235} and U^{238} with molecular weights of 235 and 238, respectively) led to the development

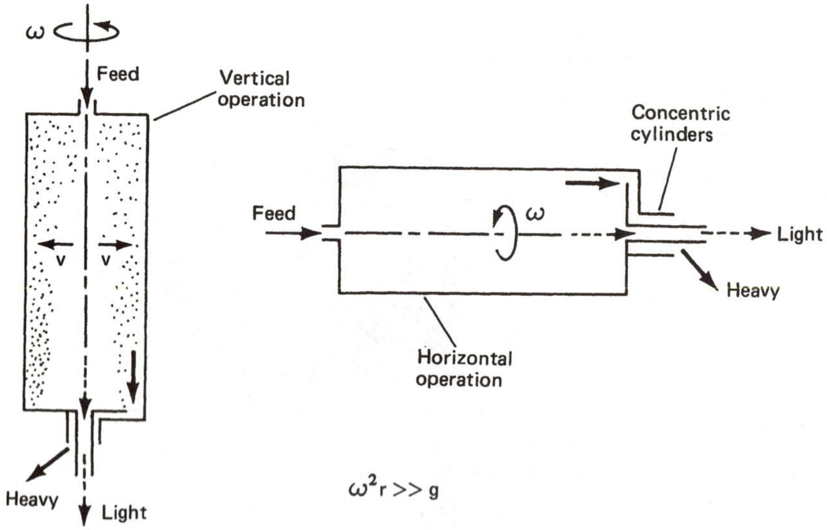

Figure 15.6: *Concurrent flow design centrifuges. Spin axis may be vertical or horizontal because $\omega^2 r >> g$, but vertical operation has structural design advantages.*

of high-capacity, continuous-process gas centrifuges. The active component (U^{235}) exists normally as a negligible percentage of the raw material, and it must be increased to about 3% of the total mass for use as a fuel in a nuclear power plant. The uranium is first processed chemically to uranium hexafluoride (UF_6), a gas. The molecular weights of the gas components ($U^{235} F_6$ and $U^{238} F_6$) are 349 and 352, respectively, a difference of less than 1%. The gas centrifuge relies on its high centripetal acceleration to concentrate the heavier gas molecules ($U^{238} F_6$) near the periphery and the lighter molecules ($U^{235} F_6$) near the rotor axis.

Figure 15.6 shows an early *concurrent flow* design in which the raw gas mixture is spun up to cylinder speed and introduced uniformly over one end surface. As the spinning mixture flows along the cylinder, U^{238} moves radially outward and U^{235} moves radially inward. Gas enriched in U^{235} is removed at the opposite central axis, and U^{238}-enriched gas is removed at the periphery. An approximately exponential radial profile for gas pressure and density exists in the centrifuge. Alloy steel materials available at that time permitted spin speeds yielding pressure ratios, axis to periphery, of up to about 1:500. The need for greater enrichment and capacity (i.e., separation efficiency) led to better materials, higher speeds, and a different design that utilized rotating fluid, secondary-flow phenomena.

Figures 15.7a and 15.7b show two *countercurrent flow* designs that form the basis for most gas centrifuges presently in use. In these designs, the enclosed gas is forced to circulate within the cylinder, moving along the central axis to one end, then outward, a reverse flow along the cylinder wall, and finally inward to complete the

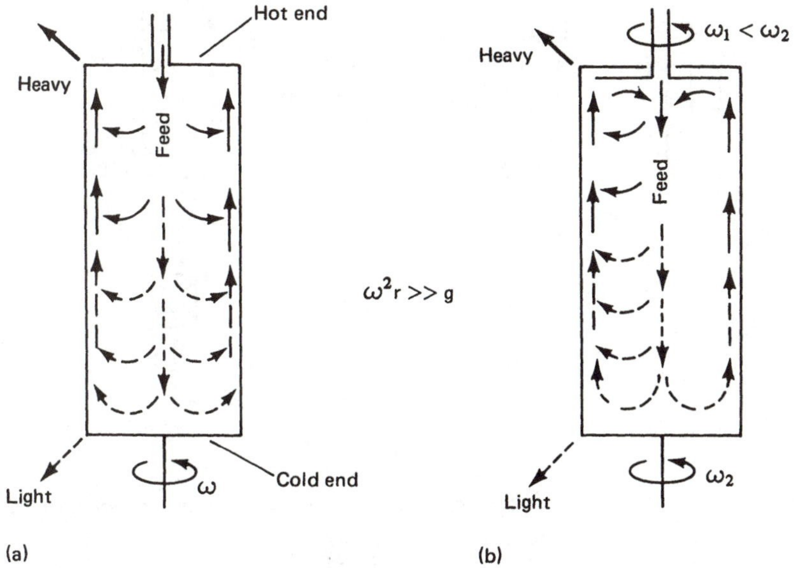

Figure 15.7: *Countercurrent designs with secondary flows (a) thermal convection, (b) end plate $\Delta\omega$. Compare (a) with an equivalent settling tank. In (a) thermal convection is driven by centrifugal force ($\omega^2 r$), not by gravity (g) because $\omega^2 r >> g$.*

flow pattern as shown. Lighter molecules constantly move inward and heavier molecules outward during the axial flows, so that the gas at the end of its passage along the cylinder wall tends to be enriched in heavier molecules, while the central axis flow at the opposite end is enriched in lighter molecules. Flow convection then forces this central lighter-gas mixture radially outward through a thin, boundary-layer flow along the end surface, increasing in pressure and density as it does so. Gas enriched in U^{235} can then be removed at the periphery. Raw gas is admitted centrally and heavier gas removed peripherally at the opposite end.

The internal circulation can be driven in either of two ways. In Figure 15.7, (a) shows thermal convection, and (b) shows secondary flow driven by $\Delta\omega$ of an end plate. In the thermal convection mode, a temperature difference is applied between the two end plates causing axial flows, hot to cold along the axis and cold to hot along the cylinder wall. Centrifugal force replaces gravity as the convective driving force. Compare this convective flow with that in a horizontal layer of fluid where one end of the fluid is heated and the other end is cooled. In the mechanically driven $\Delta\omega$ design shown (b), a disk at the raw gas input end spins slightly slower than the cylinder. Gas is forced radially inward through the boundary layer on the disk surface, because the pressure gradient (dp/dr) in the boundary layer is less than (dp/dr) in the central portion of the fluid spinning at the cylinder speed. Alternatively, a disk spinning slightly

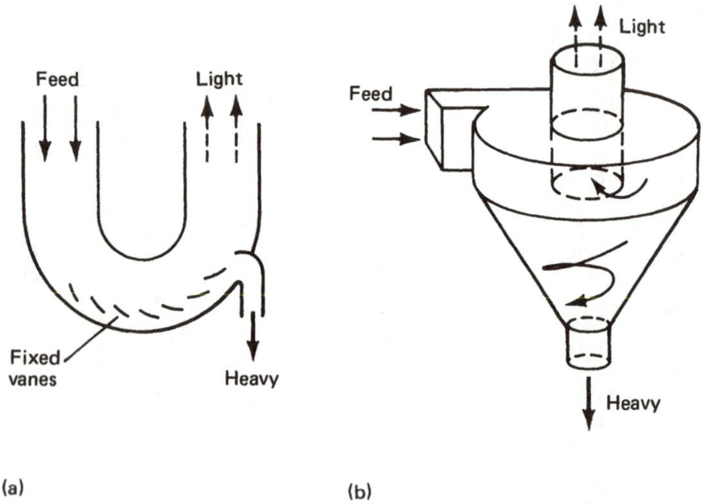

Figure 15.8: *(a) U-shaped separator is used for high volume separation of dense low-drag particles. (b) Cyclone separator can remove fine dust from air. These use* $(\omega^2 r)$ *for separation and secondary flows and (g) for collection.*

faster than the cylinder can be placed at the opposite end. See Chapters 11 and 14 for quantitative procedures.

Separation effiency (s) is given as

$$s \sim (\Delta m)\,(r\omega)^4\,L \tag{15.13}$$

where Δm is the mass difference of the molecules, $(r\omega)$ is cylinder wall peripheral speed, and L is cylinder length. Strength-to-weight ratios of available materials limit $(r\omega)$, while flexural stiffness (whirl vibrations) constrain L. Steel, aluminum, and titanium alloys have maximum $(r\omega)$ of 400 to 500 m/s. Glass fiber and carbon fiber/resin composites have increased permissible $(r\omega)$ to about 700 m/s. Because optimum design dictates that each cylinder be fairly small (say 15 cm diameter and 60 cm long), separation efficiency and useful flow rate for one unit are both also very small. Many thousands of identical units are used in a typical production facility in series (to increase separation purity) and in parallel (to increase output flow rate). Projects 15.6 and 15.7 in Section 15.5 include references and a suggested analytical/experimental study.

15.4 Cyclone Separators

In cyclone-type separators, the raw mixture of substances is introduced at high velocity into a nonmoving mechanical structure that deflects the fluid mixture into a circular path. Heavier components resist being deflected more than lighter components, and they pass preferentially toward the outside of the flow. Figures 15.8a,b show two designs. In (a) the fluid is simply forced to change direction by 180°. This design is fairly effective with a fluid-solid mixture when the solids are much denser than the fluid and have small enough drag to move radially through the fluid during the short time the fluid is in the separator. They are usually operated in an upright position so that gravity aids the collection process.

In (b) the flow closely approximates an inviscid vortex with a central sink. Angular momentum is conserved as the mixture moves radially inward. Centripetal acceleration (v_θ^2/r) and separation efficiency reach high values near the axis. Secondary flows on the conical portion help to carry the heavier components to the bottom outlet. As with the gas centrifuge, design and cost considerations favor small size for each unit, and practical facilities typically use many units in series and/or in parallel. Other variations are noted in the suggested projects in Section 15.5.

15.5 Study and Research Projects

The projects suggested here are intended for individuals or groups. Some can be completed in a week or two, others are term or longer projects. Only a few will have exact answers. References to other texts and to professional journals are included.

Project 15.1: Design and build an experiment that rotates a disk in an unbounded fluid. Preferably use a direct-current motor having a permanent-magnet field. With this motor, output moment (M) is directly proportional to input current (i), making it also useful as a device to measure moment. Additionally, M(i) will be nearly independent of angular velocity. If M(i) is not known, calibrate it by raising a weight with a thread wrapped around a spool on the output shaft. Perform your first test using water in a large container or pool (fluid nearly unbounded) and compare actual M and ω with computed M at the known ω and disk size. See (15.1). Test at various ω and with different size disks. Use a flow tracer to observe transition from laminar flow to more complex flows. Do edge effects depend on ω and/or R? Next, use a liquid with unknown viscosity and estimate the liquid's viscosity. For references see Greenspan (1968), Schlichting (1979), Van Dyke (1982), and Granger (1988).

Project 15.2: Vary the Project 15.1 experiment by placing a disk very near the rotating disk and examine variations of M(h), where h is the gap distance between the rotating disk and the stationary disk. Other variations include spinning a large smooth disk in air. How effective is it as a pump (fan)? For another variation, operate a motor in a soft vacuum (say ~10 mm Hg) and then at 1 atm. Measure the difference in current for the same angular velocity and compare it with a computed estimate for drag losses using the theory of this chapter, (15.1) and (15.10). See Szeri et al. (1983) and their references.

Project 15.3: Chapter 11 describes a numerical solution (laminar flow) for the problem of a rotating disk in a fluid at rest (Table 11.1 and Figure 11.1). Schlichting (1979) describes the same solution but with a different notation. The sixth and earlier editions of Schlichting provide more detail. See von Kármán (1921), Cochran (1934), and Sparrow and Gregg (1960) for complete details. In their solutions, the partial differential Navier-Stokes equations are first reduced to a set of four ordinary differentials as shown in Chapter 11. Various approximate and numerical methods are then used to seek solutions. What compromises (i.e., special conditions or restrictions) that limit the general validity of the solutions are explicitly or implicitly included in these solution methods? Consider a numerical solution of the original partial differential equations. How feasible is that? Must compromises again be made in order to reach useful solutions? Will the solutions be significantly different? Given such a program, would it be relatively easy to include edge effects and flow in a finite volume?

Project 15.4: Explore the literature for quantitative estimates for the onset of Taylor vortices in a cylindrical annulus. Most of this research will be found in selected journals on mathematics, fluid mechanics, and physics. Current research is typified by Murray et al. (1990), Takeda et al. (1990), Babcock et al. (1991), and Ning et al. (1991). Compare Taylor vortices to Görtler vortices. Can the equations used to estimate the onset of one be used also for the other? Examine the internal structure of the Taylor vortices. Based on how they are forced into existence and maintained, should their velocities about a vortex axis be represented best by $v_\theta = \omega r$ or by $v_\theta = \kappa/r$? Compare total vorticity in a cylindrical annulus before the onset of Taylor vortices to total cross-sectional vorticity of the set of generated Taylor vortices. Can this analysis be extended to the onset of turbulence?

Project 15.5: Flow between two concentric cylinders has become a major research tool for understanding stability/instability of fluid flows. The development of Taylor vortices and their transition to more complex forms, and finally turbulence, is used to understand bifurcation theory, chaos, and other concepts that trace the transition to turbulence. Examine the literature and write a summary report on current research activities in this field. Review articles in recent issues of applicable journals, e.g., *Journal of Fluid Mechanics*, *Physics of Fluids*, and *Journal of Fluids Engineering*. The history of this research is reviewed by Donnelly (1991). You may wish also to consult the National Science Foundation (1800 G. Street N.W., Washington, DC 20550) for information on current research supported by them.

Project 15.6: Success in the design and operation of any separation centrifuge requires knowledge of the diffusion and settling times of the components in the mixture. The centrifuge not only decreases the time scale of a separation process but can even make the process possible by dominating the normal processes that retard or prevent separation, e.g., convection, thermal diffusion, Brownian motion, etc. Familiarize yourself with the physics and mathematics of these processes, and assess the centripetal acceleration (in terms of number of equivalent earth gravities) necessary to overpower them by a factor of 100. Stokes's theory for a small sphere falling through a viscous medium is a good starting place. Next, assess the magnitude of secondary flows that occur with even small speed variations of the centrifuge. Compare the magnitude of these secondary flow or other rotating fluid phenomena with the separation process. See McCall and Potter (1973), Avery and Davies (1973), Fujita (1975), Aston (1978), and Conlisk et al. (1982).

Project 15.7: Read Avery and Davies (1973) and Conlisk et al. (1982), and then design and build a simple countercurrent centrifuge. To aid in observing and understanding details of its performance, including secondary flow characteristics, construct it from transparent plastic cylinders. Design for comparatively slow rotation speeds to separate some easily separable liquids or a liquid and suspended particles. With this design it should be possible to use inexpensive commercially available seals and bearings. Estimate the centrifuge's efficiency during the design portion of your research and then compare this estimate with your test results.

Project 15.8: Assume a cyclone separator shaped as a vertical hollow cylinder with diameter (D) and height (H), and with flat ends perpendicular to the cylinder axis. The top and bottom surfaces have central holes with diameter (d). A thin slot admits an air/particle mixture tangentially at the cylinder periphery. The bottom hole is open to an airtight container. Select an air/particle mixture for an application of interest and estimate its particle size and density distributions. Select separator design criteria to achieve about 90% separation of the particles into the container. First, assume rigid body rotation of the air/particle mixture within the cylinder to get an estimate of radial and vertical separation/diffusion times and an estimate of radial inward secondary flow on the bottom surface. How does the top surface participate? Next, assume an approximate real flow pattern. Develop a numerical solution for more detail. For more realism, have the bottom surface a cone tapering from diameter D down to the hole diameter d, and place a short tube below the top outlet hole as in Figure 15.8b. How will these affect each solution? To what extent can a cyclone separator be used as a gas/liquid or liquid/particle separator? Quantify some examples. See the book by Gupta et al. (1984) titled *Swirl Flows*.

Project 15.9: Design a new and novel centrifugal separator. Review the various ways that vortices, rotating fluids, and secondary flows originate and consider methods for using each to separate different liquids, gases, and/or suspended particles. First, examine various flow phenomena for some ideas and then some consistent structure (mechanical, electrical, magnetic, etc.) that would reduce your idea to a useful application. If your idea is new, novel, and useful, it might be a patentable invention with commercial value. Consult your library for books on inventions and patents. Achieving commercial success with an invention is usually a difficult and expensive process, but when successful, an inventor can be well rewarded.

Project 15.10: Prepare a report summarizing commercially available centrifuges and cyclone separators. Catalog by application area, size and capacity, special features, and cost. Rank the centrifuges and separators as to which would be the best buy for an application of interest. Include reliability and usefulness at slightly off-design conditions in your analysis.

Chapter 16

Wings, Lift, and Drag

16.1 Circulation and Lift (Inviscid)

Chapter 16, as with Chapters 14 and 15, introduces an application area, here basic aerodynamic theory and practice. The chapter provides an introduction to the general theory and practice of aerodynamics, but only enough to identify important specific applications of the theory presented in Part II. Ten projects, listed in Section 16.6, include references to the professional literature for readers who wish to pursue ideas in more detail.

This section will derive the expression relating lift and circulation

$$L = -\rho U \Gamma \tag{16.1}$$

where L is lift (force perpendicular to flow), ρ is the fluid density, U is the free-stream (or object) velocity, and Γ is the circulation of fluid about the object. The chapter will later discuss drag and apply the results to a variety of applications. The derivation of (16.1) is based on a constant-density fluid assumed to have zero viscosity. However, Stokes's theorem states circulation can exist only if vorticity exists, and in a constant density fluid, creation of vorticity requires viscosity. The existence of Γ (and ζ) in (16.1) appears to contradict the assumptions of $\rho = c$ and $\mu = 0$.

Although the relationship of circulation to lift may not be obvious, the relationship of a difference in pressure between the bottom and top of a wing is obvious. In addition, a force in one direction (lift) must be balanced with a net rate of change of fluid momentum in the opposite direction to satisfy Newton's law. A useful explanation of lift must integrate these sometimes contradictory ideas into a consistent and rational theme.

Equation (16.1) is based on an inviscid flow analysis. Inviscid two-dimensional (2-D) flow about a circular cylinder of radius R is obtained by combining the velocity potentials for a uniform stream, (u, v, w) = (U, 0, 0), with the velocity potential for a doublet (source and sink as their separation approaches zero). Refer to Chapter 3 and

Appendix B for a review of potential (inviscid) flow theory. The solution to inviscid flow about a cylinder in polar coordinates is

$$\phi = -U\cos\theta\left(r + \frac{R^2}{r}\right) \tag{3.16a}$$

$$v_r = U\cos\theta\left(1 - \frac{R^2}{r^2}\right) \tag{3.16b}$$

$$v_\theta = -U\sin\theta\left(1 + \frac{R^2}{r^2}\right) \tag{3.16c}$$

The Bernoulli equation ($p + \rho V^2/2$ = constant) can be applied to this inviscid flow. Use of $V^2 = v_r^2 + v_\theta^2$ permits p to be computed around the surface of the cylinder. The result shows that the flow-wise component of pressure force, integrated along the front of the cylinder, equals that integrated along the rear. Therefore, drag equals zero. This is consistent with a theorem by d'Alembert that drag about an object in a steady-state inviscid (irrotational) flow must be zero. The transverse component of pressure force can also be integrated along the top and along the bottom; again they are equal, so lift is zero. These are illustrated in Figure 16.1.

If ϕ for an inviscid vortex (see Chapters 8 and 10)

$$\phi = -\kappa\theta = -rv_\theta\theta = -\frac{\Gamma}{2\pi}\theta$$

is added, flow about a cylinder with circulation is the result, as shown in Figure 16.2 and given below.

$$v_r = U\cos\theta\left(1 - \frac{R^2}{r^2}\right) \tag{16.2a}$$

$$v_\theta = \frac{\Gamma}{2\pi r} - U\sin\theta\left(1 + \frac{R^2}{r^2}\right) \tag{16.2b}$$

Net pressure again is obtained using the Bernoulli equation, with r evaluated at the surface of the cylinder. At $r = R$, $v_r = 0$, and $v_\theta = 2U\sin\theta + \Gamma/2\pi R$. Use of $V^2 = 0 + v_\theta^2$ in the Bernoulli equation then gives

$$p = \frac{\rho U^2}{2}\left[1 - \left(2\sin\theta + \frac{\Gamma}{2\pi RU}\right)^2\right] \tag{16.3}$$

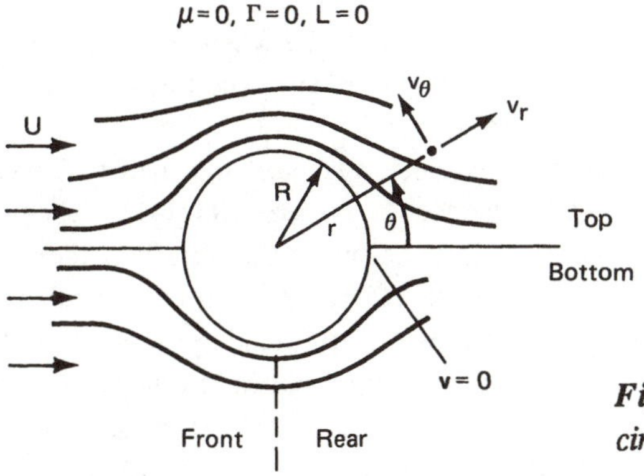

Figure 16.1: Flow about a circular cylinder.

After integrating the vertical component of pressure force from $\theta = 0$ to π and 0 to $-\pi$, the equation for net force per unit length transverse to the flow direction (called *lift*) is obtained.

$$L = -\rho U \Gamma \tag{16.1}$$

The flow is symmetric front to rear so that pressure integrated front and rear again equals zero. Chapter 9 transformed the Navier-Stokes equations from the variables (ρ, **V**) to the variables (**V**, ζ), showing that flows can always be interpreted solely as interactions between vorticity (circulation) and velocity as shown in (16.1).

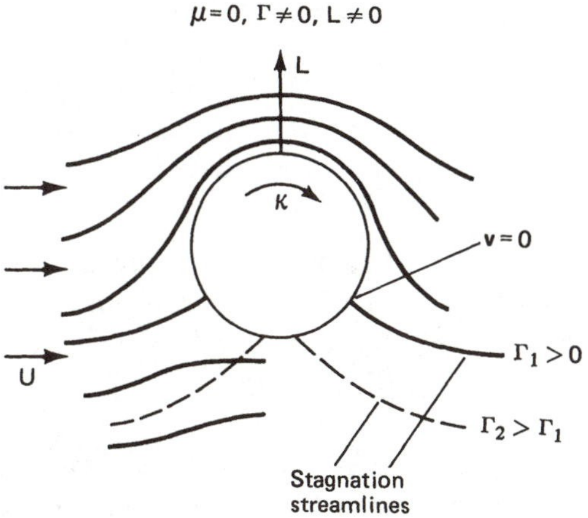

Figure 16.2: Same as Figure 16.1, except $\Gamma \neq 0$. Positive U and negative Γ produce upward (positive) lift.

16.2 Circulation and Lift (Viscous)

But as noted earlier, *How can circulation, which requires vorticity which in turn requires viscosity, be consistent with an inviscid flow?* The answer is that an inviscid vortex has a line of concentrated vorticity at its origin (hidden here inside the cylinder) as part of its definition. The theoretician simply says, *Given a line of vorticity*. . ., and then places it at the center of the cylinder where it will not interfere with the inviscid assumption. But where is a *real* object going to get vorticity and circulation without the help of a theoretician?

In 1924, an ocean-going ship (the Flettner) was built that had two rotating smokestacks (cylinders 50 ft high and 9 ft diameter)—no screws or paddle wheels, just rotating smokestacks. A viscous boundary layer forms on the surface of the rotating cylinders. Then, in what is known as the Magnus effect, air is carried in circular paths providing circulation. The spinning cylinders and their boundary layers provide the vorticity to satisfy Stokes's theorem. When the wind blows, a transverse force is created per (16.1). Note that the faster the cylinder spins, the greater the force. However, as with the sailboat, unless the wind blows, lift is zero. Chapter 10 also discusses a mechanism by which two vortices can propel each other forward in a stationary fluid, but two rotating cylinders without a wind are not an effective propulsion system since each creates only a thin layer of moving fluid. The inviscid vortex model assumes rotating fluid extending to infinity, and each of a pair of vortices is completely enveloped in the flow of the other.

Ship rudders based on the Magnus effect have been used experimentally. A spinning cylinder projects downward beneath the ship and provides a force transverse to water flow direction. By placing the *spinning rudder* behind the ship's screw, a flow velocity can be maintained with the ship motionless. A spinning sphere has a similar effect, although the mathematical model is more complicated. Baseballs, golf balls, and tennis balls provide common examples.

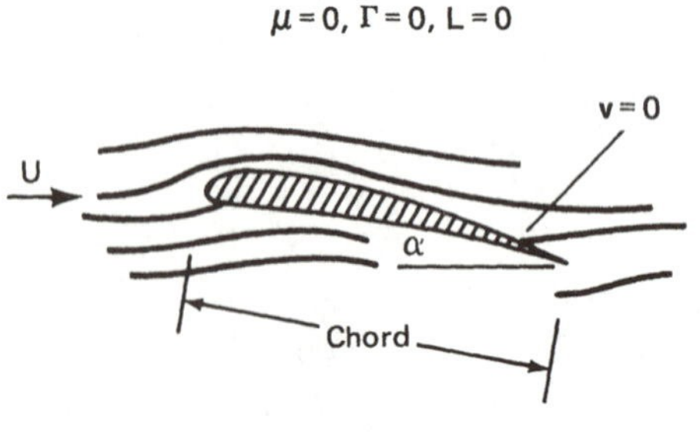

Figure 16.3: *Inviscid flow, with Γ = 0 about a wing section, has zero lift and zero drag, similar to Figure 16.1.*

One way to create circulation, then, is to spin the object, but how can a nonspinning aircraft wing generate circulation? Figure 16.3 shows an inclined wing section in an inviscid uniform stream. The inviscid fluid forms no boundary layer. The fluid can maneuver, without loss, around the leading and trailing edges. Note the location of the stagnation points. Pressure is symmetric top and bottom, front and rear, and both lift and drag are zero. However, if $\mu > 0$, viscous effects in the boundary layer, and the inability of a real fluid to reach $V = \infty$ around a sharp edge, forces flow about a real aircraft wing to be approximately as shown in Figure 16.4. Note that the fluid moves slightly faster and farther along the top than the bottom.

The integral $\int_C \mathbf{V} \cdot d\boldsymbol{\ell}$ along the top, therefore, exceeds the same integral along the bottom, with the difference being the net circulation. Integration of $\int_S \boldsymbol{\zeta} \cdot d\mathbf{S}$ over the top and bottom boundary layers will provide the same result as shown by Stokes's theorem. Equation (16.1) then provides a quantitative evaluation of lift.

As long as the boundary layer remains thin (no separation), the Bernoulli equation can be used just outside the boundary layer to estimate pressure on the wing. Integration of this pressure top and bottom will provide an estimate of lift, and integration front to rear will provide form drag. Surface friction is estimated by integrating shear stress at the surface. Form drag increases significantly if boundary layer separation occurs. Refer to Chapter 4 for a discussion of boundary layer theory. The analysis so far has assumed an infinite-span, constant-chord wing. When a wing is finite, a third type of drag, called *induced drag*, must be included and will be discussed presently.

16.3 Kutta-Zhukowskii Theory

The difficulty in the design stage is in predicting flow velocities about a practical aerofoil shape. An early solution, that provides values of inviscid flow velocities about a range of aerofoil shapes, uses the Zhukowskii transformation (also spelled Joukowski)

$$z = Z + \frac{\ell^2}{Z} \tag{16.4}$$

Here z, Z, and ℓ are complex numbers. An off-center circle, defined in coordinates of Z, is transformed point by point by (16.4) to coordinates z. In the z plane, the circle appears as an aerofoil shape with the trailing edge at $Z = -\ell$. The circle in the Z plane represents the cross section of a circular cylinder. Solutions for flows about circular cylinders in Z-frame coordinates (16.2a,b) then can be transformed point by point to the z plane also. The transformed solutions are flow solutions about the aerofoil shape. Milne-Thomson (1960) provides complete details.

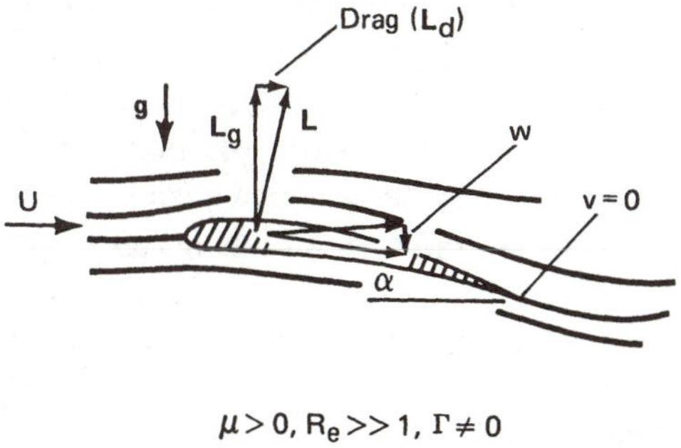

Figure 16.4: *Flow with viscosity. Zhukowskii transformation changes Figure 16.3 to this same flow pattern. Induced drag (L_d) is explained in the text.*

Figure 16.4 shows a typical Zhukowskii 2-D aerofoil. Flow of a uniform stream about a cylinder without circulation (Figure 16.1) transforms to the flow shown in

Figure 16.3. This inviscid flow solution is obviously unrealistic because the circular cylinder rear stagnation point transforms to a point on the upper rear of the aerofoil rather than at its trailing edge. The stagnation point is moved back to the trailing edge as shown in Figure 16.4 by adding circulation to the circular cylinder flow. The Kutta-Zhukowskii hypothesis states that the circulation required to move the transformed aerofoil stagnation point to the trailing edge provides the correct estimate for lift in a real flow. Predictions of this model are quite accurate, with the implication that the viscous boundary layer does indeed force the same amount of circulation to exist about a real wing. The Zhukowskii aerofoil represents an ideal flow situation. Modern aerofoils are based on additional compromises between performance and construction factors.

Lift is increased by increasing the angle of incidence α, although lift can exist at zero incidence. An object flat on the bottom and rounded on top with zero incidence has lift because the air must travel farther and faster along the top than along the bottom, and therefore $\int_C \mathbf{V} \cdot d\boldsymbol{\ell} = \Gamma \neq 0$. Increasing α increases lift by increasing Γ, up until the angle where separation on the top rear of the wing reduces lift to less than the weight of the aircraft. At this point the aircraft stalls and immediately begins to fall. Both form drag and induced drag increase as α is increased.

16.4 Finite Wings and Vortices

Figure 16.5 illustrates a real aircraft and the idealized vortex system associated with an aircraft in flight. The idealized system (upper right of figure) illustrates Helmholtz's theorem (see Chapter 10) that a vortex cannot end in a fluid. As the aircraft accelerates on takeoff, circulation around the wing increases. At the same time, flow from below the wing rolls up behind the trailing edge, creating a vortex that is equal to, but opposite, the circulation about the wing. As the aircraft speeds up, the trailing vortex is left behind. The wing carries its circulation, i.e., lift vortex, with it. The two vortices are joined by the tip vortices into a continuous pattern. The vortices left behind the aircraft eventually dissipate due to viscous effects, but as can be seen in the Figure 1.7 condensation trails, the ratio of vortex energy to viscous dissipation is large enough that the vortices persist over a considerable distance (time). Figures 1.7 and 16.5 (upper-right-hand sketch) also show a 3-D instability associated with long parallel vortices of opposite sign. Random perturbations and viscous effects sometimes cause them to touch, break, and then rejoin as toroids resembling the smoke rings of Figure 1.3.

Condensation trails observed behind large jet aircraft with their engines near the fuselage often are not well-defined wing-tip vortex patterns as in Figure 1.7. The flight altitude of jets is usually above atmospheric levels that contain enough natural moisture for condensation by the small pressure/temperature variations of wing-tip vortices, and further, large quantities of water vapor are produced as a combustion

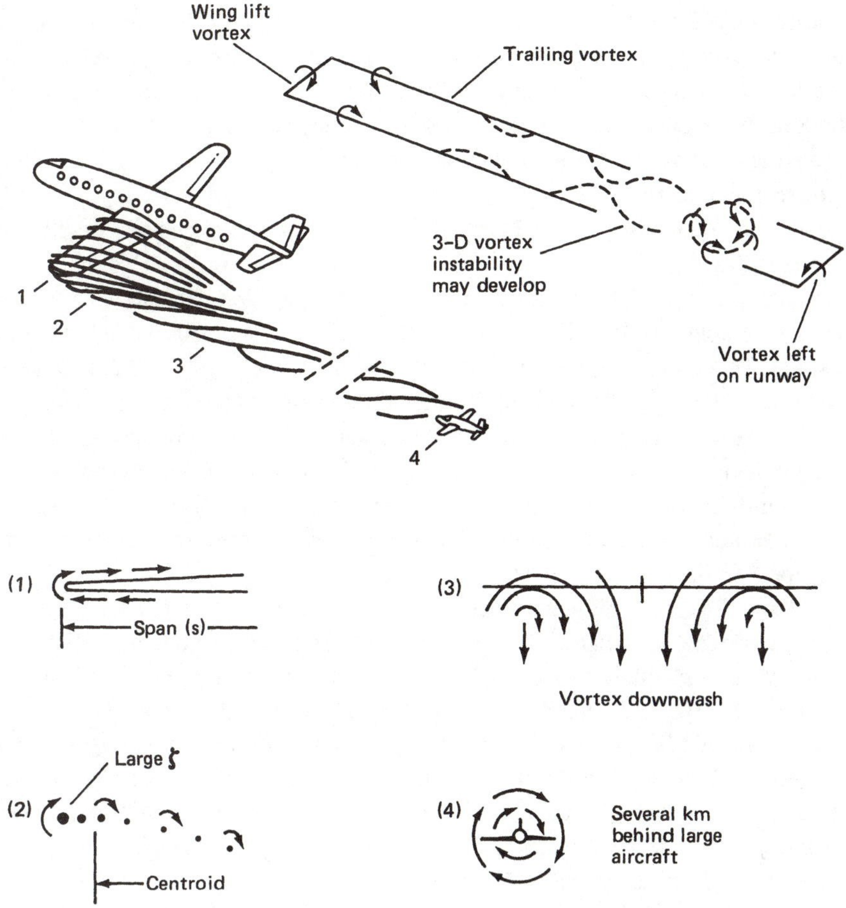

Figure 16.5: *Vortex system behind a lifting wing. See text for explanation.*

by-product in the jet engines. The vapor may leave the engines, expand, cool, condense, and become visible immediately behind the engines rather than behind the wing tips. However, wing-tip vortices, visible or not, always exist for an aircraft in flight, and even if condensate starts at an engine outlet, it tends eventually to integrate into the wing-induced vortex patterns. The well-defined vortices shown in Figure 1.7 were, however, produced by a jet aircraft.

The sketch of a large passenger aircraft details the formation of wing-tip trailing vortices. An elementary explanation is that pressure below the wing is greater than that above, so air tends to flow up around the end of the wing tips, creating a vortex whose axis is parallel to the flight direction. A more rigorous explanation notes that in a finite wing, especially a wing that gets smaller toward the tips, circulation per unit distance along the wing must get less and less toward the tips, reaching zero abruptly at the tips. Each decrease in Γ represents a decrease in vorticity ζ at that point on the wing. This vortex strength cannot simply just end at that point. Note that this explanation applies equally well to viscous and inviscid flow analyses. Use of a vortex line within the wing that decreases in strength toward each wing tip, with corresponding increases in trailing vortex strength, is the crux of *lifting line* theory as proposed by Lanchester in 1907. The theory was later formalized by Prandtl; see von Kármán (1963).

At position 1 of the wing vortex pattern of Figure 16.5, vorticity shows up as a net inward component of flow on top of the wing and an outward component on the bottom. This same transverse flow just behind the wing, at position 2, is represented by a line of vortices whose centroid is slightly inboard of the wing tip. Vortices from the left wing all have the same sense of rotation, and all interact, causing a net rotation about that centroid (see Figure 10.1). Right and left wing vortices rotate in opposite directions with down flow between them. These two parallel vortices of opposite rotation mutually cause each other to have a downward velocity as shown at position 3 in Figure 16.5 (see also Figure 10.2).

A small aircraft following a large aircraft is shown at position 4 in Figure 16.5. Trailing vortices are a hazard for smaller aircraft at distances up to 5 to 10 km behind large aircraft, especially on landing. In order to land at a reasonable speed, the pilot of the large aircraft has increased the angle of incidence and the power of his aircraft to ensure adequate lift at the lower speed. The trailing vortices are stronger than when cruising, and the small aircraft pilot must follow near them in order to land on the same runway. The large aircraft can have landed and taxied off the runway, while 5 km back one of its trailing vortices is forcing a small aircraft to unexpectedly roll over--sometimes a fatal occurrence. The experienced small aircraft pilot preferably avoids large aircraft completely, or stays above the landing path of the large aircraft, knowing that the vortices tend to sink, and stays slightly to the windward side, knowing that they blow away. They sink because they propel each other downward, and Helmholtz promises the vortices will move with the wind by his theorem, *Vortices in a fluid always remain attached to the same particles of fluid.* For various operational reasons, including safety, most large jet aircraft airports now have rules that prohibit smaller aircraft from using them or even entering their airspace.

Birds have increased their survival rate by making use of these concepts. For example, groups of birds that fly long distances do not follow in each other's downwash, but rather fly slightly behind and to the side of each other. They fly in the slight updraft just outside each other's wing tips and form the familar V formation. Individual birds, that soar while seeking their fortunes, have developed long, individually controllable, tip feathers that correspond somewhat to fingers. They apparently adjust the position and angle of incidence of each of these feathers to reach out and gain lift from their own tip updrafts as shown in Figure 1.8. An alternative explanation is that they simply reduce the magnitude of the strong vortices typically formed at wing tips. Soaring birds perform at speeds consistent with laminar boundary layers; only high-performance gliders (sailplanes) operate in this region; all powered aircraft, large and small, operate at higher speeds.

When aircraft performance is critical, modified wing tips are added to their design. For example, small aircraft in Alaska that operate from rough fields and river banks sometimes are equipped with downward tilting wing tips for slower and safer landing and takeoff speeds. Other small aircraft have added wing-tip fuel tanks shaped to reduce wing-tip vortex formation. One very large four-engine passenger jet (Boeing 747-400) was redesigned for more efficient transoceanic flight with the addition of upward tilting *winglets* at each wing tip for the same reason.

16.5 Vertical Momentum and Induced Drag

Two related topics remain

1. The fluid momentum change $d(\rho\mathbf{V})/dt$ necessary to satisfy Newton's equation $\mathbf{F} = m\mathbf{A}$
2. The connection between trailing vortices and induced drag

Fluid momentum change occurs when a narrow jet of fluid is deflected by a vane. Suppose a volume ($\forall$) of fluid with horizontal momentum $\rho\forall w$ is deflected 90° downward so that its new momentum is $\rho\forall(-w)$. Rate of change of momentum for a continuous jet of fluid replaces $\forall$ with the flow rate Q to get, for reaction force on the vane,

$$F = \rho Q w \tag{16.5}$$

The flows in Figures 16.1 through 16.4 are not jets, but fluids extending to infinity and moving relative to the cylinders or aerofoils with a uniform velocity U. The inviscid fluid of Figure 16.3 is not simply an isolated jet. The flow must react also with the remainder of the fluid. The net effect for an inviscid fluid is that it will reform as shown on the rear side and resume its uniform flow to infinity in the opposite direction, just as does flow about the cylinder in Figure 16.1. There is no net deflection up or down.

Consider now the inviscid flow with circulation in Figure 16.2. This fluid is lifted by the vortex as it approaches the cylinder and then is depressed as it leaves. Equations (16.2a,b) and (16.3), and Figure 16.2, define a flow that is deflected to some $(-\rho w)$ momentum. The momentum deflection here is an integral part of the definition of circulation in a uniform stream. As a simple model, assume w is some averaged vertical velocity for the volumetric flow rate Q of fluid affected by Γ. The vertical momentum flux becomes $\rho Q w$ as in (16.6). This is a vertical force called lift,

$$L = \rho Q w \tag{16.6}$$

where L and Q are lift and flow rate per unit length. It is the same lift given by (16.1), although in a less convenient form.

Figure 16.4 has been used to represent both a viscous flow and a Zhukowskii transformation of Figure 16.2. The Zhukowskii transformation interpretation leads also to (16.6), except that the vertical displacement of the flow field, obvious in Figure 16.2, now cannot be seen. The viscous interpretation is partially different, because a viscous flow past the same aerofoil can be deflected. The viscous fluid does not completely reform behind the aerofoil but acquires also some small net downward velocity.

The discussion of Figures 16.1 through 16.4 assumed 2-D wings, i.e., wings with constant chord and infinite span. If the wings are finite in length, or if the chord varies along an infinite span, trailing vortices occur as in Figure 16.5. These vortices cause increased *downwash* behind the aircraft and the phenomena of induced drag. It occurs whether the fluid is inviscid or viscous.

Downward deflection of air behind a real aircraft and its relation to lift can be analyzed as follows. After being deflected downward by the wing, the air has a vertical component of momentum per unit mass of $-\rho w$. This occurs at a rate equal to the aircraft velocity U. The total rate of change of z-momentum is force in the z direction,

$$F_z = L_g = \int_A (\rho w) U \, dA \qquad (16.7)$$

Figure 16.4 shows L to be total lift perpendicular to the deflected wind direction, and L_g to be its vertical component available to oppose gravity. Downward flow is assumed to be averaged over some area A. Obviously the area is approximately as wide as the wing span and from the trailing edge to some distance behind the wing. Let the wing have span s and assume a distance behind the wing as some fraction of the span, say ks. Then

$$L_g = \rho k s^2 w U \qquad (16.8)$$

Induced drag (L_d) can be related to L_g using $L_d/L_g = w/U$. Note that w behind a real wing is mostly caused by vortex-induced downwash, but partly also by direct deflection of the viscous air by the wing.

Induced drag is a minimum for a wing with spanwise elliptic lift distribution. Downwash is uniformly distributed across the span for this design. A British aircraft during World War II, the Spitfire, had an elliptical planform wing approximating this design and performed excellently. It is nearly as effective, and much less expensive, to use a tapered planform wing slightly twisted along its span to vary the local angle of attack and circulation. Additional details are included as suggested projects in Section 16.6.

16.6 Study and Research Projects

The projects suggested here are intended for individuals or groups. Some can be completed in a week or two, others are term or longer projects. Only a few will have exact answers. References to other texts and to professional journals are included.

Project 16.1: The book *Theoretical Hydrodynamics* (4th edition) by Milne-Thomson (1960) is a classic presentation of hydrodynamics, especially frictionless fluids, vortex motion, and the use of complex variables. Chapter V is devoted to the review and use of complex variables that are then applied to a large variety of fluid problems, including aerofoils. Chapter VI introduces streaming motions and the Joukowski (Zhukowskii) transformation. The topic of aerofoils and the use of the theorems of Kutta and Joukowski are detailed in Chapter VII. The notation for vector operations follows a British format (**ab** = a dot product, **a**∧**b** = a cross product, $\partial/\partial\mathbf{r} = \nabla$, and **a;b** is a dyad). Write a brief report on the use of the Joukowski transformation and its role in the development of aircraft. Include also its inherent disadvantages. Conduct a literature search to locate recent applications of this theory.

Project 16.2: Select an airfoil cross section from Fox and McDonald (1992, pp. 450–460) or from Anderson (1985, Appendix D), and estimate the total vorticity in the boundary layer for a selected angle of attack, flow velocity, and air density. An approximate calculation might assume Hiemenz flow at the front stagnation point, transition to flow about a cylinder for a short distance, and then flat plate boundary layer flow, first laminar and then turbulent. Induce turbulence, if necessary, to suppress separation. Use any reasonable approximations necessary to achieve a preliminary numerical answer. Next, use Stokes's theorem to relate this vorticity to net circulation about the airfoil, and finally, use ($L = -\rho U\Gamma$) to estimate lift. How does this compare to the actual value of lift reported in Fox and McDonald (or Anderson)? Critique this computation and suggest improvements. See Bertin and Smith (1989).

Project 16.3: Research the use of wing-tip structures on birds, and on private and commercial aircraft. Prepare a report on birds that compares their lifestyles (feeding, migration, social patterns, etc.) with their wing characteristics. See Scorer (1978). Include information on their wing-tip designs and tail functions. Next, or alternatively, prepare a report on use of aircraft wing shapes and tail assemblies that can be modified in flight. Include unusual wing-tip designs. Some wing tips turn down sharply, others include fuel tanks, and yet others tip up sharply. Examine especially the Mooney Mark 21 with its dramatic increase in performance when flying "on the step" with its tail empennage properly aligned, the Boeing 747-400 with winglets for long oceanic flights, a twin Cessna with wing-tip fuel tanks, and a new regional commuter jet with winglets as used by Comair (Cincinnati, OH). Consult manufacturers and users for data.

Project 16.4: Downwash behind an airfoil occurs for a finite wing span and is due mostly to the wing-tip vortices. Can or must downwash also occur for a wing with infinite span and uniform cross section moving in a viscous fluid (air)? What if the fluid were inviscid? Must an upward force on a wing (lift) be associated with a downward acceleration of air? If so, where is the net downward acceleration of fluid in the classic solution for flow of a perfect fluid past a circular cylinder with circulation? Some introductory texts are Shevell (1983), Anderson (1985), and Bertin and Smith (1989).

Project 16.5: Conduct a literature search (include textbooks and commercial organizations) on the availability and utility of ready-to-use computer programs for aircraft flow visualization, analysis, and design. Acquire and use one or more of these programs suitable for computers that are available to you, e.g., Roache (1982) and Olfe (1987). Many university and college departments, as well as companies active in this field, will have them in their libraries. Compare their capabilities with some of the professional computer analyses being performed at government laboratories, aerospace research and development firms, and aircraft manufacturing companies. Write or call one of each of the three types of organizations for references to some of their current activities.

Project 16.6: Locate a text on aerofoil wind tunnel experiments and construct a test aerofoil, preferably one with known flight characteristics. Use an NACA aerofoil shape or a shape scaled to a light aircraft wing. Contact the manufacturer for dimensions and flight characteristics. If a wind tunnel is not available, fabricate a simple one or use a moving vehicle. At least two people are needed if a vehicle is used, one to drive and one to experiment. Place the aerofoil in front of the vehicle in the relatively undisturbed flow. Measure lift and drag of the scale model under test conditions and then mathematically scale to full size using dimensionless parameters. Measure flow velocities very near the wing to estimate net circulation and then get a rough estimate of boundary layer thickness to match total vorticity to circulation. First, assume a linear boundary layer profile and then more realistic profiles. Fasten a scale model on either wall of a wind tunnel with the free end at the tunnel centerline. Observe and measure the wing-tip vortex. Experiment with tip winglets to modify (reduce) the vortex. See Van Dyke (1982) and Granger (1988) for photographs and ideas.

Project 16.7: Carefully observe a light, single-engine propeller aircraft on the ground and also in the air. If you are not a pilot, arrange to be flown by a qualified pilot authorized to fly simple maneuvers with a passenger. All the observations suggested here can easily be completed in a 1 hr flight. A flight instructor and light aircraft at a licensed and insured flight school should not be more than $100/hr (U.S.) and possibly much less (1993). Note the wing, stabilizer, and fin cross sections and alignment on the aircraft. Do the wing and stabilizer lie in the same or parallel planes? Is the fin perfectly aligned along the aircraft centerline or is it set on the fuselage at a slight angle? Estimate the lift forces generated by movement of the control surfaces, ailerons, flaps, elevators, and rudder. Observe that the tail surfaces are used primarily to orient the wing. Is the planform of the wing rectangular or tapered, and do the outer portions have the same angle of attack as the inner portions? On takeoff, note that the rudder must be held either left (or right) to maintain straight flight. Which way does the propeller rotate? Ask the pilot to perform a standard rate turn and observe movement of the controls. What movements are needed to initiate a turn, and what are needed to maintain the turn at a standard rate and constant elevation? The effects are more obvious for sharp turns with the wings about 45° from the horizontal. Ask the instructor to stall the aircraft after explaining the maneuver to you. What is happening to circulation and lift on the various surfaces? Try to interpret all the phenomena in terms of vorticity generation and circulation. The physics of aerobatic flight is discussed by O'Dell (1987).

Project 16.8: Conduct research on extrapolating test results on scale models to larger models and/or to full-size aircraft. As a first test, purchase a good quality scale model (molded from plastic) of an existing commercial aircraft. Obtain as large and detailed a model as is practical. Test the model in an available wind tunnel and also in a water tunnel (or any smooth flowing air stream or water flow). Using scaling laws, attempt to correlate the results of the air and water tests. If possible, obtain flight performance characteristics of the real aircraft and compare your test results, after scaling, to them. Alternatively, build two models of the same aircraft (or just a wing section) and attempt to predict results of the larger model (say, two times the size) from test results on the smaller. Most intermediate texts have good introductions on scaling, e.g., Fox and McDonald (1992), White (1986), and Roberson and Crowe (1990).

Project 16.9: Estimate (analytically, numerically, or experimentally) flow about a spinning cylinder in a fluid at rest (air or water). Next, introduce a movement of the fluid perpendicular to the rotating cylinder and estimate the new flow (Magnus effect), again analytically, numerically, or experimentally. Do your new results include both a drag force and a lift (transverse) force? Vary the cylinder rotation speed and the uniform flow speed. A small boat mainsail can be analyzed as a wing; compare propulsive effectiveness of a rotating cylinder sized to fit the same boat. The Flettner ship is discussed in White (1986) and in Granger (1988, p. 280). Also compare efficiency of a spinning cylinder in lieu of a standard rudder. In a conventional design the cylinder is placed behind the screw (propeller) and can be retracted to reduce or eliminate drag when underway. The U.S. Navy has investigated this design. Contact Naval Research Laboratories, Washington, DC. The Cousteau Society boat, the Alcyone, is sometimes said to have "rotating" sails. The rotating sails actually are constructed as masts shaped in cross section like very fat sails. Their thickness is approximately one-half that of their chord, and they use suction along their rear surfaces to prevent separation. They are rotatable only in the same sense that a conventional sail can be reoriented to meet changing winds and desired boat movements. Compare vorticity and circulation for the Alcyone sails and a conventional mainsail. A brochure is available from The Cousteau Society, 425 East 52nd St., New York, NY 10022.

Project 16.10: Lift and its relationship to vorticity and circulation is a more complicated matter for flight at supersonic speeds. Here the craft and its wings interact with the ambient air only through a shock-wave surface. Continuum mechanics apply on either side of the shock wave but not across it. See Bertin and Smith (1989) for an introduction, and see the professional literature for current developments. von Kármán (1963) summarizes the theory and traces its historical development.

Chapter 17

Turbomachinery

17.1 Definitions and Classifications

This chapter introduces the theory and practice of turbomachinery for the nonspecialist. It emphasizes those portions of the subject most relevent to a study of rotating fluids and vorticity. The project suggestions in Section 17.6 include references to the professional literature for those wishing more detail.

The word *turbomachinery* applies generally to any machine that transfers energy from a rotating device to a fluid (e.g., a fan, propellor, pump, blower, or compressor) or that transfers energy from a fluid to a rotating device (e.g., wind turbine, water turbine, steam turbine, gas turbine). Some machines, like a jet engine, include both types of devices. Reciprocating machinery with pistons and cylinders are not considered turbomachinery even though parts of the machine may rotate. Positive displacement rotating devices such as the Mazda automobile rotary engine or a gear-type pump also are not included.

Energy transfers from device to fluid or vice versa are achieved using any of several phenomena, for example

1. Centripetal acceleration of a fluid leading to a radial pressure differential, $dp/dr = V_\theta^2/r$, e.g., a centrifugal pump.
2. Deflection of fluid momentum leading to a force ($\mathbf{F} = d\mathbf{P}/dt$) at a distance $\mathbf{R}$ from the rotation axis creating a moment or torque ($\mathbf{R} \times \mathbf{F}$), e.g., an impulse turbine.
3. Lift on an aerofoil-shaped vane or blade subjected to a directed flow, $L = f(p)$ or $f(\Gamma)$, e.g., a wind turbine or airplane propellor.

Two or more principles may apply in a given machine. The fluid can be incompressible or compressible. When the fluid is compressible, reaction of the compressing or expanding gas must also be considered in an analysis.

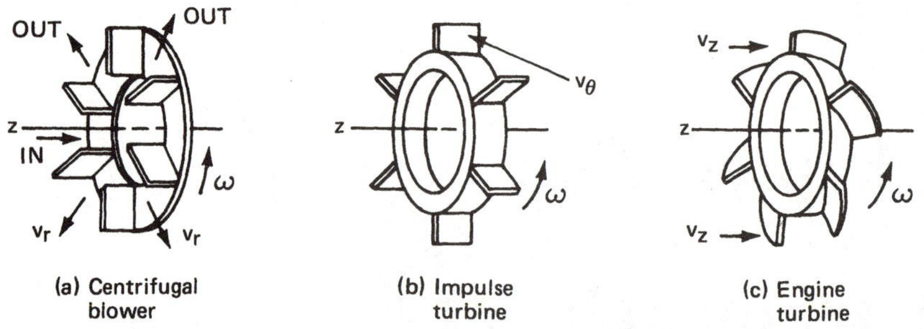

Figure 17.1: *(a) Simplified radial-flow, (b) tangential-flow, and (c) axial-flow turbomachine rotors.*

Turbomachines are often classified by fluid flow direction: radial, tangential, or axial. Figures 17.1a, 17.1b, and 17.1c illustrate these flows using cylindrical coordinates (r, θ, z), with z always aligned with the axis. A centrifugal pump or compressor (a), or a radial-flow water turbine, as used in many hydroelectric installations, are examples of radial-flow devices. Tangential-flow devices are less common, with the best example being a Pelton wheel impulse turbine (b). Aircraft propellors, ship screws, most steam turbines, and aircraft jet engines are examples of axial-flow devices (c). In these devices there may be some radial or tangential flow components, but the net mass flow rate is axial. In some mixed-flow devices the flow path is more complicated, and simple classifications are not useful. In Figures 17.1a, 17.1b, and 17.1c, the radial-flow device (a) is a blower that adds energy to the fluid, while (b) and (c), as shown, are turbines that extract energy from the fluid.

All turbomachines use either shaft torque to impart pressure or motion to a fluid, or they use a moving or pressurized fluid to create shaft torque. Therefore, regardless of whether the basic flow is axial, radial, or tangential, the device must be constructed so that the flow creates (or uses) the axial component of torque (M_z). Any component of $\mathbf{R} \times \mathbf{F}$ not aligned with the axis does no work, is resisted by the structure as a nonuseful load (stress), and typically leads to energy or power loss. The power-producing motion of the fluid is usually analyzed as an irrotational (inviscid) flow. Fluid rotation (vorticity) typically appears as secondary smaller-scale dissipative fluid motions.

Figure 17.2a shows a tangential-flow impulse device. In this type of device, fluid energy is completely converted to flow velocity at a nozzle. Note that the fluid can enter the machine from any direction but is reoriented to a tangential direction by nozzles at the point of use. This tangential momentum is deflected by the rotor vanes to produce a turning moment at the shaft. Pelton wheel output torque is maximum with the shaft stopped and zero when vane speed ($r\omega$) equals fluid speed (V). Maximum power for this device is obtained when vane speed equals one-half the speed of the entering fluid. Sometimes, as shown in Figure 17.2b, two rows of rotor vanes

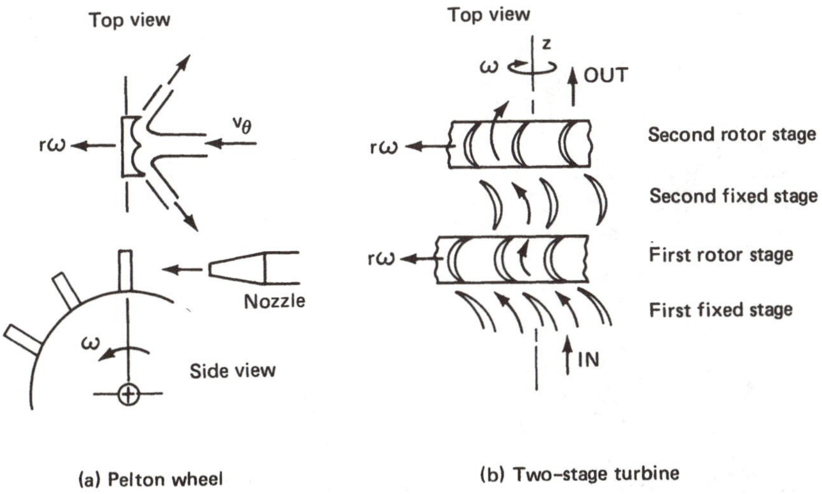

Figure 17.2: *Two examples of tangential-flow devices. (a) is a Pelton wheel impulse turbine, and (b) is a more efficient two-stage turbine that combines tangential and axial flow.*

are used with an intermediate set of stationary vanes that serve as nozzles for the second set of rotor vanes. Although the basic principle of the device in Figure 17.2b is the same as that of Figure 17.2a, it is common also to refer to the Figure 17.2b device as an axial-flow machine because net flow is axial.

Figures 17.3a and 17.3b show axial-flow devices. Axial-flow devices typically operate using lift (pressure) produced by flow over and around aerofoil-shaped vanes. An aircraft propellor (a) provides the simplest model. Here energy flows from a rotating shaft to a fluid. Analysis proceeds exactly like the analysis of an aerofoil in Chapter 16, including blade shapes, flow separation, tip vortices, etc. Both the propellor blades and the axial-flow turbine vanes are twisted along their lengths so that their effective angles of attack are constant for all distances (r) from the axis. The horizontal axis wind turbine (b) is similar to an aircraft propellor except the angle of attack is adjusted so that lift is tangential rather than axial. In all but the simplest applications, the blade pitch (angle of attack) is manually or automatically adjusted to the optimum angle of attack for different aircraft or fluid forward velocities. Propellor tip speeds sometimes approach sonic speed, and air then should not be assumed incompressible.

Figures 17.4a and 17.4b show two variations of complex axial-flow devices: (a) a commercial turbofan jet aircraft engine and (b) a military turbojet engine with afterburner. See also Figure 1.10. Turbofan jets are designed for optimal performance at high subsonic speed and at high altitudes. The forward part of the engine (not the fan) is a radial-flow compressor with alternating rows of stator vanes and rotor vanes.

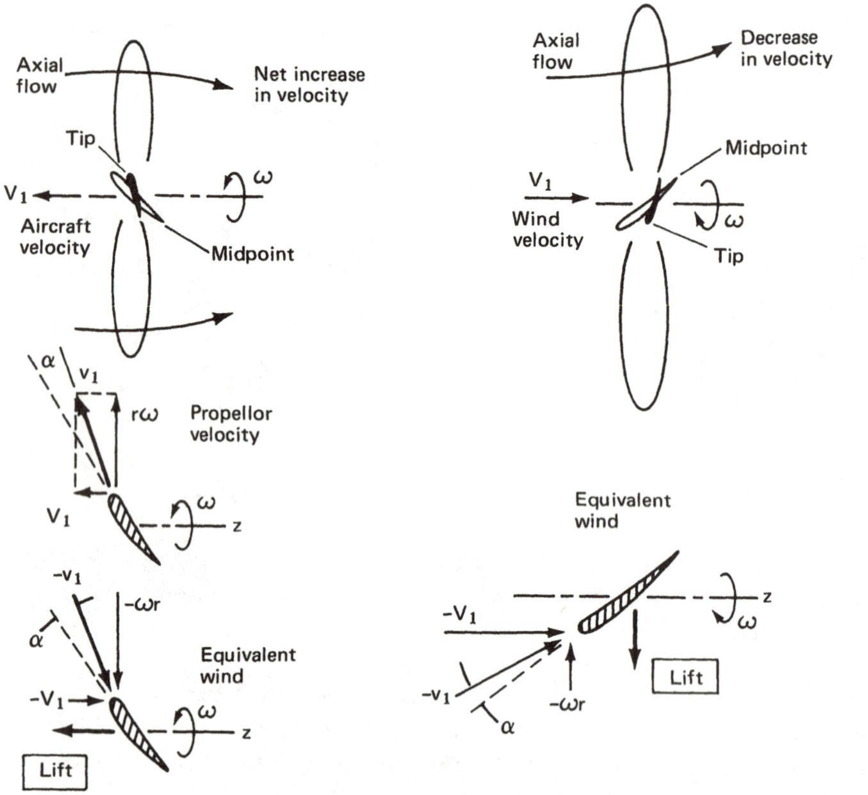

(a) Aircraft propellor (b) Wind turbine

Figure 17.3: *Basic axial-flow devices. Lift on a propellor blade (a) is axial (forward) to provide thrust, while in a wind turbine (b), lift is tangential to provide a turning moment. Blades must be twisted to have effective angles of attack (α) at each radial distance (r).*

The annular cross-section area is reduced as the air is adiabatically compressed. Fuel is injected and burns continuously in the combustion portion. The greatly increased internal energy of the burned air-fuel mixture is expanded through a series of stator and rotor vanes in the turbine.

Energy of the hot gas is used in three ways:

1. To power the compressor section rotor
2. To power the turbofan to get thrust from bypassed air, often using a separate axle inside the main compressor-turbine axle
3. To obtain thrust by ejection of the exhaust gas at high velocity.

The turbofan component provides high efficiency at subsonic aircraft speeds.

High-Mach-number military jet aircraft engines, as in Figure 17.4b, do not use a turbofan, but do usually include an afterburner section. For rapid acceleration and brief periods of increased velocity, the pilot can cause fuel to be injected into the afterburner section. The fuel combines with bypass air and with the remaining oxygen in the exhaust gas, again increasing gas internal energy. In this mode, the rear portion of the engine has many characteristics of a rocket. A fan and an afterburner are incompatible; engines never have both, and many engines have neither. Stationary gas turbomachines generate rotor power, e.g., drive an electric generator, and attempt to extract all the energy available in the exhaust gas.

Flow conditions at each stator stage and at each rotor stage are too complex to be modelled as simple momentum change vanes or as simple aerofoil vanes. Reference sources that address specific gas compressor and gas turbine design features are included at the end of this section. References are also included for steam turbines that use high-pressure, high-temperature (superheated) steam to generate power, most often for operating electric generators. Aircraft engines compromise fuel efficiency in favor of small size and low weight. Weight and size are not as critical in megawatt-sized electric generating systems, but fuel efficiency over a lifetime of several decades is critical. These steam turbines have many more stator and rotor stages, and each stage has more vanes along with other features in an attempt to reach the very highest efficiencies.

Figure 17.5 shows (a) a radial-flow centrifugal pump and (b) a radial-flow turbine. See also Figure 1.9. Flow in a radial-flow turbine is analyzed much like that in a tangential- or axial-flow device. The vanes deflect the fluid momentum and/or obtain lift from the moving fluid. A centrifugal pump can be analyzed in the same way at high flow rates but not at zero or reduced flow rates. Output pressure of a centrifugal pump at zero flow rate is obtained solely from translation of fluid elements in circular paths as the fluid follows the circular motion of the rotor. Pressure in the fluid then is obtained by integrating $dp/dr = \omega^2 r$, with ω the angular velocity of the rotor. As output flow rate increases, the fluid traverses the radial distance of the rotor without completely achieving the tangential speed of the rotor.

In Figures 17.5a and 17.5b, $r\omega$ is tangential speed of the rotor at each value of r, and V_r is radial speed of the fluid. The quantities v_1 and v_2 are fluid velocities relative to the rotor. The rotor vanes shown for the centrifugal pump align at each end with these velocities to avoid losses due to separation and turbulence. In some designs the "shockless entry" is preserved but the outlet portion of the blade is swept forward to yield increased tangential fluid velocity and pressure at high flow rate. Note that the pump will not function efficiently as a turbine and vice versa.

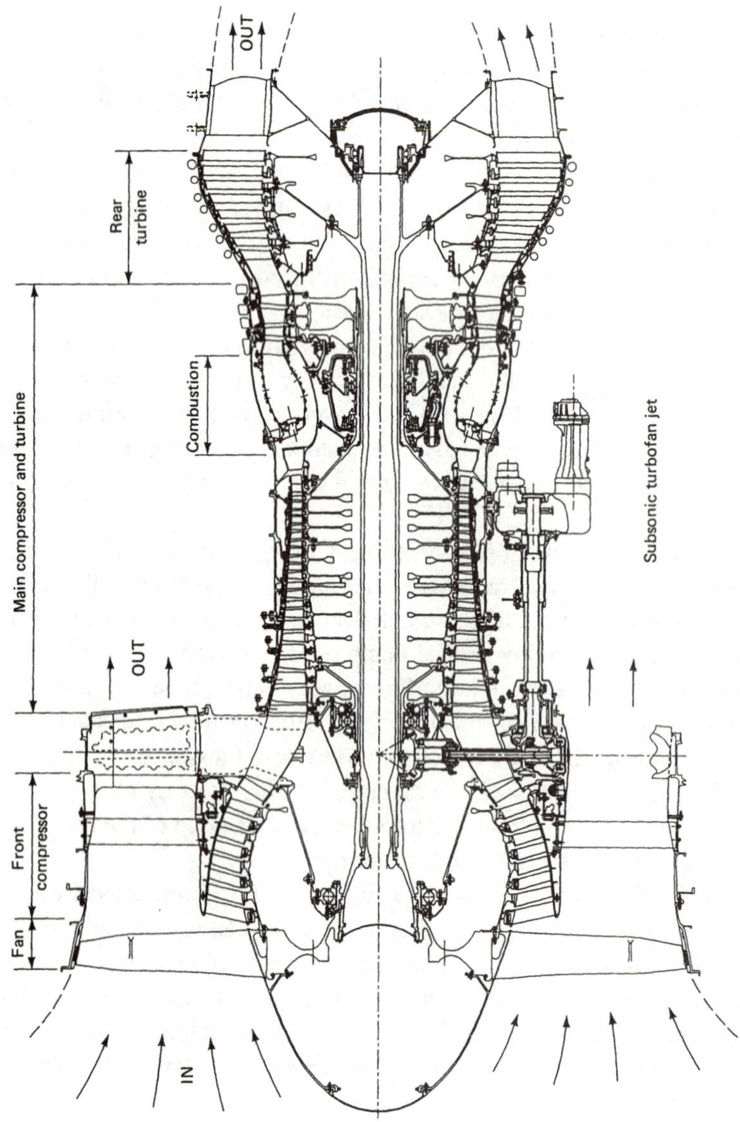

Figure 17.4: *(a) A subsonic turbofan jet engine is an axial-flow turbomachine designed for commercial aircraft propulsion. Drawing courtesy Pratt & Whitney.*

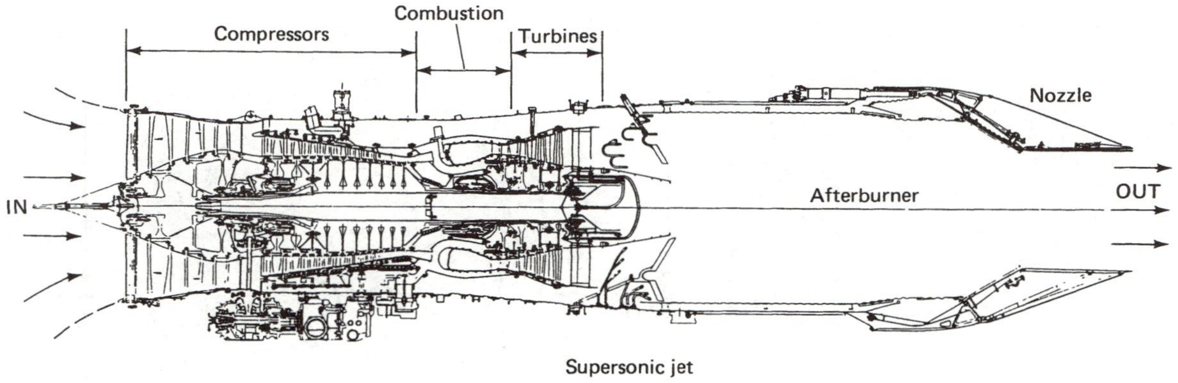

Figure 17.4: (b) Military supersonic jet engines do not use a turbofan but do typically include an afterburner section. Many jet engines use neither a turbofan nor an afterburner. Drawing courtesy Pratt & Whitney.

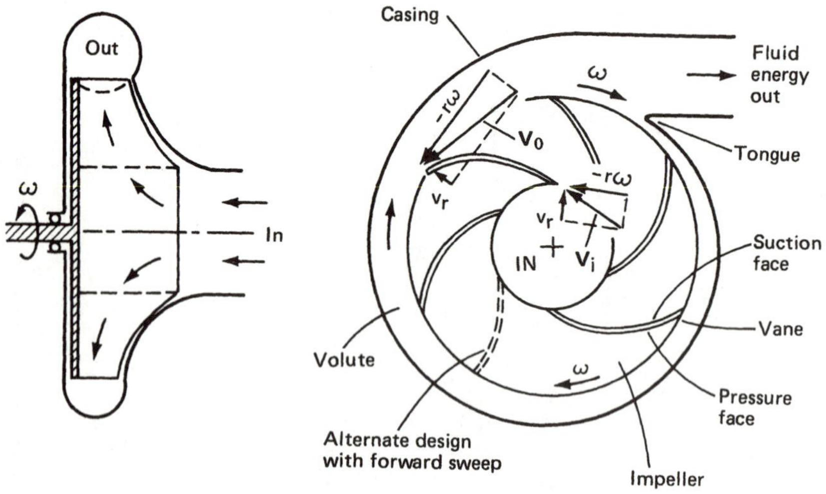

(a) Centrifugal pump

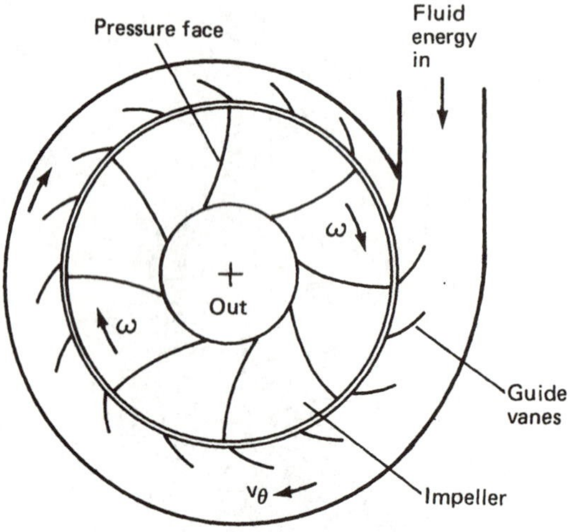

(b) Radial-flow turbine

Figure 17.5: *A simplified radial-flow pump (a) and a radial-flow turbine (b). The pump uses input torque to create (p) using ($r\omega^2$). The turbine uses input ($\mathbf{V}$) to create torque using ($\mathbf{R} \times \rho\mathbf{V}$). See also Figure 1.9.*

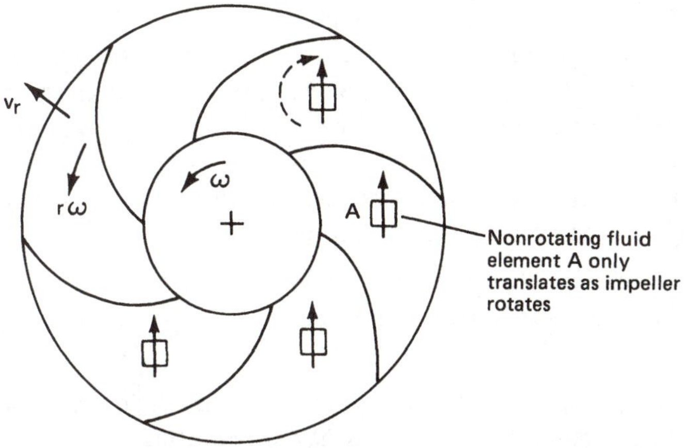

Figure 17.6: *Relative circulation of an inviscid fluid in an impeller. Fluid elements (A) have inertially fixed directions as they translate, but they rotate clockwise relative to an observer on the impeller.*

17.2 Internal Flow Characteristics

Figure 17.6 illustrates circulation in the spaces between vanes of a centrifugal pump. The phenomenon is easily understood by remembering that a nonrotating, inviscid, constant density fluid cannot acquire rotation (vorticity). For example, if a stationary cube is filled with a $\mu = 0$, $\rho = c$ fluid and then rotated, the fluid only deforms. It does not rotate with the cubical container, although it will translate with it, of course. If the cube is returned to its initial angular orientation, each element of fluid returns to the exact position within the cube it initially had.

Each space between vanes in a rotor is also a translating and rotating container, and fluid in that space will attempt not to rotate also. This is seen in the rotating frame as a negative rotation. All real fluids, however, have viscosity, and rotation (vorticity) is transferred from the rotor to the fluid by viscous diffusion, secondary flows, turbulence, etc. If pump flow-through rate is zero, real fluid within the impeller soon spins up to the rotor's angular velocity. Only limited spin-up can occur at high flow rate, because the fluid then passes through the impeller too fast for the spin-up mechanisms to be completely effective. Be careful here to distinguish translation of fluid elements from fluid rotation (vorticity) as illustrated in Example 8.1.

The limiting cases of fluid motion passing through a radial-flow impeller are as follows.

Case 1. Fluid velocity V_r at each r has exactly the velocity necessary to keep even with swept-back vanes designed to match that $V_r(r)$. The fluid, assumed inviscid and with no entering V_θ, then passes radially through the impeller without interacting with the vanes. This can be called the *zero energy transfer* or *free-flow* condition. Each type of rotor (radial, tangential, or axial) can be designed for some speed and flow rate for which no energy is transferred from rotor to fluid or vice versa. In many centrifugal blowers (high flow rate and pressure), the vanes are straight and exactly radial as in Figure 17.1a. In this design, or if the vanes are swept forward, there is no free-flow condition with the impeller spinning.

Case 2. If V_r is less than the free-flow condition, the device is a pump because the impeller vanes then force the fluid to be compressed and/or to acquire some V_θ.

Case 3. If V_r is greater than the free-flow condition, the fluid presses against the inside concave side of the vane, and the device is a turbine.

Case 4. If output flow is stopped, $V_r = 0$, and the fluid in each space is forced to translate in curved paths as the rotor turns. If the fluid is (assumed) inviscid, it translates without acquiring vorticity. Any real fluid, however, acquires vorticity, and the fluid spins up to the rotor's angular velocity (i.e., rigid rotation mode).

When a radial-flow device is to be used as a turbine, the flow direction is usually reversed as shown in Figure 17.5b. The function $V_r(r)$ is proportional to r^{-1} in a constant thickness radial-flow device, but $V_r(r)$ can and usually is modified so that V_r is nearly constant or even increases with r by narrowing impeller width as r increases. All turbomachine elements have flow rates equivalent to those of Cases 1 through 4. For example, when an airplane maintains its forward speed by losing elevation, there is some propellor speed where the airspeed relative to the propellor is aligned with the blade profile at each r. Neither drag nor lift occur, as in Case 1. An increase in propellor speed produces lift (forward thrust), and a decrease in propellor speed produces drag. The effects, however, are not linear with speed.

Fluid flow within turbomachinery typically exhibits the entire range of rotating fluid phenomena. Such phenomena may severely distort the flow field and cause energy losses; for example, secondary vortex flows have been estimated to cause 10% or more of the efficiency loss in jet turbine engines. They should be analyzed with the same care and attention given to the main flow field and the thermodynamic variables. Figure 17.7 shows some of the secondary flows and vortices that form in an axial-flow turbine or compressor. Shown are wing-tip vortices, Couette flow (scraping) vortices between the vane tips and the stationary housing, (cascade) secondary flow vortices due to curvature of the space between vanes (as in river meander), Görtler vortices, and relative circulation due to rotor speed (as in Figure 17.6). Longitudinal swirl in the noncircular flow path, boundary layer separation, and other phenomena can also occur but are not shown. See Project 17.6 in Section 17.6.

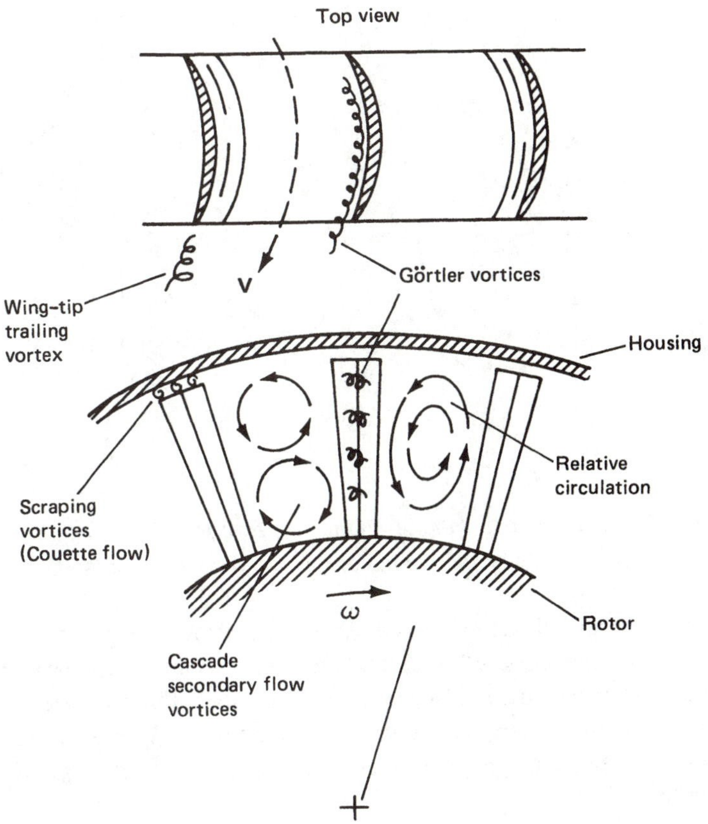

Figure 17.7: *Some of the secondary flows and vortices in an axial-flow compressor turbine rotor. Mechanical features and vortices are shown exaggerated for illustration.*

17.3 Torque and Power Output

For all turbomachinery, shaft moment (torque) is related to rate of change of moment of momentum ($\mathbf{M} = \dot{\mathbf{H}}$ in its simplest form) using

$$\mathbf{M} = \frac{\partial}{\partial t} \int_{\mathcal{V}} (\mathbf{R} \times \rho \mathbf{V})\, d\mathcal{V} + \int_S (\mathbf{R} \times \rho \mathbf{V})\, \mathbf{V} \cdot d\mathbf{S} \tag{17.1}$$

Here $(\mathbf{R} \times \rho\mathbf{V})$ is $\mathbf{H}$ of a fluid element in volume $\mathcal{V}$, which is bounded by surface S as shown in Figure 17.8. Obviously $\int_{\mathcal{V}} (\mathbf{R}\times\rho\mathbf{V})\, d\mathcal{V}$ is total $\mathbf{H}$ in that volume, and $\partial/\partial t$ denotes any changes in total $\mathbf{H}$ while the fluid is inside the volume.

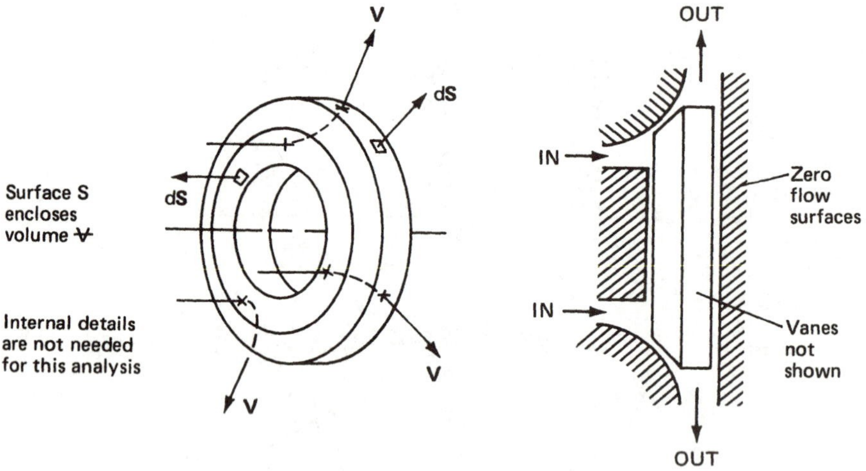

Figure 17.8: *Typical control volume shape for turbomachine analysis. Flow lines for a mixed-flow device with axial input and radial output are shown.*

The last term vectorially describes the flux of **H** passing through any portion of the closed surface S. The quantity $\mathbf{V} \cdot \mathbf{dS}$ gives the component of **V** perpendicular to the surface (defined at each point of the surface as an outward-directed element $dS = \hat{\mathbf{n}}ds$). The net effect of this term is to give $\partial/\partial t$ of the flux of $(\mathbf{R} \times \rho\mathbf{V})$ into or out of the volume. The two terms together then define total change of **H** per unit time $(\dot{\mathbf{H}})$ to obtain $\mathbf{M} = \dot{\mathbf{H}}$.

Most turbomachines are operated in a steady-state mode, and (17.1) reduces to

$$\mathbf{M} = \int_S (\mathbf{R} \times \rho\mathbf{V})\, \mathbf{V} \cdot \mathbf{dS} \tag{17.2}$$

In some applications transient performance is critical, for example, a military jet aircraft or a gas-turbine powered automobile, and the entire equation must be retained. Equation (17.2) can be simplified further when only output torque (moment) along the z axis is desired, i.e., M_z. In this case, $(\mathbf{R} \times \rho\mathbf{V})$ reduces to $\rho R V_\theta$, and $\mathbf{V} \cdot \mathbf{dS}$ reduces to $(V \cos\alpha\, dS)$, where α is the angle between the flow direction and a normal to the control surface across which the fluid is entering or leaving. With this convention (17.2) can be written as

$$M_z = \int_S RV_\theta(\rho V \cos\alpha\, dS) = \int_S RV_\theta\, d\dot{m} \tag{17.3}$$

This equation is valid for all steady-state turbomachines; axial, radial, and centrifugal. For many analyses the integral can be replaced by a summation

$$M_z = \int_{S_2} RV_\theta\, d\dot{m} - \int_{S_1} RV_\theta\, d\dot{m} \tag{17.4}$$

Assuming R, V_θ and $d\dot{m}$ are averaged over (or are nearly constant over) each of S_1 and S_2, then

$$M_z = \dot{m}\left(R_2 V_{\theta 2} - R_1 V_{\theta 1}\right) \tag{17.5}$$

Turbomachine power (in or out) is $P = M_z\omega$.

$$P = \dot{m}\omega\left(R_2 V_{\theta 2} - R_1 V_{\theta 1}\right) \tag{17.6}$$

The pressure rise is often quantified as *head*, the height of a column of fluid of density ρ, computed as

$$H = \frac{M_z\omega}{\rho g Q} = \frac{P}{\rho g Q} \tag{17.7}$$

where Q is volumetric flow rate. Head also has equivalent units of energy per unit weight, equal to (E/g). The energy transfer per unit mass (E) from fluid to rotor, or the reverse, is

$$E = P/\dot{m} = \omega\left(R_2 V_{\theta 2} - R_1 V_{\theta 1}\right) \tag{17.8}$$

Work (energy) provides an important basis for analysis. Elevation changes within turbomachines are almost always negligible, leaving for the turbine energy equation

$$e_1 + \frac{p_1}{\rho_1} + \frac{V_1^2}{2} + Q_h = e_2 + \frac{p_2}{\rho_2} + \frac{V_2^2}{2} + W \tag{17.9}$$

where e is internal energy, Q_h is heat transferred to or from the fluid, and W is work done on or by the fluid. If the fluid is incompressible, e and Q_h are not needed, and the equation reduces to the simple Bernoulli equation. Viscous loss can be included in Q_h and/or W. In a compressible fluid machine, viscous loss can add to Q_h and then to e as increased temperature, with some recovery as useful work. Additional performance equations expressed in a variety of dimensionless variables are included in Section 17.5.

17.4 Flow Calculations

Analysis of fluid velocities entering and leaving spinning rotor blades is a critical part of turbomachinery flow calculations. Complete analysis requires knowledge of fluid velocities both relative to the stator and to the spinning rotor at each point in the flow. The specific notation developed in Chapter 6 for analysis of positions, velocities, and accelerations relative to a fixed frame (**R**, **V**, **A**), and the same quantities relative to a rotation frame ($\mathbf{r}$, $\mathbf{v}$, $\mathbf{a}$), is used here

$$\mathbf{R} = (X, Y, Z) \text{ Cartesian, or } (R, \theta, Z) \text{ cylindrical} \quad (17.10a)$$

$$\mathbf{r} = (x, y, z) \text{ Cartesian, or } (r, \theta, z) \text{ cylindrical} \quad (17.10b)$$

$$\mathbf{V} = (U, V, W) \text{ Cartesian, or } (V_r, V_\theta, V_z) \text{ cylindrical} \quad (17.10c)$$

$$\mathbf{v} = (u, v, w) \text{ Cartesian, or } (v_r, v_\theta, v_z) \text{ cylindrical} \quad (17.10d)$$

Also for each analysis here, the machine is assumed to remain in a steady-state condition, i.e., constant angular velocity of the rotor and constant fluid input/output flow rates. Typical velocity diagrams are shown in Figures 17.9a and 17.9b. An aircraft propellor is shown, but the illustration applies also to an axial-flow compressor as in a jet engine. The coordinate system is fixed in the aircraft, and $\mathbf{V}_1$ is airspeed in a wind tunnel or negative velocity of the aircraft in flight. For given engine rpm, ω is constant, and $\omega r \hat{\boldsymbol{\theta}}$ is the tangential velocity of a blade element at radius r.

All velocity diagrams make use of the relative velocity equation of Chapter 6,

$$\mathbf{V} = \mathbf{V}_0 + \boldsymbol{\omega} \times \mathbf{r} + \mathbf{v} \quad (6.2)$$

By selecting the origin of the rotating frame and the fixed frame on the rotation axis, $\mathbf{R}_0$ and $\mathbf{V}_0$ are zero, $\mathbf{r} = \mathbf{R}$, and

$$\mathbf{V} = \boldsymbol{\omega} \times \mathbf{r} + \mathbf{v} \quad (17.11)$$

with subscript 1 for the inlet condition and subscript 2 for the outlet condition. In Figure 17.9 the inlet velocity diagram (a) is prepared by first drawing in the known vectors $\mathbf{V}_1$ and $r\omega\hat{\boldsymbol{\theta}}$, and then adding in the unknown vector $\mathbf{v}_1$ to complete the triangle. The outlet diagram (b) is prepared by first drawing in the known quantities and then completing the triangle. When drawing the velocity diagrams, first draw in all known vectors, keeping track of how they join together as specified by (17.11), and then close the triangle to complete the analysis. Sometimes additional information is needed. The volumetric flow-through rate is often needed to deduce the output triangle. For example, in Figure 17.9 it is often reasonable to assume that air is incompressible, so that the axial-flow velocity at the front of the propellor blade ($-V_{z1}$) is equal to ($-V_{z2}$) at the rear of the propellor blade based on conservation of mass (also here $v_{z1} = v_{z2} = V_{z1} = V_{z2}$). When there is significant expansion or compression of the fluid between blade inlet to blade outlet, thermodynamic analyses are needed to relate these values.

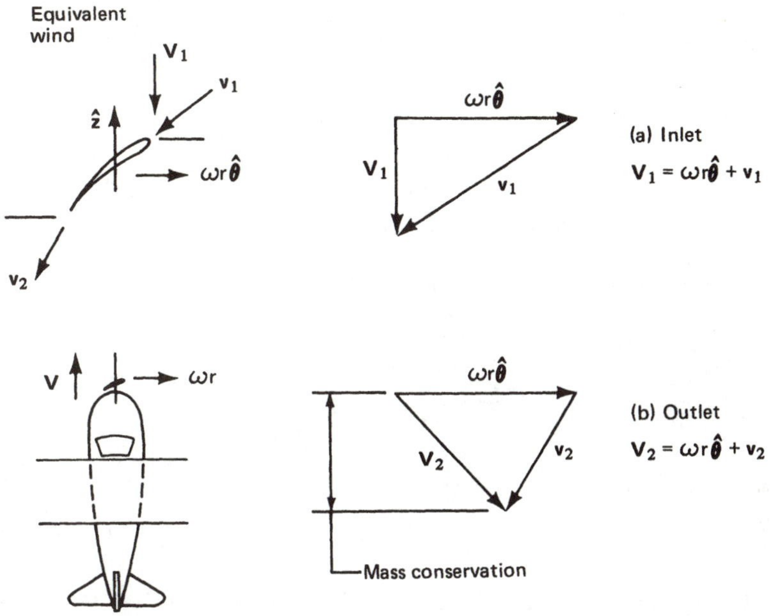

***Figure 17.9:** Aircraft propellor analyzed as an axial-flow pump, (a) inlet velocity diagram and (b) outlet velocity diagram.*

Remember that because $\boldsymbol{\omega} \times \mathbf{r}$ and $\mathbf{v}$ are added in (17.11), they are drawn head to tail, and $\mathbf{V}$ must be drawn to represent the sum of the two. The diagram can have two different appearing, but exactly equivalent, forms as in Figures 17.10a, 17.10b, and 17.10c. The inlet and outlet diagrams of Figure 17.9 are on the left (a) in Figure 17.10, with the alternative form (b) in the middle. Each represents (17.11) exactly. The forms on the left represent $\mathbf{V} = \boldsymbol{\omega} \times \mathbf{r} + \mathbf{v}$, and those in the middle represent $\mathbf{V} = \mathbf{v} + \boldsymbol{\omega} \times \mathbf{r}$. Some authors combine the inlet and outlet triangles as shown at (c); this may be useful for professionals, but it makes explanation more difficult.

The angles β_1 and β_2 in analyses of existing machines are simply the leading and trailing edge angles of an existing vane or propellor, and the inlet and outlet angles for $\mathbf{v}_1$ and $\mathbf{v}_2$ are chosen to match these to avoid turbulence and/or separation at the leading and trailing edges. However, in the design stage, these angles are not known and must be selected by the designer to match anticipated or desired performance criteria for the machine as in the following.

Consider the velocity triangles for a centrifugal pump for water. Figures 17.11a, 17.11b, and 17.11c illustrate a design analysis; see also Figure 17.5a. Assume a high-volume, low-pressure system is desired; impeller vanes are to be tapered from 15 mm thick at the vane inlet diameter of 10 cm to 8 mm thick at the vane outlet diameter of 20 cm; and a flow rate of 5 liters/s is needed. Inlet is axial and will be

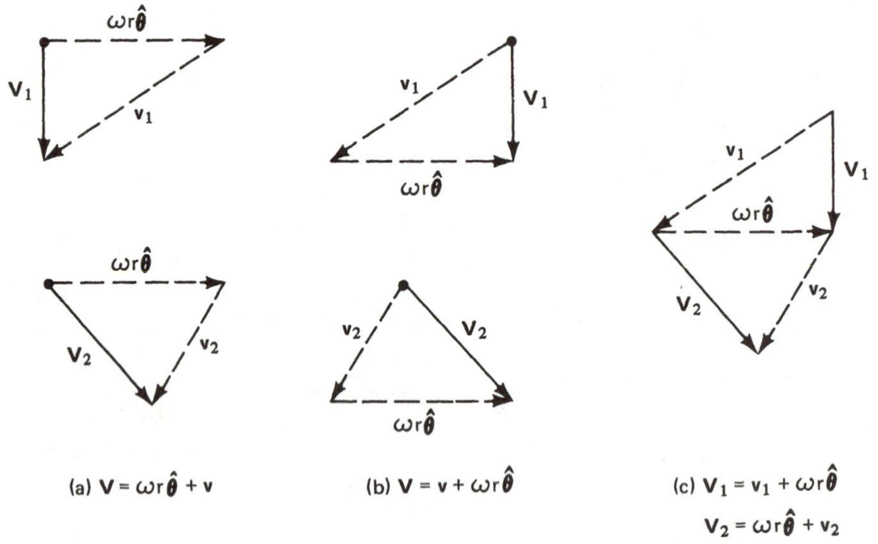

Figure 17.10: *Alternative forms of velocity diagrams. Any arrangement that preserves the vector relationship is correct, (a)* $V = \boldsymbol{\omega} \times \boldsymbol{r} + v$ *as in Figure 17.9, (b)* $V = v + \boldsymbol{\omega} \times \boldsymbol{r}$*, or (c) some combined form.*

deflected by a cone on the axis to reorient to a radial direction. Commercial availability and cost indicates an inlet pipe diameter of 8 cm to be practical, provided pipe flow Reynolds number does not exceed 10^5.

First, check whether the desired flow rate and pipe size provide an acceptable Reynolds number [all values in centimeter-gram-second (cgs) units].

$$\overline{V} = Q/A = 99.5 \text{ cm/s}$$
$$R_e = Ud/\nu = 8 \times 10^4, \text{ and is acceptable.}$$

This flow first passes through the 10 cm diameter by 1.5 cm thick inlet to the vanes and is moving radially. Therefore, $V_1 = V_{r1}\,\hat{r}$, with $V_{r1} = Q/A_1 = 106.1$ cm/s. The question now is, *What rotor angular velocity and blade angles are consistent with this inlet velocity?* The blade entry should be *shockless*, i.e., the leading edge should be tangent to $\mathbf{v}_1$. In Figure 17.11 an inlet velocity diagram is drawn (a) showing all known quantities. The diagram shows that either ω or β_1 can be estimated and the other computed. Try $\beta_1 = 65°$. Use ($\tan \beta_1 = \omega r_1/V_{r1}$) to get $\omega = 45.5$ rad/s = 435 rpm. Easily obtainable induction motors operate typically at 1728 rpm full load, and 4:1 gear reducers are readily available for $\omega = 432$ rpm. The values $\beta_1 = 65°$ with $\omega = 45.5$ rad/s is used for a first iteration. For later use, $\omega r_1 = 227.5$ cm/s, and $\omega r_2 = 455$ cm/s.

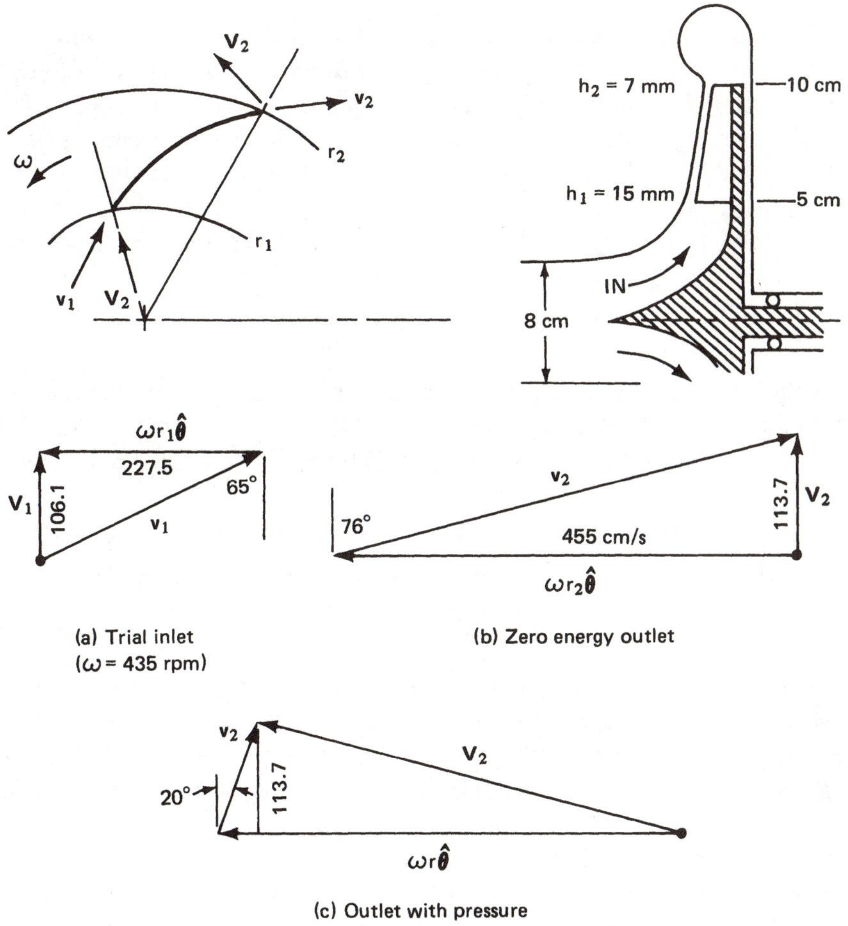

Figure 17.11: *Radial-flow (centrifugal) pump, (a) inlet velocity diagram, (b) outlet velocity diagram for zero energy transfer, and (c) outlet velocity diagram for positive pressure increase.*

The vanes must be steeper than the zero energy transfer slope of Case 1 of Section 17.2 in order for the machine to function as a pump at the given revolutions per minute. Calculate the exit angle for the zero transfer condition as a limit for β_2. For mass conservation, the radial component of $\mathbf{V}_2$ must be $V_{r2} = Q/A_2 = 113.7$ cm/s. In Figure 17.11 the outlet diagram (b) is drawn showing known and desired quantities, and their relationships. The relative velocity $\mathbf{v}_2$ is drawn tangent to the trailing edge for a smooth exit, and the absolute velocity $\mathbf{V}_2$ is drawn radially, i.e., for zero energy transfer the fluid is simply passing through the impeller without any interaction. For this case, $\beta_2 = \arctan(\omega r_2/V_{r2}) = 76°$, or an angle with the periphery of 14°.

No pressure is generated in the zero energy transfer design; the fluid moves as a line source flow between r_1 and r_2. In order for pressure to be generated the absolute velocity of the fluid must acquire a tangential component. As an iteration, try $\beta_2 = 20°$. Again have v_2 parallel to the trailing edge. Using the new outlet velocity diagram (c) with $\beta_2 = 20°$ and $V_{r2}=v_{r2} = 113.7$ cm/s, obtain $v_2 = v_{r2}/\cos\beta_2 = 121.0$ cm/s. The trigonometric relationship ($a^2 = b^2 + c^2 - 2bc \cos A$) provides $V_2^2 = v_2^2 + (\omega r_2)^2 - 2 v_2 \omega r_2 \cos 70°$ or $V_2 = 429$ cm/s. The tangential component of V_2 is

$$V_{\theta 2} = (V_2^2 - V_{r2}^2)^{1/2} = 414 \text{ cm/s}$$

Analysis of theoretical moment, power, or energy transfer to/from the fluid requires only ω, $\dot{m}$, R_1, R_2, $V_{\theta 1}$, and $V_{\theta 2}$ per (17.5), (17.6), and (17.8). The problem statement provides $\dot{m} = 5000$ g/s, i.e., 5 liters/s of water, and M_z, P, and E are

$$M_z = 5000(10 \times 414 - 5 \times 0) = 2.07 \times 10^7 \text{ dyne-cm} = 2.07 \text{ Nm}$$

$$P = M_z\, \omega = 2.07 \times 10^7 \times 45.5 = 9.42 \times 10^8 \text{ ergs/s} = 94.2 \text{ W}$$

$$E = P/\dot{m} = 2.12 \times 10^8/5000 = 1.884 \times 10^5 \text{ ergs/g}$$

If the output flow also flows through an 8 cm diameter pipe, then the energy equation, E = W of (17.8), simplifies to $\rho E = p_2 - p_1 = \Delta p$, and the pressure rise within the pump is $\Delta p = \rho E = 1.884 \times 10^5$ dyne/cm^2 $= 1.884 \times 10^4$ pascals = 2.73 lb/in^2, enough pressure to raise water 6.1 ft or about 2 m. Head can also be computed directly using (17.7).

All these values are marginal to justify the cost of the machine. Examination of the equations indicates increases of β_2 and ω are needed to achieve increases in Δp for the given flow rate and overall size of the machine. Practical values of β_2 are often smaller, i.e., the outlet vane tips have a radial direction like those in Figure 1.9 or are even tilted forward slightly. An increase in ω will require a shallower inlet vane angle (larger β_1). Increases in ω can be achieved better by preconditioning the inlet flow at r = 5 cm by adding vanes fixed to the stator that add whirl to the inlet fluid as in Figure 17.12a, or by continuing the impeller vanes to the inlet pipe diameter, e.g., see Figure 1.9.

A second iteration for the design using stator-induced inlet whirl, larger ω, and a radial output ($\beta_2 = 0°$) is summarized using the diagrams in Figures 17.12b and 17.12c. Let $\beta_0 = 60°$, $\beta_2 = 0°$, use a 2:1 gear reducer for the motor, and compute a shockless entry for β_1. Then $\omega = 1728/2 = 854$ rpm = 90.5 rad/s, and $\omega r_1 = 452.5$ cm/s. V_{r1} remains the same (106.1 cm/s) for the same Q and r_1, and $V_{\theta 1} = 106.1 \tan 60° = 184$ cm/s. From the triangle $v_{\theta 1} = (452.5 - 183.8) = 268.7$ cm/s, and $\beta_1 = \arctan(268.7/106.1) = 68.5°$. Note that the impeller inlet angle is even a little shallower than for the first iteration, but ω is twice as large. With a radial output,

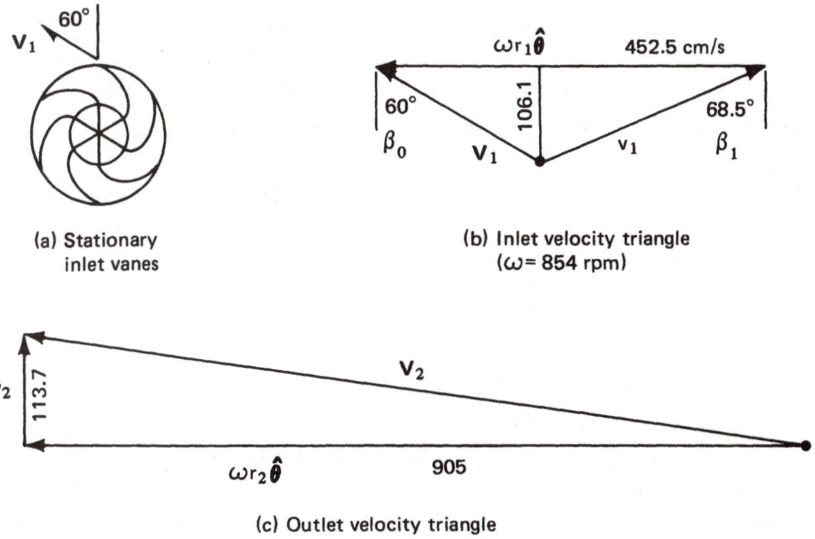

Figure 17.12: *Second iteration of Figure 17.11 pump showing (a) pre-whirl stator vanes, (b) inlet velocity diagram, and (c) outlet velocity diagram.*

$v_2 = v_{r2} = V_{r2} = 113.7$ cm/s from mass conservation as before, but now $V_{\theta 2} = \omega r_2 = 90.5 \times 10 = 905$ cm/s. For reference, $V_2 = 912$ cm/s at an angle of 7.2° from the tangential direction. This output flow direction nearly coincides with the volute direction and helps to reduce losses due to excessive turbulent motions in the volute.

Now recompute M_z, P, E, and Δp. The control volume now extends to the beginning of the stator pre-whirl vanes because that is now part of the analysis; however, at that point the flow is straight with no whirl. Accordingly

$$M_z = 5000(10 \times 905 - 0) = 4.53 \times 10^7 \text{ dyne-cm} = 4.53 \text{ Nm}$$
$$P = M_z \omega = 4.10 \times 10^9 = 410 \text{ W}$$
$$E = P/\dot{m} = 8.19 \times 10^5 \text{ ergs/g}$$
$$\Delta p = 8.19 \times 10^5 \text{ dyne/cm}^2 = 8.19 \times 10^4 \text{ pascals} = 118.8 \text{ lb/in}^2$$

Pressure can also be estimated as the difference in output fluid kinetic energy at r_2, with and without the impeller adding energy. Outlet velocity at r_2 based on mass conservation is $V_{r2} = 113.7$ cm/s, but because of the impeller, output velocity is $V_2 = 912$ cm/s. Assuming this energy is returned to pressure in the 8 cm diameter output pipe, the Bernoulli equation yields $\Delta p = (V_2^2 - V_{r2}^2)\rho / 2 = 8.25 \times 10^5$ dyne/cm^2.

These flow analyses provide β_0, β_1, and β_2, but not the intermediate vane contours nor the optimum number of vanes. The intermediate contours for an incompressible fluid are selected to have gradual or uniform changes in flow direction and pressure increases. This usually will minimize internal losses, impeller stresses, and also manufacturing cost. Lazarkiewicz and Troskolanski (1965) discuss impeller pump design features. The study and research projects in Section 17.6 provide additional references. When the fluid is compressible, e.g., a gas turbine or steam turbine, the internal energy of the fluid must be assessed as it expands or compresses while moving through the stator and rotor vanes. The analysis obviously is much more complicated and typically requires use of a computer for a useful design.

17.5 Performance Criteria

Performance criteria for turbomachines are measured using dimensionless numbers. All turbomachines have a characteristic diameter (D), an angular velocity (ω), a volumetric flow rate (Q), and an energy transfer per unit mass of fluid (E). These are functions only of the dimensions length and time, and therefore the four variables in two dimensions can be combined into two dimensionless variables. Several forms are possible; the two most useful are the flow coefficient (Q_c) and the head coefficient (H_c) defined by

$$Q_c = \frac{Q}{\omega D^3} \tag{17.12}$$

$$H_c = \frac{gH}{\omega^2 D^2} = \frac{E}{\omega^2 D^2} \tag{17.13}$$

where H is energy per unit weight and has units of length. The flow coefficient is often represented in professional literature by the letter (ϕ) and the head coefficient by the letter (ψ). This notation is not used here because of the possibility of confusion with the velocity potential and the stream function. Because Q_c and H_c are dimensionless, they can be combined in terms of powers of each to yield additional useful parameters. Study projects in Section 17.6 include references that present this theory more fully.

These concepts can be illustrated and applied better when explicit design and flow values are used in the nondimensional process. For example, Q can be made dimensionless by dividing by an area times a velocity, e.g., the output flow area (A_2) and the tangential speed of the rotor at r_2 (ωr_2) as defined in the last section. Then

$$Q_c = \frac{Q}{A_2(\omega r_2)} = \frac{V_{r2}}{\omega r_2} \tag{17.14}$$

and

$$H_c = \frac{gH}{(\omega r_2)^2} \tag{17.15}$$

Other useful dimensionless parameters include a power coefficient (P_c) where P is made dimensionless by $\rho Q(\omega r_2)2$

$$P_c = \frac{P}{\rho Q(\omega r_2)^2} \tag{17.16}$$

and a specific moment (M_s) made dimensionless with $\rho r_2 A_2 (\omega r_2)^2$

$$M_s = \frac{M_z}{\rho A_2 \omega^2 r_2^3} \tag{17.17}$$

Efficiency is the ratio of output to input power, is dimensionless, and is always less than one for real machines. Pump efficiency (η_p) is measured as the ratio of fluid power output (P_f) to mechanical power input (P_m) computed as

$$\eta_p = \frac{P_f}{P_m} = \frac{\rho g Q H}{P_m} = \frac{Q_c H_c}{M_s} \tag{17.18}$$

Turbine efficiency (η_t) is computed as

$$\eta_t = \frac{P_m}{P_f} = \frac{P_m}{\rho g Q H} = \frac{M_s}{Q_c H_c} \tag{17.19}$$

Specific speed (ω_s) represents rotor angular velocity when the machine is delivering unit volume flow rate at unit head. It is usually derived by combining powers of Q_c and H_c, and is given here in conventional form for a pump

$$\omega_s = \frac{Q_c^{1/2}}{H_c^{3/4}} = \frac{\omega Q^{1/2}}{(gH)^{3/4}} = \frac{\omega Q^{1/2}}{E^{3/4}} \tag{17.20}$$

For given Q and H, ω_s is proportional to ω. A specific diameter can be derived as

$$D_s = \frac{H_c^{1/4}}{Q_c^{1/2}} = \frac{D(gH)^{1/4}}{Q^{1/2}} \tag{17.21}$$

where again, for given Q and H, D_s is proportional to D.

Section 2.5 summarizes the advantages of using dimensionless variables over dimensional variables. Performance data and design criteria useful for any size, speed, flow rate, head, etc., of geometrically similar machines can be presented in a concise format, and inexpensive models can be tested to aid in the design of full-size machines. For example, various turbomachines are known to operate most effectively and efficiently within a limited range of specific speeds. Csanady (1964) notes that radial-flow centrifugal pumps are generally used in the range ($0.2 \leq \omega_s \leq 1.0$), while mixed- and axial-flow pumps span the range ($1.0 \leq \omega_s \leq 5.5$). Csanady uses D_2 rather than r_2 when computing specific speed. The second iteration for the pump designed in the last section, using Csanady's equation for specific speed, yields $\omega_s \approx 0.24$, which is within the lower normal range for a centrifugal pump. A Cordier diagram [again see Csanady (1964)] is a relationship between ω_s and D_s for effective and efficient operation. At $\omega_s = 0.24$ as obtained previously, D_s per Csanady's Cordier diagram should be about $D_s = 12$. After conversion to dimensional variables, the preferred D of 28 cm can be compared to D = 20 cm of the example. The reader may wish to redesign the pump by first selecting desired Q and H; next, iterating using Q_c, H_c, ω_s, and D_s for efficient pump performance; and then transforming to dimensional variables for internal flow calculations to complete the design process.

See the suggested projects for broader and more advanced discussions. When using published graphs and tables, note that authors sometimes present their data and the previous parameters in inconsistent, but commonly used, dimensional units, e.g., gallons per minute (gpm), head in feet of water, speed in revolutions per minute (rpm), pressure in pounds per square inch (psi), etc. Their presentations are useful provided the same units are used for comparison and/or design.

17.6 Study and Research Projects

The projects suggested here are intended for individuals or groups. Some can be completed in a week or two, others are term or longer projects. Only a few will have exact answers. References to other texts and to professional journals are included.

Project 17.1: Measure a propeller at a local airfield or contact a manufacturer for specifications. Compare forward velocity of the aircraft and propeller angular velocity, and examine relative velocities along the blade. Because the propeller accelerates the flow, the mean flow diameter behind the propellor should be smaller than that in front. See, e.g., Fox and McDonald (1992, Example Problem 11.12). However, air in the boundary layer on the propeller spins with the propeller, is subject to centrifugal "force," and should increase the flow diameter. What is happening? Integrate secondary and other flows in the vicinity of the propeller surface into the flow field outside the boundary layer. How should a propellor be shaped to give a forward lift force, comparable to cantilever beam strength, as a function of radius? For a variable pitch propellor, examine performance at take-off and during cruise. For additional references see Kückemann and Weber (1953), Shevell (1983), Anderson (1985), and Clark and Scott (1986).

Project 17.2: Analyze relative velocity, torque, and power output for an impulse turbine. Build a simple, single-stage impulse turbine and operate with water from a faucet. Design a nozzle to get maximum fluid velocity and design several different turbine blade shapes. Glue or bolt the blades (e.g., sheet metal or cast epoxy) to your turbine wheel. Calculate predicted torque and power as functions of fluid velocity and wheel speed, and compare with measured amounts. How can you improve performance while using less water? Does vorticity or secondary flow phenomena appear to have significant effects? If you are in a class, have a contest to see who can design the most efficient turbine blade shape.

Project 17.3: Centrifugal pumps are often designed for moderate to high pressure, and low to moderate flow rate, by staging several impeller/volutes in series, with outputs going to the inputs of the next. Design and build a centrifugal pump (impeller and volute) for low pressure (several pounds per square inch) and large flow rate (several gallons per second). As a suggestion, consider an eight-vane vertical-axis impeller 18 inch outside diameter and 5.5 inches high, with water flowing by gravity into its center through a 12 inch diameter hole. Students at the University of California, Santa Barbara, built a similar impeller into a rectangular water tunnel 18 inches high, 28 inches wide, and 6 feet long with a platform 6 inches above the bottom. The volute (scroll) fit between the platform and tank bottom, and varied in width from zero to a cross section equal to about the area of the inlet hole. The impeller and volute were bounded top and bottom by the platform and the tank bottom, and are at one end of the tank. Vanes at the far end deflected the water above the platform for return to the gravity-fed impeller. A honeycomb was placed after the vanes to straighten the flow and reduce large-scale turbulence. Vanes were added as necessary above the impeller to prevent formation of a vortex and cavitation. A one horsepower motor (with speed control) and a speed reducer to give angular velocities of zero to 90 rpm was adequate to give flow speeds over one foot per second in the test section. Make the platform and at least one wall from transparent plastic, and observe the interior flow at slow speed. Observe especially relative velocities, vortex formation, and secondary flows. Lazarkiewicz and Troskolanski (1965) and Lobanoff and Ross (1985) provide theory and design data for this and other commercial impeller pumps. For introductory references on turbomachinery see Logan (1981) and Balje (1981).

Project 17.4: Conduct a literature search and prepare a report on hydroelectric water turbines. Some countries obtain a major portion of their electric power from hydroelectric installations, e.g., Norway and New Zealand. In your library, search for the topics of water turbines, hydroelectric power, Tennessee Valley Authority (TVA), and Hoover Dam, among others. For references see Creager and Justin (1963), Olson (1980), Roluti (1985), and Inversion (1986). Also search relevant professional journals. These can be identified by examining the reference lists in the books and articles. The largest such turbines are mixed-flow devices with very sophisticated shapes and sometimes adjustable geometries to achieve very high efficiencies.

Project 17.5: Flow through channels are an important aspect of any pump or turbine. This includes, first, flow through stationary inlet and outlet passages, including transition portions designed to modify the flow field in some way. A second flow is through channels and passages that are rotating, and a third flow is through the active portions of the pump or turbine, e.g., the impeller or the rotating blades of a turbine. Apply the material of Chapter 14 to these flows and write a report on your research. For some specific references, see Csanady (1964), Langston et al. (1977), Herring et al. (1979), Tabakoff et al. (1980), Hamed and Baskharone (1980), Murakami et al. (1980), and Morris (1981). See Tritton (1978), Howard et al. (1980), and Bertoglio (1982) for discussions of turbulence in the rotating frame. Search more recent literature for additional references.

Project 17.6: The vortex and secondary flows that occur within a turbomachine generally represent lost energy, whether the machine is operating as a pump or a turbine. Analysis of these loss mechanisms and the implementation of design features to reduce them is an important area of research and development. Csanady (1964) reviews the topic in his Chapters 7 and 8. Some of the many journal references include Langston (1980), Stenning (1980), Bario et al. (1982), Gregory-Smith (1982), Sieverding and Van Den Bosche (1983), Moore and Adhye (1985), Sieverding (1985), and Chen and Dixon (1986). Can you suggest design modifications to reduce some of these losses or to reorient the flow to reclaim the energy? Particulate laden flows, whether of steam that contains condensed water or gases that contain combustion products or random particles, can cause severe damage to high-speed parts and increase any other losses. See Tabakoff et al. (1982).

Project 17.7: Thermodynamics and compressibility are distinguishing features of steam and gas turbines. Steam turbines drive the generators in all large fossil fuel and nuclear power plants, and power all large ocean-going vessels. Gas turbines, as jet engines, power almost all modern commercial aircraft. Gas turbines also provide turbochargers for automobiles and power for auxiliary generators. Power and efficiency are directly related to operating temperature, and a great deal of research is directed to developing materials capable of maintaining strength at high temperatures. Efficiency and power can also be improved by reducing flow losses. Select and investigate an application of rotating flow theory to improve efficiency of a gas turbine (or jet) of interest. Introductory books on gas turbines include Cohen et al. (1973), Boyce (1982), Dixon (1984), and IMechE (1986).

Project 17.8: Gas turbines typically are light compared to their power output, making them excellent for aircraft applications. Steam turbines (and their boilers) are much more massive, but their efficiency makes them ideal for power generation and for ship propulsion. Steam power and steam turbines were a major part of most mechanical engineering programs in the 1940s and 1950s with the advent of high-temperature (~1000°F) and high-pressure (~1000 lb/in^2) designs. Academic programs now often emphasize gas turbines, but steam turbines continue to represent a major and important role in commerce. As a study project summarize the critical design features affecting flow through steam turbine stator/rotor wheel assemblies. Church (1950) is a classic text. More modern reviews (conferences) include IMechE (1979), Bellanca (1990), and IMechE (1990).

Project 17.9: Most intermediate texts on fluid mechanics have chapters on, and assign projects on, the use of computers to solve for, and graphically display, solutions to flow problems. The inherent complexity of flows of real fluids in real applications restrict such examples to a limited number of classic solutions to carefully restricted flows. However, realistic solutions to real flow problems are also feasible and have been routinely achieved, starting in the late 1970s using supercomputers. See Litt et al. (1989). For this project write a report on recent successes in the use of supercomputers to analyze internal flow details of turbomachines. Your research should include a literature search in a technical university library, or equivalent, and telephone and/or mail requests for information from turbomachinery research and development organizations (e.g., Pratt & Whitney and General Electric) and also from manufacturers of supercomputers (e.g., Cray). Discuss your research interest with the library's technical reference librarian. It will be helpful to examine relevant journals (e.g., *Journal of Engineering for Gas Turbines and Power*, *International Journal of Computers and Fluids*, and *Journal of Fluids Engineering*) for the last decade. Locate relevant articles and use their lists of references to expand your research. Also use the texts and articles referenced in Projects 17.1 through 17.8 and their references. You will note that many engineering studies are found only in industrial/government reports (e.g., NASA) and in conference proceedings. These are often in large libraries or may be available from the sponsoring organization.

Project 17.10: Create a new type of turbomachine or an improvement to an existing design. Your concept may have commercial utility, in which case you should very cautiously consider patenting it, selling or licensing the manufacturing and sales rights, and/or producing it yourself. Consider using residual or waste energy and consider applications to a wide variety of fields (e.g., medicine, home utility, toys, etc.).

Chapter 18

Liquids in Precessing Spacecraft

18.1 Rigid-Body Dynamics

The equations of motion of a rigid object can be separated into three-dimensional (3-D) translation of the mass center and 3-D rotation about the mass center. When formulated in this manner, problems of rotation can be analyzed independent of translation. Difficulties arise when analyzing nonrigid objects, such as a flexible space structure, or an object containing nonrigid components, such as fluids. Some of these difficulties are introduced in earlier sections, especially the Chapter 11 discussion of secondary flows and spin-up. See also Example 7.4.

Figure 18.1 shows a rigid object, having a spherical cavity filled with liquid, and spinning about an inertially fixed line passing through the cavity center. Assume spin-up has occurred, and that the object and the liquid are spinning at constant rate ω. In this mode, the angular momentum **H**, and the rotational kinetic energy E, of both the object and the liquid are constant also. Next, assume the rotation vector of the object is impulsively changed to a new fixed direction. The liquid mass will attempt to continue rotating in its original direction. It will change only after interaction with the cavity wall imparts the new motion of the cavity wall.

In the case of a spherical cavity, the cavity wall has only tangential motion relative to the liquid so viscous stresses provide the only interaction forces. Stress on the liquid can be computed as the product of the liquid viscosity and the velocity gradient normal to the wall. Secondary flows are initiated that transfer wall momentum to the fluid, until finally the fluid spins in the same direction as the cavity. The problem is an extension of the basic spin-up problem of Chapter 11 where both the fluid and the cavity were initially at rest, and then the cavity was impulsively accelerated to ω^c = constant. During fluid spin-up, secondary flows were created within the fluid, and no unique ω^f could be defined. Because of the no-slip condition, the fluid adjacent to the wall had ω^c = constant, and until spin-up was

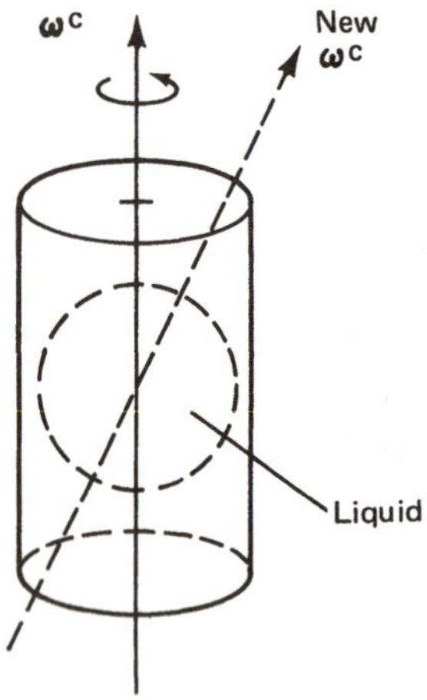

Figure 18.1: *Spinning object with enclosed liquid.*

complete, there was a region at the axis where $\omega^f = 0$. If the cavity and fluid have already reached some uniform angular velocity, and the magnitude of $\boldsymbol{\omega}^c$ is impulsively changed, the problem is similar. Secondary flows are again initiated until finally the central portion of the fluid also has the new $\boldsymbol{\omega}^c$.

This new problem is different in that the direction, rather than the magnitude of ω^c, changes. The angular velocity of the cavity relative to the fluid (ω^{cf}), with the fluid assumed rotating as a rigid sphere, is obtained using ($\omega^{cn} = \omega^{cf} + \omega^{fn}$), where the superscript n refers to Newtonian (inertial) space,

$$\boldsymbol{\omega}^{cf} = \boldsymbol{\omega}^{cn} - \boldsymbol{\omega}^{fn} \tag{18.1}$$

The cavity angular velocity relative to inertial space (ω^{cn}) then remains constant, but there is relative motion within the liquid. Neither $\boldsymbol{\omega}^{fn}$ nor $\boldsymbol{\omega}^{cf}$ can now be defined because the liquid is no longer in a rigid-body rotation mode.

This problem is a common occurrence when maneuvering space vehicles that are already spinning. Some are rotated to provide a type of gyroscopic stability; some, such as the Apollo vehicles, are slowly rotated to prevent the solar side from overheating; and others rotate so they keep a specific side directed toward the object they are orbiting about. For example, if a satellite orbits the earth once every two hours, it must also rotate once every two hours if it is to keep a camera always pointed toward the earth. All such vehicles carry large amounts of liquid fuel for making orbital and attitude corrections, with the result that each vehicle attitude correction requires a new spin-axis direction for the liquid fuel also. Orbital corrections move the mass center of the vehicle to a new position in space, and if a tank is only partly filled, the correction induces transverse oscillations in the liquid. Sloshing induced by translation is covered in other references. See especially Abramson (1966).

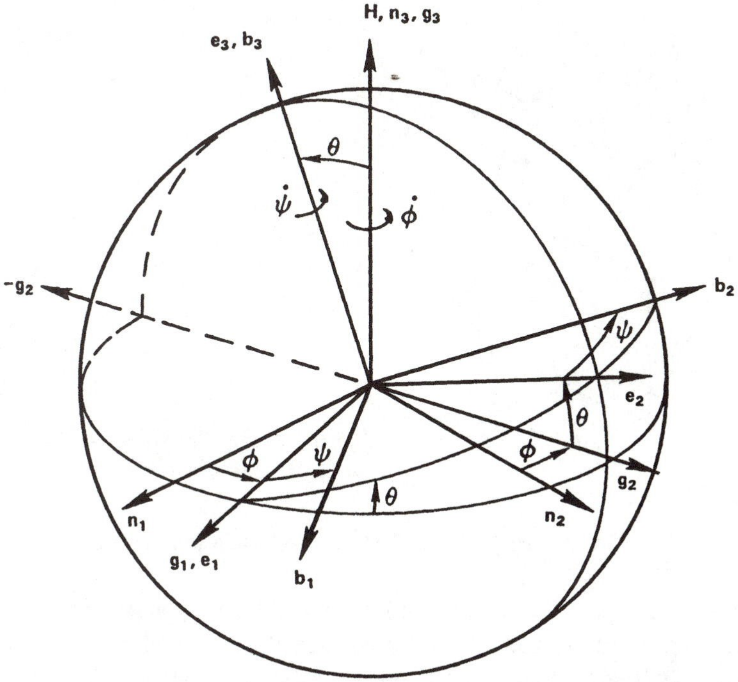

Figure 18.2: Euler angle coordinate system (Vanyo and Likins 1972).

A more general, more complex, and more serious problem of rotating space vehicles containing liquids is related to the *natural* or *free* precession of the vehicle. In this motion, the direction of the spin axis of the vehicle (and cavity) *continuously* changes direction. Understanding the phenomenon requires a brief review of rigid-body dynamics. Assume an inertially axisymmetric rigid body with moment of inertia I_3 about its spin axis and with moment of inertia I_1 about a transverse axis through its mass center. Its general motion in torque-free space is spin about the 3 axis at a rate $\dot{\psi}$, while the 3 axis rotates in a conical path about an inertially fixed line passing through the mass center. Euler angle coordinates (ϕ, θ, ψ) are shown in Figure 18.2 and sketches of the motion in Figures 18.3 and 18.4. The half coning angle is θ, and the rate at which the 3 axis cones about the inertially fixed direction is $\dot{\phi}$. Depending upon the author, this motion is called *precession*, *nutation*, and/or *wobble*. Because the object was assumed torque free, its angular momentum (**H**) is a constant and can be represented by an inertially fixed axis as shown. Figure 18.3 shows the *prograde* ($\dot{\phi}\dot{\psi} > 0$) motion of an inertially *prolate* ($I_1 > I_3$) object, and Figure 18.4 shows the *retrograde* ($\dot{\phi}\dot{\psi} < 0$) motion of an inertially *oblate* ($I_3 > I_1$) object. The oblate object is in a minimum energy condition when rotating about I_3, and the prolate object is in a minimum energy condition when rotating about I_1. In each case, rotation about the maximum moment of inertia is the minimum energy condition and

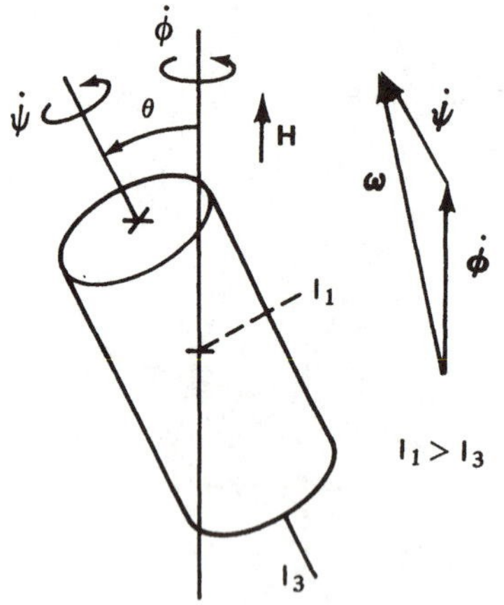

Figure 18.3: Prolate object and prograde motion.

is the end condition for an energy dissipating system. Example 7.4 provides additional details.

Analysis is augmented by use of the Euler variables ϕ, θ, and ψ, which provide a useful method for describing an arbitrary orientation of one 3-D reference frame relative to another 3-D frame. Figure 18.2 illustrates these variables. Assume first an inertially fixed frame n represented by the axes (n_1, n_2, n_3). Next, assume the g frame is initially coincident with the n frame but is then rotated a positive angle ϕ about the n_3 axis. The n_3 and g_3 axes continue to coincide after this simple 1-D rotation. Now assume an e frame, originally coincident with g but shown after rotated an angle θ about the g_1 axis. This is the second 1-D rotation, here with the g_1 and e_1 axes parallel (coincident). A third 1-D rotation displaces the b frame an angle ψ relative to e with e_3 and b_3 the parallel axes. Euler proved that these three rotations are adequate to express the 3-D angular displacement of any reference frame relative to any other reference frame, here the b frame relative to the n frame. If the b frame represents a solid body and n an inertial frame, with (n_1, n_2, n_3) and (b_1, b_2, b_3) specified, then the orientation of b relative to n is completely defined given (ϕ, θ, ψ). The rate of change of this orientation is the angular velocity of b relative to n, which is completely specified by $(\dot{\phi}, \dot{\theta}, \dot{\psi})$ as follows:

$$\boldsymbol{\omega}^{bn} = \boldsymbol{\omega}^{be} + \boldsymbol{\omega}^{eg} + \boldsymbol{\omega}^{gn} \tag{18.2}$$

$$\boldsymbol{\omega}^{bn} = \dot{\psi}\,\hat{e}_3 + \dot{\theta}\,\hat{g}_1 + \dot{\phi}\,\hat{n}_3 \tag{18.3}$$

Direction cosine matrices (see Example 6.1 and Appendix A) can then be constructed for converting each term to a common reference frame. This yields $\boldsymbol{\omega}^{bn}$ in components in the desired reference frame (n, g, e, or b).

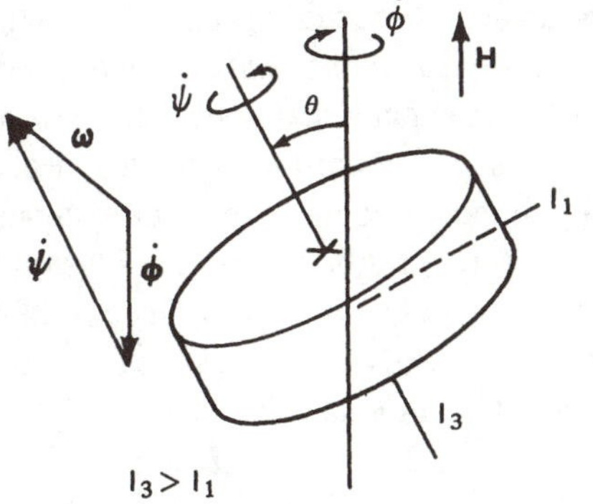

***Figure 18.4:** Oblate object and retrograde motion.*

If these details are applied to the solution of a torque-free, inertially axisymmetric rigid body, and the equations simplified, the following result becomes apparent:

$$\dot{\psi} = \dot{\phi}\left(\frac{I_1}{I_3} - 1\right)\cos\theta \tag{18.4}$$

with $\dot{\phi}, \dot{\psi}$, and θ constants. The conclusion is that a torque-free, inertially axisymmetric rigid body can *only* have motions described by (18.4).

Consider a thrown football (American version). An expert player throws the ball so it spins about its long axis with imperceptible wobble. This represents the case $\theta \sim 0$, and $\dot{\phi}$ cannot be distinguished from $\dot{\psi}$. In this case, ω^{bn} is a vector with fixed direction, and it coincides with the direction of **H**. Most amateurs who throw a football cause it to have a distinctive wobbling motion as it moves through the air. In this case, θ has an easily observed value, the football spins at $\dot{\psi}$ about its long axis, and the long axis cones about a fixed direction at the rate $\dot{\phi}$ as in Figure 18.3. The $\dot{\psi}$ motion can be made more apparent if the ball is painted with longitudinal stripes. Note in this mode that ω^{bn} also wobbles; only **H** is constant. A sphere has $I_1 = I_3$ and (18.4) becomes indeterminate. This indicates that a sphere cannot wobble. Therefore, baseballs, golf balls, tennis balls, etc., can only spin about an inertially fixed direction (if torque free of course). In practice, air drag (torque) modifies these results.

18.2 Energy Dissipation and Stability

Most spacecraft are nearly rigid, nearly inertially axisymmetric (by design), and operate in a nearly torque-free space, and therefore have motions approximated closely by (18.4) over useful time scales. The interest here is the matter of useful time scales and the fact that an inertially prolate body spinning with $\theta \sim 0$ is in a state of *maximum* rotational kinetic energy. If the body is slightly nonrigid and $\theta > 0$, it will dissipate energy internally due to material hysteresis and/or viscous dissipation. The rotational kinetic energy of the body, a function of θ, is dissipated into heat energy and radiated away. When θ reaches 90°, the body is spinning about a major axis of inertia and is in its *minimum* energy state. It is then in a stable condition of pure spin about a transverse axis. Note that an oblate body spinning about its axis of symmetry is already in its minimum energy configuration. Any perturbations here will damp out if energy dissipation occurs. It will return to spin about its axis of symmetry--although the direction of its spin axis may have changed relative to inertial space, since the perturbation may represent a small impulsive external torque.

Rotational kinetic energy of a torque-free rigid body as a function of its half coning angle θ is derived as

$$E = \frac{1}{2} I_3 \, \omega_0^2 \, (\sigma \sin^2\theta + \cos^2\theta) \tag{18.5}$$

where ω_0 is its initial spin rate about the I_3 axis with $\theta = 0$, and σ is the ratio I_3/I_1. Auxiliary equations provide the proper division of ω_0 to $\dot{\phi}$ and $\dot{\psi}$ as follows:

$$\dot{\phi} = \sigma \omega_0 \tag{18.6}$$

$$\dot{\psi} = (1 - \sigma)\, \omega_0 \cos\theta \tag{18.7}$$

In (18.5), I_3, σ, and ω_0 are constants. Therefore, E is only a function of θ, so that if $\dot{E}$ occurs,

$$\dot{E} = (\sigma - 1) I_3 \, \omega_0^2 \, (\cos\theta \sin\theta)\, \dot{\theta} \tag{18.8}$$

The critical question is, *Given $\theta > 0$, is the system unstable so that θ continues to grow?* The answer to this is, *Yes, θ for a prolate object will grow if $\dot{E} \neq 0$.* This leads to the more precise question, *Will θ remain small for a long enough time for the space vehicle to perform a useful function?* This is resolved by examining (18.8) for the limit of θ small. Equation (18.8) can then be set in the form

$$\frac{\dot{\theta}}{\theta} = \frac{(\dot{E} / \theta^2)}{(\sigma - 1) I_3 \omega_0^2} \tag{18.9}$$

If $\dot{E}/\theta^2 \approx$ constant, as can be shown to often be approximately true for θ small, then (18.9) can be integrated to a form

$$\theta = \theta_0 e^{t/\tau} \tag{18.10}$$

with τ a characteristic time given by

$$\tau = (\sigma - 1)I_3\omega_0^2 / (\dot{E} / \theta^2) \tag{18.11}$$

The question now reduces to, *Is τ large enough, i.e., will θ remain small for a long enough time to complete that phase of the mission?*

18.3 Solution: Analysis and Computation

Three solution techniques are generally available for consideration for any physical problem including this one, (1) analysis, (2) numerical computation, and (3) experimentation. Analysis and numerical (computer) methods have been partially successful in a limited number of these types of problems, especially when the cavity is completely filled and of simple shape (spheres, ellipsoids, and circular cylinders). Equation (18.4) provides the motion of the cavity wall as boundary condition. Response (usually V) of the internal fluid is the solution. Steady state or at least oscillatory modes are typically obtained. Usually, however, approximations must be made that limit the utility of the solution to specific ranges of Reynolds number or other criteria.

Persons, motivated to understand motions in the earth's liquid core as affected by forced- and free-precession of the mantle, have attempted to solve the full Navier-Stokes equations, even including magnetohydrodynamic effects. These attempts are briefly reviewed in the next chapter. They tend to be subject to two main criticisms, (1) too many restricting assumptions and conditions, and (2) inaccessibility of the basic problem for input data and/or confirmation of the solution.

Persons, motivated to design a space vehicle and its mission, are usually under schedule pressures to obtain approximate, or at least useful, performance data. They also make assumptions, but unlike persons analyzing the earth's core, they are plagued by the knowledge that their estimates all too soon will be compared to reality. Inability to predict the effect of internal energy-dissipation mechanisms (especially liquids) has been one cause of satellite mission failure, beginning with a failure of the first U.S. satellite and continuing up to the most sophisticated communication satellites.

An early analytical approach had used Stokes's solution for an oscillating flat plate. This approach wrapped the "flat plate" around the fluid to the shape of the cavity wall and used a component of ω^{bn}, often $\dot{\phi}$, as the frequency of oscillation. In this analysis, all $\dot{E}$ occurs in the boundary layer. The ease with which the method gave answers often tempted analysts to apply it to inappropriate parameter regions. Sloshing modes in partially filled containers have sometimes been approximated as one or more oscillating rigid masses and/or pendulums with modest success. Following the idea of replacing the fluid with an equivalent rigid mass viscously coupled to the cavity wall, another approach, valid for a filled sphere, replaced the interior sphere of liquid by a rigid spherical mass with the same density as the fluid, and coupled it to the cavity wall by a boundary layer of the given fluid. Motion of the cavity wall was stated as (18.4). The transient and steady-state (limiting) motions of the interior rigid sphere were obtained in closed analytical form. See Vanyo and Likins (1972).

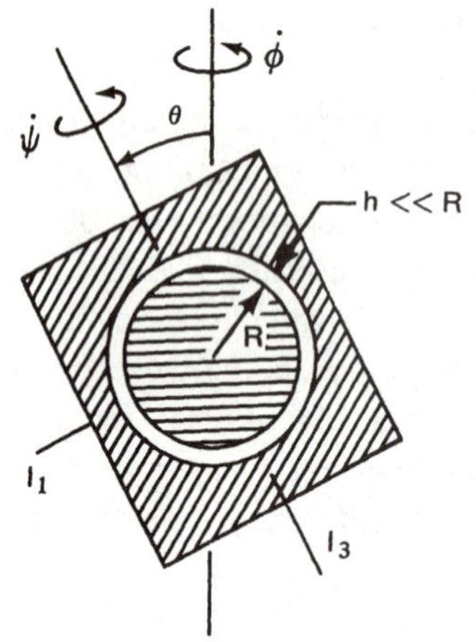

Figure 18.5: *Precessing rigid sphere used as a model for net fluid motion.*

Figure 18.5 shows the rigid sphere model. The derivation assumes the cavity is given an impulsive acceleration to an arbitrary set of values of $\dot{\phi}$, $\dot{\psi}$, and θ represented by $\boldsymbol{\omega}^c$, with the angular velocity of the rigid interior sphere ($\boldsymbol{\omega}^s$) the solution. If $h << R$, the interaction between the cavity and sphere can be approximated as Couette flow. After integrating over the spherical surface, the moment that the cavity applies to the sphere is

$$\mathbf{M}^{cs} = \frac{8\pi\mu R^4}{3h}(\boldsymbol{\omega}^c - \boldsymbol{\omega}^s) \tag{18.12}$$

The sphere becomes a spinning rigid body with motion described by

$$\mathbf{M}^{cs} = \dot{\mathbf{H}}^s = \mathrm{I}\dot{\boldsymbol{\omega}}^s \tag{18.13}$$

The two expressions for $\mathbf{M}^{cs}$ are equated yielding a vector first order differential equation in the variable $\boldsymbol{\omega}^s$. After expansion to three scalar equations, solutions are obtained for ω_1^s, ω_2^s, and ω_3^s in g-frame components. These yield the solution $\boldsymbol{\omega}^s(t)$, which, in the limit as $t \rightarrow \infty$, gives a steady-state solution. Energy dissipation is given as

$$\dot{E} = \mathbf{M}^{cs} \cdot (\boldsymbol{\omega}^c - \boldsymbol{\omega}^s) = \frac{8\pi\mu R^4}{3h(1+\zeta^2)}\dot{\psi}^2 \sin^2\theta \tag{18.14}$$

where $\zeta = 5\nu/hR\dot{\phi}$, and h is approximated by $h = [2\nu/(\dot{\phi} + \dot{\psi}\cos\theta)]^{1/2}$. This solution was verified to be accurate within several percent by laboratory $\dot{E}$ experiments over a parameter range typical of many space vehicle motions. Details of these and other approaches are explored in the Section 18.5 projects.

No analyses, applicable to the complexity of real space vehicle tank designs, have been successful in the absence of experimental or flight data. The same result has generally been true for numerical attempts. Most of the difficulty has been the presence of fluid instabilities (including inertia wave phenomena), transition to turbulence, and/or the churning effect of baffles and other propellant management devices.

18.4 Solution: Scaled Experiments

Consider an actual satellite of the early 1980s typical of the complexity of a real problem and its economic consequences. A communication satellite (Figure 1.11) with a value of $100,000,000 was to be placed in orbit. It was to be placed first in a near-earth circular orbit and then boosted along an elliptical path to arrive at its final 22,300 mile high geosynchronous orbit. During transition from its low circular orbit to its elliptical transfer orbit, it was to be spun at 40 rpm about its long axis (minimum axis of inertia). The ratio of its inertias was $I_1/I_3 = 5.62$ based on the rigid portion of the vehicle; however, at that time the vehicle was to be 55% liquid by mass. The liquids (fuel and oxidizer) were in two spherical tanks, each tank 0.84 m diameter and 83% filled with liquid. Each tank also contained an assembly of thin vanes whose purpose was to feed liquids to the outlet ports, using surface tension, when in its final (zero-g) orbit.

Figure 18.6a is a schematic drawing of the vehicle, and Figure 18.6b shows fuel location and vanes while the vehicle spins in zero-g. Energy is inversely proportional to moment of inertia for a given angular momentum. If $I_1/I_3 = 5.62$, then $E_3 = 5.62\ E_1$, and an unusually large amount of energy has to be dissipated by the liquids before the vehicle can tumble to spin about its transverse axis. Assume an estimate of $\tau \gtrsim 1$ hour as acceptable for the transfer orbit condition, i.e., given some small value of θ at orbit inception, the value of θ must remain moderately small for 1 hour.

Two professional organizations completed numerical predictions. One estimated $\dot{E}$ using a spherical boundary layer (related to the Stokes's solution) and computed $\tau \sim$ days. The other estimated $\dot{E}$ based on the full Navier-Stokes equations but did not fully include turbulence and/or fluid churning. They estimated $\tau \sim$ hours. Both predictions indicated mission success. Satellite design and mission plan proceeded on that basis.

A decision was made to check the predicted behavior of the satellite with a simple test. A scale model of the vehicle, using 5 inch diameter tanks to model the 0.84 m diameter tanks of the real vehicle, with matching Reynolds numbers and other

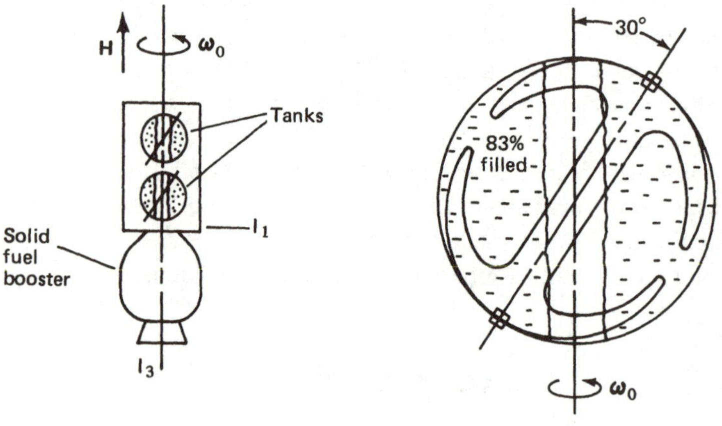

(a) Satellite and booster (b) Tank details

Figure 18.6: *(a) Communication satellite with booster rocket and liquid fuel and liquid oxidizer tanks. (b) Spinning fuel tank in zero-g with propellant management vanes.*

parameters, was spun up to an appropriate angular velocity and dropped from a one-story-high building. During its one second free fall, internal energy dissipation caused its coning motion (θ) to increase. With this estimate of $\dot{\theta}$, it was possible to estimate $\dot{E}$ using (18.9), and finally to estimate τ using (18.11). The result was a predicted τ for the real vehicle of $\tau \sim 1$ min. This value was confirmed using an apparatus that measured $\dot{E}$ as the electric power necessary to maintain $\dot{\phi}$, $\dot{\psi}$, and θ constant for the 5 inch diameter scale model tanks under conditions that scaled the laboratory results to the predicted flight conditions.

The satellite manufacturer accepted the experimental results, modified the mission plan to anticipate a rapid tumble to spin about a major axis (called *flat spin*), and as a space first, provided means to recover from this mode to complete a successful orbit insertion. Telemetry data from the actual flight indicated flat spin was achieved in 5 minutes. Actual τ was within 20% of both the drop test τ and the laboratory $\dot{E}$ test for τ. The best numerical estimate was off by a *factor* of 20 and the other by a *factor* of 2000. As a positive statement, the success in recovering from the predicted flat-spin condition quickly set the stage for deliberately planning for this mode of operation on many succeeding flights.

Figure 1.12 shows a later (1992) communication satellite attached to its last stage solid-fuel booster during the prelaunch assembly procedure. Fuel tanks in this satellite were designed to avoid the high energy-dissipation rates of the Figure 1.11 satellite. To verify the design, the manufacturer sponsored $\dot{E}$ laboratory tests to predict the stability/instability of the satellite during the spin portion of the launch trajectory.

Measured $\dot{E}$ were very small for the anticipated manuevers, indicating stability, i.e., flat spin would not occur. Telemetry data during launch verified the prediction.

These examples show that caution must be used in all fluid analyses, whether rotating or not. Some observational or experimental evidence is always needed before risking a great deal solely on calculated predictions. Even with experiments, there are difficulties in scaling from known data to anticipated phenomena.

Continuing with the example, energy dissipation $\dot{E}$ is a function of the variables

$$\dot{E} = \dot{E}(t, \omega_o, R, \rho, \mu, g, \theta, s, B, G) \tag{18.15}$$

where t = time, ω_o = initial angular velocity, R = tank radius, ρ = fluid density, μ = fluid viscosity, g = gravity and/or acceleration, θ = half coning angle, s = surface tension, B = a vane deflection parameter, and G represents the geometry of the tank and vanes. Because there are three physical dimensions, mass, length, and time, these can be reduced to a set of dimensionless parameters three less in number $(10 - 3 = 7)$

$$E/\rho\omega_o^3R^5 = f(\omega_o t\,, \omega_o R^2\rho/\mu\,, \omega_o^2 R/g\,, \theta\,, s/\rho g R^2, B/\rho R^6\omega_o^2\,, G) \tag{18.16}$$

where the left-hand side is dimensionless energy dissipation, and the quantities in the parentheses are, respectively, dimensionless time, Reynolds number (R_e), Froude number (F_r), half coning angle, Bond number, a vane stiffness parameter, and a parameter representing the geometry.

Progress was furthered by assuming a steady-state solution existed, by neglecting surface tension relative to other phenomena, matching the angle θ, and satisfying G by building the scale models to be geometrically similar to the vehicle tanks. The assumptions on surface tension and steady state were substantiated as approximately correct by comparison to known analytical solutions that had been verified by experiments and/or observations. The result, where f represents dimensionless energy dissipation and K the stiffness parameter, is

$$f = f(R_e, F_r, K) \tag{18.17}$$

Three variables (ρ, μ, ω_o) remained that could be varied to scale laboratory (R_e, F_r, K) to actual vehicle (R_e, F_r, K).

In this application, no formal or explicit use was made of rotating fluid theory. However, knowledge of that theory aided in the selection and understanding of the dimensionless parameters used to scale the experimental results. It may be that lack of use of the relevant theory, and its implications, were contributing factors in the inaccuracy of the numerical predictions. Presumably, if ever it is possible to predict such results analytically or numerically, the methods will include all significant rotating fluid phenomena. Further details and additional readings are included in Section 18.5.

18.5 Study and Research Projects

The projects suggested here are intended for individuals or groups. Some can be completed in a week or two, others are term or longer projects. Only a few will have exact answers. References to other texts and to professional journals are included.

Project 18.1: Chapter 6 illustrated a technique for transforming vector equations from an inertial (Newtonian) reference frame to a reference frame rotating at constant rate and direction. More generally, a reference frame might be rotating with variable rate and/or direction. Using the coordinate systems and notation of Figure 18.2, place the equation $\mathbf{F} = 12\hat{\mathbf{n}}_3$ successively in the g, e, and b reference frames. Try it next with the equation $\rho[\partial\mathbf{V}/\partial t + (\mathbf{V} \cdot \nabla)\,\mathbf{V}] = \mathbf{f} + \rho\mathbf{g} - \nabla p + \mu\nabla^2\mathbf{V}$, with $\mathbf{f}$ any forces not otherwise included. Refer to Appendix A.3 for help with direction cosine matrices. See also Likins (1973).

Project 18.2: Derive (18.4) that relates $\dot{\phi}$, $\dot{\psi}$, and θ for an inertially axisymmetric object with axial and transverse moments of inertia I_3 and I_1. This scalar equation is obtained by carefully examining the vector equations of motion of an arbitrary rigid object. The solution is outlined in Likins (1973), using the notation of this text, in Goldstein (1965), and in Thomson (1963). Note that the Thomson text uses an older engineering convention where $\dot{\phi}$ and $\dot{\psi}$ are exactly reversed in meaning. Vanyo (1973) outlines a derivation where the equation $\mathbf{M} = \dot{\mathbf{H}}$ is placed in the e frame, and the question is asked, *What is a necessary condition for M = 0?* Show that, when (18.4) is satisfied, the dynamic unbalance (i.e., moment) of a rigid object with $\dot{\phi}$, θ, I_3, and I_1 as given, but with $\dot{\psi} = 0$, is exactly cancelled by the gyroscopic moment due to forced precession ($\dot{\phi}$) of the spinning object with $\dot{\psi}$, θ, I_3, and I_1 as given.

Project 18.3: Derive (18.14) for the motion of a rigid sphere, inside and viscously coupled to, a precessing spherical cavity. The solution is outlined in more detail in Vanyo and Likins (1972). The same paper outlines a computer solution for a similar, but more complicated, problem of a series of n concentric shells nested one within the other, with the outside (largest) shell having a forced motion given by (18.4). The solution for 50 shells using FORTRAN, matrix inversion, and the University of California, Los Angeles, mainframe (IBM 7094) in 1967 took 15 minutes of central processing unit (CPU) time. How long does your desktop take today? Note that secondary flow effects are neglected in the solution approximation. Can you modify the solution to include such effects?

Project 18.4: Consider a computer numerical solution of the Navier-Stokes equations for motion within a constant density and viscosity liquid that fills a spherical cavity whose motion is given by (18.4). As of early 1993, the problem has not been solved. Start with a literature search using two techniques: (1) a computer search of the literature using key words, and (2) by identifying the journals most likely to publish the subject and reading their table of contents. Start with the most recent issue and go back at least 10 years. The two techniques never produce the same list of relevant publications. For this project examine initially the *Journal of Fluid Mechanics* and relevant AIAA and ASME journals (American Institute of Aeronautics and Astronautics; American Society of Mechanical Engineers).

Project 18.5: One early attempt at estimating $\dot{E}$ in a liquid-filled, precessing, spherical cavity made use of Stokes's second solution (see Section 4.1). Those analysts assumed that precession was similar to a lateral oscillation, at rate $\dot{\phi}$, of the cavity relative to the liquid. Their computations for $\dot{E}$ were low by about two orders of magnitude relative to later experiments. Repeat the computation by integrating over the area of the sphere. Note that the computation ignores that the central liquid is rotating. What rotational flow phenomena (Chapters 6 through 13) might be included to improve the analysis?

Project 18.6: Equation (18.4), and Projects 18.2 through 18.5, assume steady-state values of $\dot{\phi}$, $\dot{\psi}$, θ, I_3, and I_1. Such steady-state conditions can be controlled in the laboratory, when desired, and would apply to some very massive rigid space vehicles containing small containers of liquid. More important applications involve semirigid vehicles containing large quantities of liquid (usually fuel). See Abramson (1966). The moving liquid can dominate the motion of the vehicle. The interaction can be estimated using the moment applied to the vehicle by the internal liquid or by computing the rate at which energy is dissipated within the liquid. Examine the literature for examples (see Project 18.4). Compare their utility. Suspect any solutions based on an inviscid assumption for the liquid. Why is it impossible for a contained inviscid liquid to cause a progressive change in the vehicle's attitude? Consider d'Alembert's paradox and whether it is possible to dissipate the rotational kinetic energy of the vehicle using an inviscid fluid. See Application Example 7.4.

Project 18.7: Turbulence and transfer of momentum and energy by inertial wave phenomena in the rotating liquid are important factors in promoting energy dissipation in precessing cavities containing liquids. Sloshing phenomena is an additional factor complicating the problem when the cavity is only partially full. Consider the possibility of analytical, numerical, and/or even crude empirical approximations that include the influence of turbulence, inertial waves, and sloshing. Some references are Abramson (1966), Greenspan (1968), Johnston et al. (1972), Vanyo and Likins

(1972), Tritton (1978, 1985), Koyama et al. (1979), Bertoglio (1982), and Vanyo (1991).

Project 18.8: Space vehicle fuel tanks are never exactly spherical, and they typically include baffles and other propellant management devices. Further, they are usually only partially filled. A great deal of time, effort, and money are spent seeking solutions to these very difficult real problems. Analyses and numerical solutions are not usually reliable, and earth-based experiments, scaled to the planned vehicle size in a space environment, are needed. Examine the literature for examples of six approaches:

1. Observation of the actual satellite assembly mounted on a spherical air bearing as it spins, precesses, and changes attitude, i.e., $\dot{\theta}$ (Peterson 1976)
2. Observation of a scaled tank suspended in a gimballed assembly that permits θ to change (D'Amico 1984)
3. Drop tests to observe $\dot{\theta}$ with the scaled spinning and precessing tank in free-fall (Reiter and Lee 1966; Harrison et al. 1983; Guibert 1987)
4. Observation of spin decay of a gun-fired liquid-filled projectile (Kitchens et al. 1978)
5. Measurement of moment applied to a tank of liquid during spin and precession (Zedd and Dodge 1985)
6. Measurement of energy dissipation within the liquid during controlled spin and precession of the container (Malkus 1968; Vanyo and Likins 1971; Garg et al. 1986)

Discuss advantages and disadvantages of each approach. Note that none of them are easy or inexpensive, but several might be developed in simplified form in a typical university laboratory.

Project 18.9: Read and discuss the novel by Arthur C. Clarke (1973) titled *Rendezvous with Rama*. It is a fascinating, well-written suspense story of an encounter with an alien space vehicle. The vehicle is a spinning cylinder, many kilometers large, with a small world including a "lake" on the inside of the cylindrical surface. He correctly assesses many phenomena due to rotation. Does he miss any?

Project 18.10: After reading Chapter 19, compare the extent and type of research used to understand the problem of motions in the earth's liquid core with the research available for space vehicle liquid-fuel analyses.

Chapter 19

The Earth, Sun, and Moon

19.1 Introduction

This chapter and the remaining three chapters discuss the three major bodies of fluid on and in the earth. They are (1) the liquid core, (2) the oceans, and (3) the atmosphere. The liquid core is by far the most massive. It cannot be studied directly; however, even though it is about 2880 km below the earth's surface, much has been learned about it. For example, it extends out to 0.55 of the earth's radius and contains 0.33 of the earth's total mass. It is probably composed mostly of molten iron with some amount of nickel and silicates. A much smaller inner core with a radius of 1215 km is known to be solid. Motions within the liquid core and their relationship to other geophysical phenomena are the subject of much current research. The oceans are the next most massive. They cover 0.71 of the earth's surface, but because their average depth is only about 3.73 km, they contain only 2.3×10^{-4} of the earth's mass. The atmosphere, with an average depth (for weather phenomena) of approximately 10 km, has only 0.88×10^{-6} of the earth's mass. The oceans and the atmosphere may contain only insignificant portions of the earth's total mass, but, because of their importance to human activities, both the oceans and the atmosphere are also studied intensively and extensively.

The energy source for atmospheric flows is solar radiation. Oceanic flows are driven mostly by viscous drag of the prevailing winds. Motions in the liquid core, if any, are probably driven by thermal or compositional convection and/or precession of the mantle. The flows are complicated by the gravitational influence of the sun and moon, the equilibrium shape of the spinning earth, and other lesser phenomena. Because the earth is rotating, all three bodies of fluid rotate and all exhibit rotating flow phenomena. The study of all (atmospheric, oceanic, and liquid-core motions) begin with an understanding of earth-sun-moon dynamics.

	Sun	Earth	Moon
Diameter (km)	1.4×10^6	1.3×10^4	3.5×10^3
Mass (kg)	2.0×10^{30}	6.0×10^{24}	7.4×10^{22}

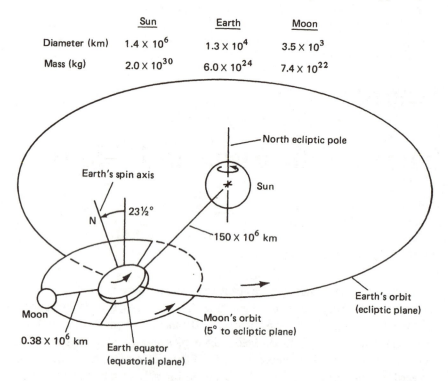

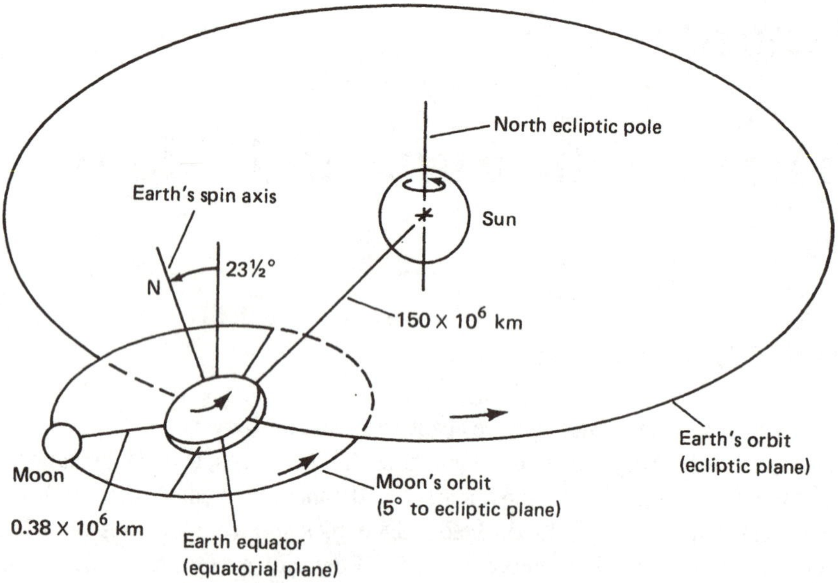

Figure 19.1: *Geometry of the earth-sun-moon system. Not to scale. Refer to Appendix E for additional data.*

19.2 Earth and Earth-Sun-Moon Dynamics

The solar system, including the earth, sun, and moon, is on the order of four billion (4×10^9) years old and has reached a nearly steady-state condition, at least on time scales relevant to human activities. The earth spins on its axis one revolution per solar day and orbits around the sun in 365 1/4 days. The earth's spin axis equator is inclined 23 1/2° to the plane of its orbit. The moon orbits about the earth, going from full moon to full moon in 29 1/2 days. Its orbital plane is inclined to the earth's orbital plane by 5°, and it always keeps the same side facing the earth. See Figure 19.1. The earth's orbital plane is often called the *ecliptic plane* because the moon must be passing through this plane for solar or lunar eclipses to occur.

The sun (a massive ball of hot gas) spins about a fixed axis with its equator nearly aligned with the earth's orbital plane. At its equator the sun rotates at about 25 days/revolution, with the poles spinning slightly slower. All of these motions, and the orbits of all the planets, are counterclockwise when viewed from a position in space north of the earth. It is probable, based on this and other evidence, that the solar system formed as a contraction from a much larger mass of gases and particles with

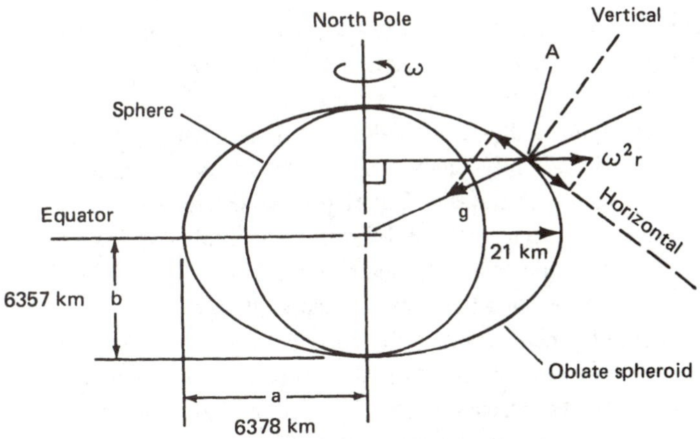

Figure 19.2: *The oblate spheroidal earth compared to a sphere. Not to scale. An element of air or water moving with the earth's west-east velocity at A is in horizontal equilibrium on the spheroidal surface.*

approximately the same mass and angular momentum as the present solar system. See Example 12.4.

Because the earth spins and is not perfectly rigid, it long ago deformed to an oblate spheroidal shape with an equatorial bulge. This shape represents an equilibrium between gravity and centripetal acceleration. The earth's ellipticity (i.e., its nonsphericity) is computed as

$$\epsilon = \frac{a - b}{a} \tag{19.1}$$

where a is the equatorial radius and b is the polar radius. Its ellipticity is about 1/298 at the surface. As shown in Figure 19.2, the equatorial radius is 6378 km, about 21 km larger than the radius to either pole. This represents a 21 km high "mountain" of material along the equator (Mount Everest is 8.8 km high). If the earth should stop spinning but keep its present shape, objects, air, and water would slide north or south down this "equatorial mountain" toward the poles.

An object stationary on the earth moves west to east at exactly the earth's tangential speed, and the horizontal component of its centripetal acceleration exactly compensates for the horizontal component of gravity. Horizontal means tangent to the actual spheroidal surface. If an object has a westward velocity relative to the earth's surface, its centripetal acceleration is inadequate to compensate for gravity, and on a smooth, equilibrium-surface earth, it would begin to slide down toward a pole. An eastward motion would cause the object (say a fluid element) to slide up the 21 km high bulge to the equator. If the equations of motion are set in a frame rotating with the earth's angular velocity, these relative east-west velocities show up in the Coriolis term ($2\boldsymbol{\omega} \times \mathbf{v}$). See Figure 7.1 and discussion. This explanation is used later to help understand north-south Coriolis acceleration of air or water that is being forced to move east or west relative to the earth's surface.

The interior of the earth, out to slightly more than half its radius (3485 km), is a liquid with a density slightly more than molten iron with some nickel and other materials. Analysis of earthquake waves propagated through the earth show the boundary between the liquid core and the rigid mantle to be very distinct. These analyses use the fact that a solid can transmit both longitudinal (pressure) and transverse (shear) waves, but a liquid can transmit only pressure waves. Discrepancies in the transmission of shear waves to the opposite side of the earth are caused by the liquid core. The mantle-core boundary has a spheroidal shape with an ellipticity of about $\epsilon = 1/400$. The boundary surface appears to be slightly irregular on the order of several kilometers. At the very interior of the core, the material again becomes rigid, either due to composition differences or to a metallic phase change. Motions of the core liquid relative to the mantle may be responsible for the earth's magnetic field, for generating and/or conducting heat outward, and for helping to activate extremely slow convection movements in the mantle. In this regard, the mantle is rigid (more so than steel) for short-term phenomena (seismic waves) but "creeps" slowly (large-scale radial and lateral convection) over time spans of hundreds of millions of years.

Interactions of the earth with the moon and the sun cause additional phenomena of interest. Figure 19.3 illustrates how the moon (and the sun) periodically apply moments to the spinning earth. The earth reacts as a gyroscope responding to a transverse moment according to the equation $\mathbf{M} = \dot{\mathbf{H}}$. This equation says that $\dot{\mathbf{H}}$ of the earth must equal in magnitude and direction the applied moment. The moon's applied moment is approximately that of a rectified sine wave with period 29 1/2 days, and the sun's that of a rectified sine wave with period 365 1/4 days. Together they cause $\mathbf{H}$ of the earth to change direction by about 50 seconds of arc per year, a pulsing motion but always in the same direction. It takes about 26,000 years for one complete conical precession (at 23 1/2°) of the earth's spin axis about the north ecliptic pole direction (perpendicular to the earth's orbital plane).

19.3 Motions in the Liquid Core

Debates have arisen whether forced precession of the mantle induces motions in the liquid core. Note the similarity of this problem to the motion of liquid fuels in a spinning and precessing satellite as analyzed in Chapter 18. Within the last half century various (and sometimes contradictory) analyses have indicated that

1. Liquid in the core follows the mantle exactly, just as if it were frozen solid and the earth were rigid throughout. This is explained to be a result of inertial waves (Chapter 13) "pressing against" the 1/400 ellipticity of the mantle-core boundary. Linearized analyses, assuming E_k and $R_o \ll 1$ and $\dot{\phi} \ll \dot{\psi}$, support this

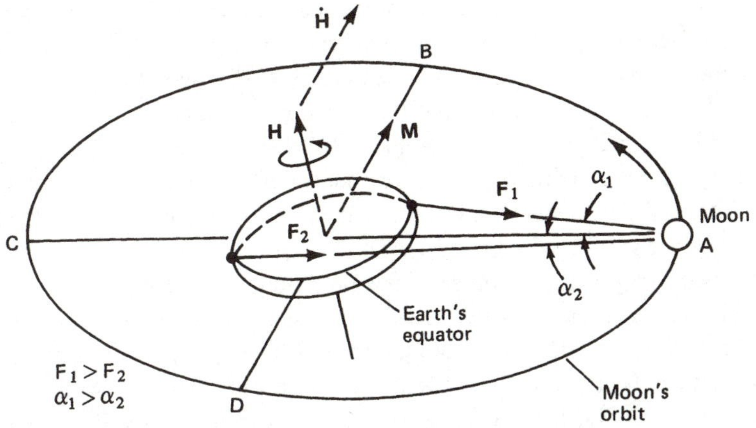

***Figure 19.3:** Lunar torque (M) on the earth due to differences in lunar attraction on the near portion of the equatorial bulge (F_1) and the far portion (F_2).*

conclusion. See Chapter 18 for definition of variables. However, most such phenomena are highly nonlinear, and details such as $\dot{\phi} = \dot{\phi}(t)$ due to pulsations of the lunar-solar torques were not considered.

2. Spin of the liquid core is completely independent of the precession of the mantle, and the resulting differential motion drives the "dynamo" responsible for the earth's magnetic field. The model has been supported by some researchers and criticized by others.

3. The liquid core responds to mantle precession somewhat like the rigid-sphere model shown in Figure 18.5. This model is consistent with some observations, e.g., dynamo dissipation and earth heat flux values, westward drift of the magnetic field, and scaled laboratory experiments.

4. Core motions consist of buoyancy-driven convective flows, restricted to those motions permissible in a rotating fluid. The solution indicates a series of helical columns with their axes parallel to and surrounding the earth's spin axis. These motions could be combined with several of the models.

5. Mantle-core coupling is affected by various mechanisms, e.g., topological coupling (bumps on the boundary surface), viscous coupling, magnetic field coupling, etc.

These analyses have been associated with, and often motivated by, parallel attempts to understand the earth's magnetic field (geomagnetism). To a good first approximation the observed geomagnetic field is bipolar with its north pole near the earth's north polar spin axis and its south pole near the earth's South Pole. The history of the geomagnetic field is often preserved in minerals (rocks) if they contain even minute amounts of ferromagnetic materials. For example, as sediment with ferromagnetic particles settles to an ocean or lake bottom, the particles orient preferentially to the ambient magnetic field. In the case of molten lava, ferromagnetic particles imprint and preserve the ambient field as the lava cools and hardens. Analyses of such rocks indicate the earth has had a magnetic field for most or all of its history, and that the field direction has reversed at nearly periodic intervals (similar to features of the sun's observed magnetic field).

It is generally agreed that motions in the molten and conducting liquid core form a *geodynamo* that maintains the earth's field. The field would dissipate quickly (about 15,000 years) if not maintained. Core motions must supply energy for maintaining the geomagnetic field as well as energy for observed heat flux from the earth, volcanoes, continental drifts, etc. The driving forces for core flow motions must of course correspond to reasonable physical mechanisms.

Decay of residual radioactive elements had been considered as the energy source, but this is now known to be improbable because the radioactive elements locate preferentially in minerals concentrated mostly in the outermost crust of the earth, not in its interior. Luni-solar precession was proposed in the 1960s, the idea attracted some attention, errors were discovered in the early analyses, other analyses suggested the concept was improbable, and other mechanisms were sought.

During the 1970s and 1980s, accretion of the solid inner core was proposed and widely accepted as a probable energy source. The accretion mechanism releases thermal energy at or near the surface of the solid inner core due to solidification of the molten material through a phase change or potential energy loss of heavier materials depositing from higher regions. Figure 19.4a shows the inner core, liquid core, mantle, and crust of the earth, and Figure 19.4b shows the ring of helical flow patterns dictated by Coriolis effects as the liquid attempts to convect radially outward. Laboratory experiments confirm these axially aligned helical flow patterns in the presence of thermal convection.

The luni-solar precession idea was reintroduced in the early 1990s based on analytical and experimental evidence learned in the aerospace industry. Chapter 18 discusses precessing communication satellites containing large mass fractions of liquid (~ 60%). The analogy to the earth is apparent in that the earth is a precessing solar satellite containing 33% liquid by mass. Figures 19.5a, 19.5b, and 19.5c illustrate application to the earth's core. The rigid-sphere model for the liquid core is, of course, only a first-order approximation to an actual flow, but it does provide the critical requirements for an earth-core flow model. Experiments also confirm this flow pattern. Some combination of convection and precession is possible, both being dominated by Coriolis and secondary flow features. Project 19.8 suggests further readings.

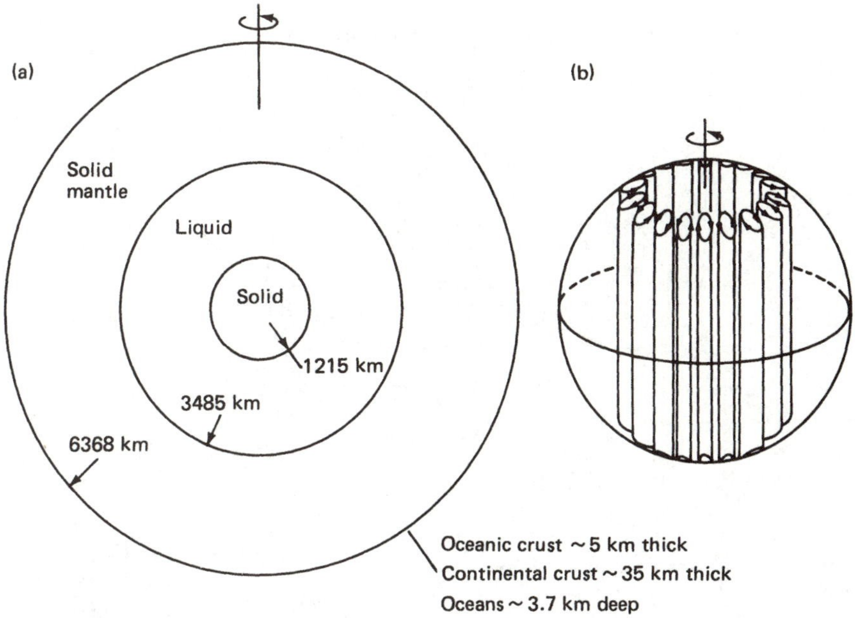

Figure 19.4: *(a) Cross section of the earth approximately to scale. A heat-producing inner core will cause convection in the liquid core, but, because of Coriolis effects, the convection will occur as axially parallel columns rather than as radial patterns. (b) View of the liquid core showing convective cells (Busse 1970b).*

19.4 Chandler Wobble

The earth also wobbles (precesses) due to its nonspherical shape in the same manner as the communication satellites of Chapter 18. This *torque-free* precession was first predicted by Euler, but its magnitude was too small to be detected with the instruments available in the middle 1700s. Near the end of the 1800s Seth Chandler correctly identified an apparent periodic variation of about 7 m in the latitude of observatories as a wobbling motion of the earth. Often called *Euler-Chandler wobble*, it is nearly consistent with Euler's result,

$$\dot{\psi} = \dot{\phi}\left(\frac{I_1}{I_3} - 1\right)\cos\theta \qquad (18.4)$$

Note that $\dot{\phi}$, $\dot{\psi}$, and θ of *torque-free* precession do not have the same values nor the same forcing mechanism as the $\dot{\phi}$, $\dot{\psi}$, and θ used for *forced* precession of the earth. Euler estimated I_1/I_3, assuming a constant-density, rigid earth with ellipticity of 1/298, and computed a 305 day period. Chandler's observed value of 427 days is due to the fact that the earth's density is not constant, and the earth is not rigid.

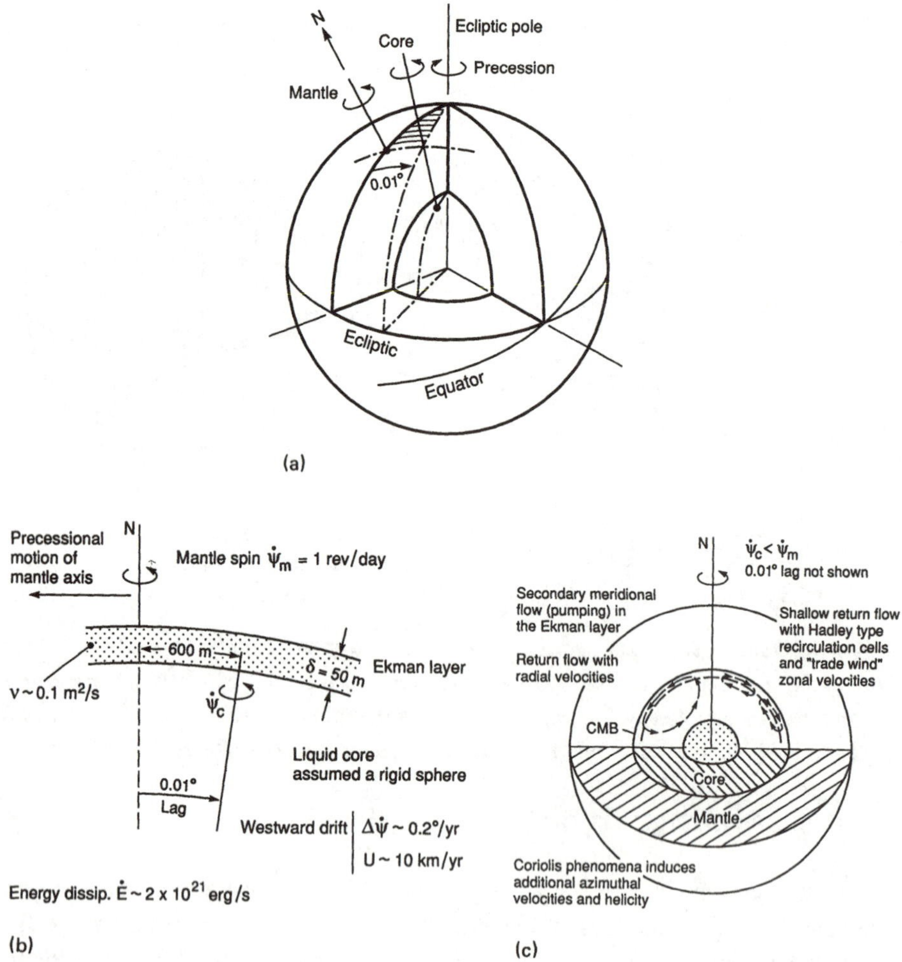

Figure 19.5: *(a) Illustration of the net liquid core lagging in precession behind the mantle. (b) Parameter values consistent with a rigid-sphere model and known earth phenomena. (c) Radial flow may be generated by Ekman layer pumping (secondary flows) at the mantle-core boundary and mass continuity requirements (Vanyo 1991).*

Equation (18.4) requires $\dot{\phi}$ (precession) be much larger (1 revolution/day) than $\dot{\psi}$ (spin of −1 revolution/427 days). The half coning angle is observed to be ~ 0.1 seconds of arc. In summary, because of free precession, the earth spins at rate (−1 revolution/427 days) and precesses at rate (1 revolution/day) at a coning angle of 0.1 seconds of arc. Because this coning angle is so small, it is conventional to refer to the algebraic sum of $(\dot{\phi} + \dot{\psi})$ as the earth's rotation rate of ~1 revolution/day. Forced precession analyses use this rotation rate of 1 revolution/day as spin $(\dot{\psi})$ and a forced precession rate (1 revolution/26,000 yr) at the observed half coning angle of 23 1/2°.

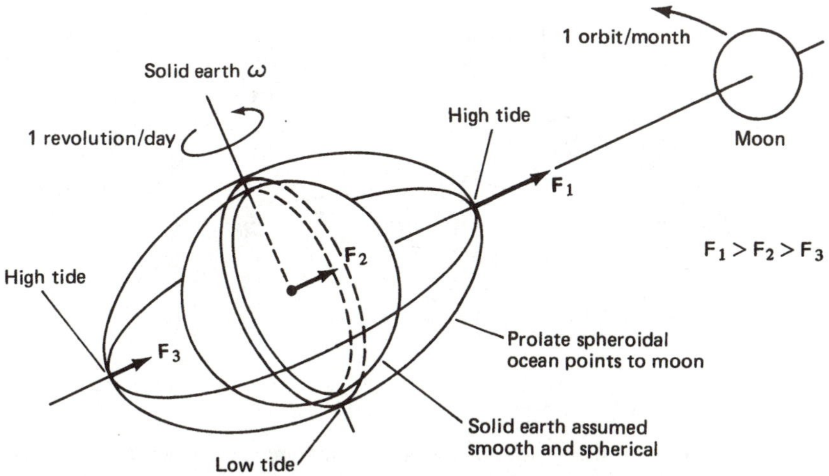

Figure 19.6: *Idealized and exaggerated production of oceanic tidal bulges due to differential gravitational forces. Oceans on the earth are constrained to basins, and tides on the open ocean are less than 1 m high.*

The energy minimum condition for an oblate object is spin about its symmetry axis without wobble. Wobble (precession) causes energy dissipation in the nonrigid earth, and therefore the earth should have converged long ago to pure spin about its symmetry axis. Instead, its wobble has been maintained at roughly the same level, with occasional abrupt changes. The mechanisms responsible for energy dissipation and for energy maintenance are topics of current research. Among the many mechanisms proposed for energy dissipation are the atmosphere, the oceans, material hysteresis in the solid mantle, and the liquid core. Mechanisms proposed for energy maintenance include atmospheric and oceanic disturbances induced by earth-sun-moon interactions, earthquakes and other solid earth movements excited by internal energy sources, and motions of the liquid core. All of these, except for solid earth hysteresis, are rotating fluid phenomena. Core motions are related to the analyses of Chapter 18. Atmospheric circulation patterns are discussed in Chapter 20 and oceanic flows in Chapter 21.

19.5 Tides and Earth-Sun-Moon Dynamics

An additional phenomena of interest is that of lunar-solar induced tides. There are atmospheric tides, oceanic tides, and solid earth tides, with oceanic tides being the most important and most easily observed. They are all produced as shown in Figure 19.6. The gradient of a gravity field causes any object in the field to be slightly stretched in the radial direction. By way of a simple explanation, the ocean on the side of the earth nearest the moon is attracted more strongly than the center of the earth and

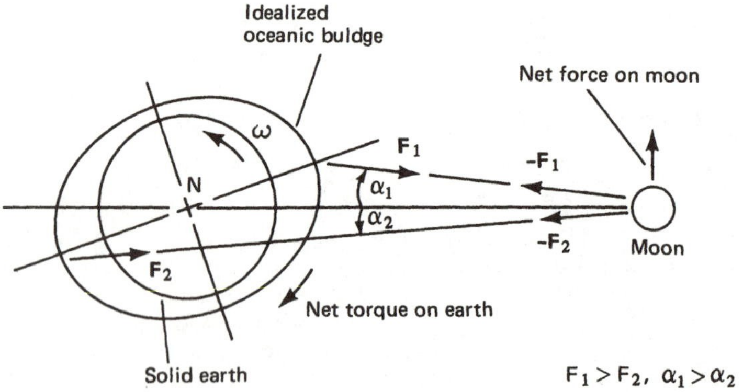

Figure 19.7: *View looking at an idealized earth-moon system from above the North Pole. Drag of the rapidly spinning earth carries the oceanic bulge forward of the earth-moon line.*

is pulled away from it, but also the center of the earth is pulled more strongly than the ocean on the far side of the earth and so is pulled away from it. The shape of an ocean covering a spherical nonspinning earth would appear as a prolate spheroid aligned with the moon's direction as shown in Figure 19.6. The sun's attraction to the earth is much stronger than the moon's attraction, but the gradient of its field is less. As a consequence, the moon is about twice as effective as the sun in producing tides. When the moon and sun are aligned (full and new moon), their effects are additive, and tides reach their maximum and minimum.

Because the earth spins faster than either the moon's orbital rate or the earth's orbital rate, the continents actually rotate into or out of the elevated or depressed regions of ocean, although an observer on the shore sees the reverse, i.e., an incoming or outgoing motion of the tides. The oceans are pretty much trapped into separate oceanic basins because of the continents, and idealized tides are not completely realized. Instead, the tidal forces tend to produce a swirling-sloshing type of motion within each oceanic basin. This motion is consistent with Coriolis effects on a rotating earth as will be shown in Chapters 20 and 21.

An interesting detail is the slight advance (relative to the earth-moon line) of the tidal peaks. These are caused by the earth's spin as shown in Figure 19.7. This slight deviation of the oceanic peaks from alignment with the moon causes the moon to exert a slight gravitational torque on the earth that slows the earth's spin rate. The opposite and equal effect is a slight force applied to the moon in a forward direction. The consequence is that our days are getting longer, and the moon is leaving us. During a person's lifetime these differences can only be detected with precise instruments, but over much longer periods the differences are impressive. For example, there is independent analytical (Munk 1966) and observational (Vanyo and

Awramik 1985) evidence that 850 million years ago the earth was spinning fast enough to have about 435 days/year and the moon was closer.

19.6 Study and Research Projects

The projects suggested here are intended for individuals or groups. Some can be completed in a week or two, others are term or longer projects. Only a few will have exact answers. References to other texts and to professional journals are included.

Project 19.1: Much of the material presented in Chapter 18, including the projects, is useful preparation for Chapter 19 and these projects. For example, a communication satellite containing a large mass fraction of liquid fuel is very much like the earth, which contains a large mass fraction of liquid core. Examine the differences and similarities. Examine also the type of research in each field. In what ways do the research emphases appear to differ?

Project 19.2: The precession discussed in Chapter 18 is usually called *free precession* (i.e., moment free) to distinguish that from *forced precession* (i.e., due to an applied moment). Engineers and geophysists also sometimes use the words *torque*, *wobble*, and *nutation*. For this discussion, torque means the same thing as moment; the words wobble, nutation, and free precession mean the same thing; and luni-solar precession is a type of forced precession. Assume a constant-density oblate-spheroid earth, and estimate $\dot{\phi}$ and $\dot{\psi}$ using (18.4). The angle θ is so small that Euler could not detect it, i.e., $\dot{\phi} + \dot{\psi} \approx 1$ revolution/day. Chandler did measure θ, but Chandler's wobble period differed from Euler's estimate. Recompute $\dot{\phi}$ and $\dot{\psi}$ using (18.4), and estimate earth's axial moment of inertia (I_a) and transverse moment of inertia (I_t) using data from Jordan and Anderson (1974). First assume the liquid core is rigid; then assume it is isolated from the mantle, i.e., its I_a and I_t do not contribute to the earth's I_a and I_t. For other references see Munk and MacDonald (1975) and Lambeck (1980).

Project 19.3: Estimate the earth's forced precession by computing the lunar and solar torques applied to the earth's equatorial bulge. Get a crude estimate by assuming a spherical earth with a cylindrical band of material around the equator, then try a numerical solution using data from Jordan and Anderson (1974). The angle θ here is known from observation, i.e., $\approx 23° 27'$. Note that the lunar and solar torques are functions of time. Again use $\dot{\phi}$, $\dot{\psi}$, and θ as variables, but remember they have different values than those for Euler-Chandler wobble.

Project 19.4: Refer to Application Example 7.4 which illustrates that an energy-dissipating, inertially oblate object should relax to pure spin about its axis of maximum moment of inertia. Why has Euler-Chandler wobble not gone to zero? As of early 1993, there was no agreement on what continues to supply the rotational kinetic energy needed to maintain Euler-Chandler wobble. Earthquakes and atmospheric perturbations are often suggested. Search the literature for these estimates. Analyze the possibility of interactions between luni-solar forced precession and Euler-Chandler wobble (torque-free precession) acting through motions in the liquid core. See Lambeck (1980), Melchior (1986), and Jacobs (1987).

Project 19.5: Could the earth "flip over" like the mechanism in Application Example 7.4 because of a sudden change in its inertia ratio? Suppose all the continents drifted quickly to within the Arctic Circle and the Antarctic Circle, and then all the oceans froze on top of them. Estimate the earth's I_a and I_t for that configuration, and verify that then $I_t > I_a$. Next, estimate the probability that this could happen; consider the speed of mantle convection and continental drifts, isostasy (adjustment of the "rigid" mantle to this new set of stresses), and the thermal flux needed to freeze all the oceans at the poles. Some references are Gold (1955), Warlow (1978), Slabinski (1981), Andrews (1985), and Peltier (1986).

Project 19.6: The sun rotates about an axis perpendicular to the plane of the solar system. Rotation is observed from motion of sunspots (relatively cool spots of concentrated magnetic field structures) and from doppler shifts of light between the east and west edges. At the equator, the period of rotation is about 25 days. This smoothly transitions to a period of about 29 days near the poles. Movement of fluid from the polar to the equatorial regions, or the reverse, should cause the equator to rotate slower, not faster, than the poles. Theories have been advanced, but the matter may not yet be resolved. What mechanism(s) might cause this anomalous behavior? Some references are Babcock (1961), Aller (1963), Busse (1970a), and Durney and Sofia (1987).

Project 19.7: Review the literature on convective motions in a self-gravitating, rotating, spherical (spheroidal) mass heated internally. The analyses and experiments apply to stars (the sun) and to planets (for the earth, both the liquid core and the mantle). Estimate the relative importance of centripetal and gravitational acceleration for typical objects. Examine variations needed to achieve experimental results, e.g., a rotating flat circular pan heated from below, a space laboratory experiment substituting a central electrostatic force for gravity, or use of centrifugal force to simulate motions in the equatorial plane. How well do these experiments model essential features of the physical problem? Note, in particular, the theory and experiments for convective patterns in a rotating filled sphere that appear to indicate that any radially induced motion must lead to patterns of helices parallel to the rotation axis. Should this pattern apply to both extremely slow and fast mantle convections? Some important references are Roberts (1968), Busse (1970b), Carrigan and Busse (1983), Hart et al. (1986), Loper (1989), Zhang (1992), and Cardin and Olson (1992).

Project 19.8: Rotation and precession of deformable bodies and bodies containing liquids began with studies by Plateau (1863), Hough (1895), and Poincaré (1910). The study was continued by Bondi and Lyttleton (1953), Roberts and Stewartson (1965), and Malkus (1968), who emphasized application to the earth's liquid core as a possible energy source for a geodynamo. Precession of the earth as a basis for core motions necessary for a geodynamo came under criticism. See Busse (1968), Loper (1970, 1975), and Rochester et al. (1975). As noted in the text, the concept has been reexamined by Vanyo (1984, 1991) and by Vanyo et al. (1992). Examine Malkus (1968) and succeeding papers, and assess the evidence and arguments. Compare this with your analysis of Project 19.7. Could both phenomena occur simultaneously? What observations of earth phenomena, if any, might help to distinguish which, if either, of the phenomena is operative? See, for example, Melchior and Ducarme (1986).

Project 19.9: A geodynamo energized by motions in the earth's molten and electrically conducting liquid core is generally accepted as the mechanism responsible for maintaining the earth's magnetic field. An equivalent mechanism has been proposed for the sun's magnetic field structure. Other more rapidly rotating stars are known to have even more intense magnetic fields. Readers interested in magnetohydrodynamics may wish to examine introductory books and survey papers on the topic: Aller (1963), Busse (1978), Priest (1982), Jacobs (1987), Roberts (1988), and Priest and Hood (1991).

Project 19.10: Select and research a topic of interest by physically examining recent issues of journals that include articles relevant to that topic. First, read the table of contents; if any titles appear interesting, read the abstracts; next, if the abstract is relevant to your interest, note the reference and/or check out that volume; finally, read, summarize, and integrate all chosen material. Journals of interest here include: *Astrophysical Journal*, *Geophysical and Astrophysical Fluid Dynamics*, *Geophysical Journal of the Royal Astronomical Society*, *Journal of Fluid Mechanics*, *Journal of Geophysical Research*, *Physics of the Earth and Planetary Interiors*, *Solar Physics*, and occasional articles in *Nature* and in *Science*.

Chapter 20

Atmospheric Circulation

20.1 Introduction

This chapter and Chapter 21 apply rotating fluid theory to large-scale motion of fluids on the exterior of the rotating earth, i.e., the atmosphere and the oceans. Because the atmosphere and the oceans interact in many ways, parts of each chapter apply to and aid in understanding important features of the other. Precession of the earth's spin axis was a factor in earth core motions but is of negligible importance here because other phenomena, much larger in magnitude and changing at a much more rapid rate, determine the motions. Chapters 20 and 21 begin with qualitative descriptions of atmospheric and oceanic phenomena and end with more quantitative analyses.

As with the other chapters in Part III, this chapter is intended to be only an introduction to a topic for the nonspecialist, and further, it only covers those portions of that topic most relevant to rotating fluids. Readers interested in broader or more advanced coverage of atmospheric phenomena should supplement this discussion with a specialized text or monograph, e.g., Palmén and Newton (1969), Holton (1979), Atkinson (1981a), Houghton (1986), and Brown (1990). Projects in Section 20.6 suggest additional readings and research.

The energy for atmospheric motions is supplied by solar thermal differentials. Heat flow from within the earth is negligible. Major thermal differentials occur because of

1. The earth's shape and orientation (larger *insolation*, i.e., solar radiation, per unit area at the equator than at poles)
2. Surface reflectivities and heat capacities (land, water, and clouds)
3. Inclination of the spin axis relative to the earth's orbital plane (seasonal variations)
4. The earth's spin rate (day-night variations)

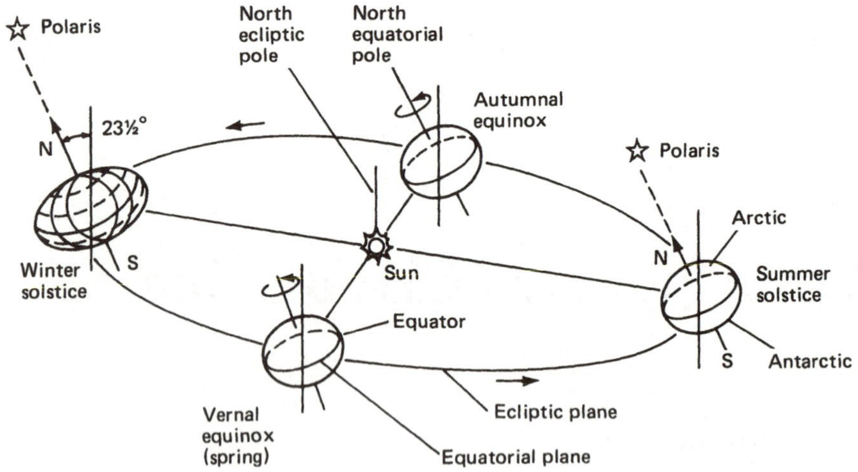

Figure 20.1: *Earth rotation, inclination, orbital motion, and seasons. Equinox means "equal day and night." Solstice means "sun stationary." Winter, summer, spring, and autumn here refer to Northern Hemisphere events.*

The fact that water occurs profusely in all three forms (ice, water, and vapor) introduces its latent heats of evaporation and solidification as major energy storage and transport mechanisms.

The energy for driving oceanic currents basically is also solar, but it operates through indirect mechanisms. Surface oceanic currents are maintained by the winds, which, as noted, are maintained by solar energy. The less energetic, slow, deep currents of the ocean are driven partly by temperature and saline density gradients and partly by the surface currents.

Figure 20.1 reviews the basis for the seasons. Some atmospheric phenomena occur on time scales which are short relative to the length of seasons; however, many major atmospheric circulation patterns persist as nearly permanent flow patterns. The oceans store energy and momenta which are large compared to seasonal variations, and therefore oceanic circulations are even more permanent. Oceanic heat capacities are so large that relatively small variations in oceanic flows are capable of causing major atmospheric variations. The discussions of this chapter and Chapter 21 relate to large-scale circulations, and will assume seasonal and daily variations have been summed to yield long-term averages. The result is fairly accurate.

Figure 20.2 shows a cross section of a quadrant of the earth. Surface irregularities are seen to be negligible compared to the earth's radius. All irregularities, and the combined average depth of the oceans and height of the *troposphere*, would fit within the width of the line in Figure 20.2 representing the quadrant. The nonsphericity of the earth is negligible for most atmospheric and oceanic flows, but it is significant for

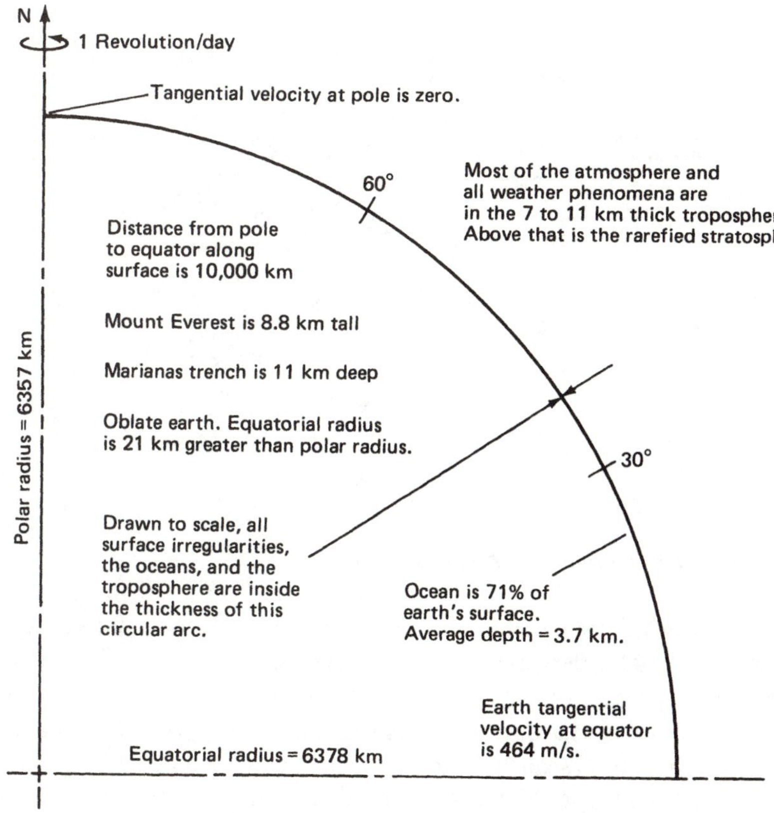
N
1 Revolution/day
Tangential velocity at pole is zero.
60°
Most of the atmosphere and all weather phenomena are in the 7 to 11 km thick troposphere. Above that is the rarefied stratosphere.
Distance from pole to equator along surface is 10,000 km
Mount Everest is 8.8 km tall
Marianas trench is 11 km deep
Oblate earth. Equatorial radius is 21 km greater than polar radius.
30°
Polar radius = 6357 km
Drawn to scale, all surface irregularities, the oceans, and the troposphere are inside the thickness of this circular arc.
Ocean is 71% of earth's surface. Average depth = 3.7 km.
Earth tangential velocity at equator is 464 m/s.
Equatorial radius = 6378 km

Figure 20.2: *Scale drawing of the earth illustrating height of the atmosphere, depth of the ocean, and surface irregularities compared to its radius.*

some Coriolis phenomena. Because the earth's oblateness is determined as the resultant of gravitational and centripetal accelerations, the earth's shape is such that the tangent to the oblate surface at any point defines the horizontal for fluid analyses. The earth has already adjusted its shape to compensate for centripetal effects, and they do not usually need to appear again in the fluid equations.

Note also the tangential speed of the earth's surface at the equator. If an object with this velocity were suddenly transported to the region of a pole, it would have a velocity greater than the speed of sound in air at sea level. Nomenclature and the earth-fixed coordinate system of latitude and longitude are reviewed in Figure 20.3.

Because the equatorial surfaces of the earth are nearly perpendicular to an earth-sun line, each square meter at the equator intercepts the solar radiation (*insolation*) available at the earth's orbital distance from the sun (~1 kW/m^2). Away from the equator, surfaces are progressively more inclined, so each square meter of the earth at higher latitudes intercepts only a portion of the available energy (~ $\cos \phi$), until at the poles, the intercepted energy on the average approaches zero. The temperature

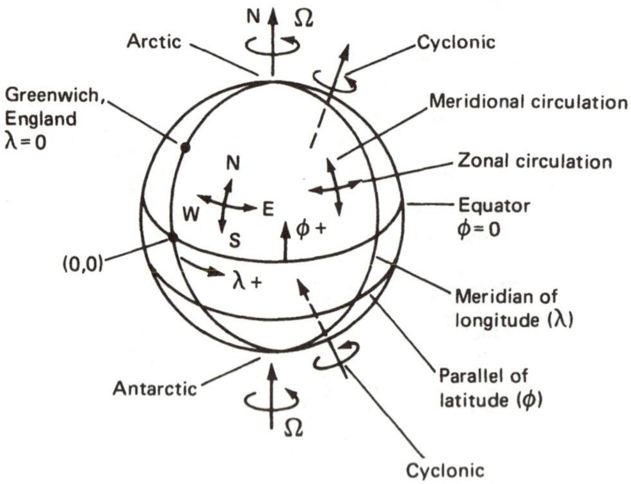

Figure 20.3: *Earth surface coordinate system.*

difference between the poles and the equator, moderated only slightly by winter-summer and day-night differences, causes constant net energy flows away from the equator toward the poles.

Air at the equator is heated, rises, and moves toward the poles. As air leaves the equator it loses heat by radiation to space. As it cools it sinks. The image of a simple convective cell is misleading, however, for three reasons:

1. *Flow instability.* The distance from the equator to either pole is 1000 times larger than the height of the troposphere (10,000 km vs. 10 km).
2. *Mass conservation and polar convergence.* Distance around the equator is about 40,000 km, and this flow volume must move to progressively smaller polar regions.
3. *Coriolis phenomena.* As air moves north or south it moves to regions where the earth's surface has smaller tangential velocities, and vice versa for flows moving toward the equator.

20.2 Convective Flow Patterns

As seen in Figure 20.2, it is unlikely that an upper heated layer of air could move all the way from the equator to the pole without interacting with the adjacent lower cool layer moving toward the equator. Figure 20.4 shows a laboratory simulation of the problem. Here the ratio of horizontal convection to fluid depth is only 20:1 rather than the 1000:1 ratio of the earth's atmosphere. Even so, it is apparent that the heated upper level will tend to interact with the oppositely flowing lower level to form

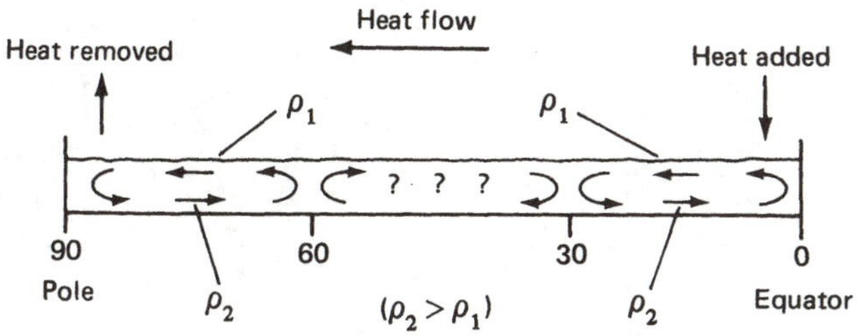

Figure 20.4: *Laboratory model of atmospheric meridional (north-south) circulation, including thermal and stratification effects, but without polar convergence or Coriolis phenomena.*

smaller cells, each cell transferring energy only a portion of the distance. Figure 20.4 shows a cell going from 0 to 30 (later representing degrees of latitude) and a cell from 60 to 90. If a cell of the same rotation sense is placed between 30 and 60, its vertical flows will oppose those of the other cells. For its vertical flows to agree with the other cells it must rotate in the opposite direction, but then the hotter fluid is on the bottom, and the region is thermally unstable to convection.

Prediction of atmospheric convective cells and their interaction is very difficult. Atmospheric circulation cells are discovered by observation--not by prediction. As shown in Figures 20.5a and 20.5b, their size and rotation sense have been discovered to be almost like Figure 20.4. Strong meridional (north-south) circulations exist between the thermal equator (near the geographic equator) and latitudes of about 30° north and south. These are named *Hadley cells* and are nearly closed cells in the sense that only small amounts of air are exchanged with neighboring cells. Very weak reverse-flow cells exist between latitudes of 30° and 60°. These are called *Ferrel cells* and are regions of atmospheric instability. Most of the poleward flow of energy in this region is a result of flow eddies of random size and flow direction. The arctic and antarctic cells form polar fronts extending equatorward to latitudes of about 60°. As will be seen later, this limit is subject to wide variations during winter, especially in the Northern Hemisphere which has 67% of the earth's land surface.

Considering Figure 20.2, it is surprising that the atmosphere subdivides into only three cells per hemisphere. Each cell has a ratio of horizontal distance to tropospheric height of over 300:1. The Hadley cells and polar cells maintain their integrity because the flow velocities normally are low, and the stratification due to density differences is normally strong and stable.

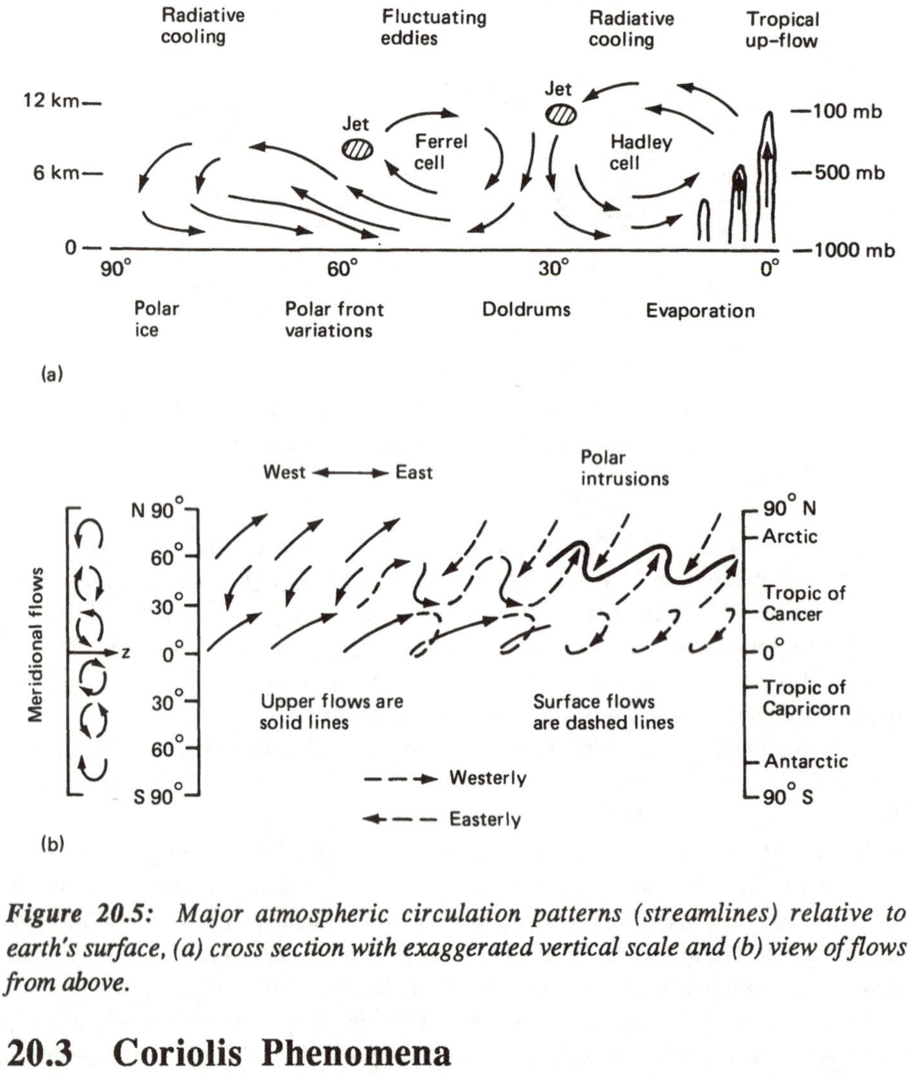

Figure 20.5: *Major atmospheric circulation patterns (streamlines) relative to earth's surface, (a) cross section with exaggerated vertical scale and (b) view of flows from above.*

20.3 Coriolis Phenomena

Coriolis phenomena are never erratic in their operation. All flows (atmospheric or oceanic) going poleward respond alike, whether they are part of large-scale smooth-flowing cells or erratic eddies, and all flows going equatorward respond alike, but oppositely to the poleward flows.

The Navier-Stokes equation for acceleration relative to a coordinate system rotating at $\mathbf{\Omega}$ = constant was derived as

$$\frac{D\mathbf{V}}{Dt} + \mathbf{\Omega} \times (\mathbf{\Omega} \times \mathbf{R}) + 2\mathbf{\Omega} \times \mathbf{V} = \frac{1}{\rho}\mathbf{f} - \frac{1}{\rho}\nabla p + \nu\nabla^2\mathbf{V} \tag{20.1}$$

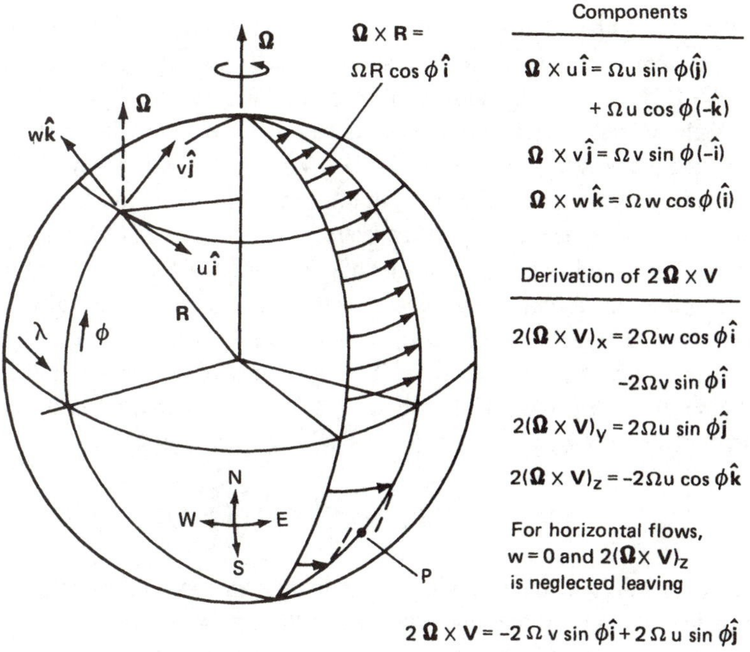

Figure 20.6: *Atmospheric and oceanic Coriolis phenomena. Analyses, valid for local regions, approximate (v_λ, v_ϕ, v_r) with (u, v, w) in rectangular coordinates (x, y, z). See text for explanation.*

In (20.1) all variables and derivatives are relative to the rotating coordinate system. Spherical coordinates (λ = longitude, ϕ = latitude, r = radius), appropriate to a nearly spherical surface, are locally orthogonal, and the typical atmospheric or oceanic analysis often makes use of this by approximating (20.1) using rectangular coordinates centered at the location of interest. It is convenient then to approximate variations in (λ, ϕ, r), at that location, by (x, y, z), and (v_λ, v_ϕ, v_r) by (u, v, w). Then x is east, y is north, u is a flow toward the east, and v is a flow toward the north. For nearly horizontal flows z and w are usually neglected. Note that this approximation neglects many terms unique to a general spherical coordinate system formulation of the equations. Because of this, the resulting equations can only be integrated over regions that are nominally flat, say a region no larger that 1000 km in diameter.

In Figure 20.6 consider first a quantity of air in the upper atmosphere at some latitude ϕ. An arbitrary velocity in the atmosphere or ocean (**V**) can be resolved into components (u, v, w) with directions of ($\hat{\mathbf{i}}$, $\hat{\mathbf{j}}$, $\hat{\mathbf{k}}$) as shown. Then

$$\mathbf{\Omega} \times \mathbf{V} = \mathbf{\Omega} \times (u\hat{\mathbf{i}} + v\hat{\mathbf{j}} + w\hat{\mathbf{k}})$$

$$= \mathbf{\Omega} \times u\hat{\mathbf{i}} + \mathbf{\Omega} \times v\hat{\mathbf{j}} + \mathbf{\Omega} \times w\hat{\mathbf{k}} \qquad (20.2)$$

Each of these is computed as shown in Figure 20.6 using the locally flat (Cartesian) coordinate system. They are then recombined to show the approximate λ, ϕ, and r components of $2\mathbf{\Omega} \times \mathbf{V}$, the Coriolis term. A typical analysis of large- and moderate-scale atmospheric and oceanic motions assumes flows are locally horizontal, so that $v_r \approx w = 0$, and the vertical component of $2(\mathbf{\Omega} \times \mathbf{V})$ is neglected. Only $(-2\Omega v \sin\phi)$ in the east-west direction and $(2\Omega u \sin\phi)$ in the north-south direction remain.

Note that $(\mathbf{\Omega} \times u\hat{\mathbf{i}})$ is a centripetal acceleration and is directed toward the earth's axis (not its center). It has components in both the ϕ and r directions. The horizontal component represents a north-south phenomena caused by east-west flows, and the vertical component (typically neglected) represents a vertical phenomena caused by east-west flows. The term arising from $(\mathbf{\Omega} \times v\hat{\mathbf{j}})$ represents an east-west phenomena caused by north-south flows. The desired equation that approximates the spherical components of $2\mathbf{\Omega} \times \mathbf{V}$ by locally flat (Cartesian) coordinates is

$$2(\mathbf{\Omega} \times \mathbf{V}) \approx (-2\Omega v \sin\phi)\hat{\mathbf{i}} + (2\Omega u \sin\phi)\hat{\mathbf{j}} \tag{20.3}$$

Terminology unique to conventional atmospheric and oceanic analyses is needed to be consistent with the professional literature of these fields.

1. Atmospheric analyses use the convention of early mariners to name a wind by the direction *from which it comes*, e.g., a westerly or west wind blows from the west toward the east (+u). Oceanographers use a convention that names a flow of water by the direction *to which it is going*, e.g., an eastward flow moves toward the east (+u). Figure 20.7 summarizes the conventional nomenclature for wind flow directions.

2. *Geostrophic* motions are flows dominated by the interaction of horizontal pressure gradients and the Coriolis term. *Ageostrophic* refers to all other flows. Later discussions will expand on the significance of geostrophic vs. ageostrophic.

As with Chapter 6, it is important to distinguish between (1) acceleration relative to inertial space and (2) acceleration relative to a rotating frame. The first involves real forces, and the second need not. Figure 20.6 shows the variation of the earth's tangential velocity $(\mathbf{\Omega} \times \mathbf{R})$ as a function of latitude. Consider point P which has some tangential velocity. A parcel of air, stationary relative to the earth at some location nearer the equator, has a larger tangential velocity than the earth at point P. If it is moved toward point P, without changing its tangential velocity, it will be a westerly (i.e., be moving from the west toward the east relative to the earth at point P). In both the Southern Hemisphere and the Northern Hemisphere, air transported toward the poles become westerlies. If air is transported from either pole toward the equator, it moves to regions where the tangential velocity of the earth is greater than that of the air, so relative to the earth, that air becomes an easterly (flows to the west). In these cases no transverse force is applied, so the air has no transverse acceleration

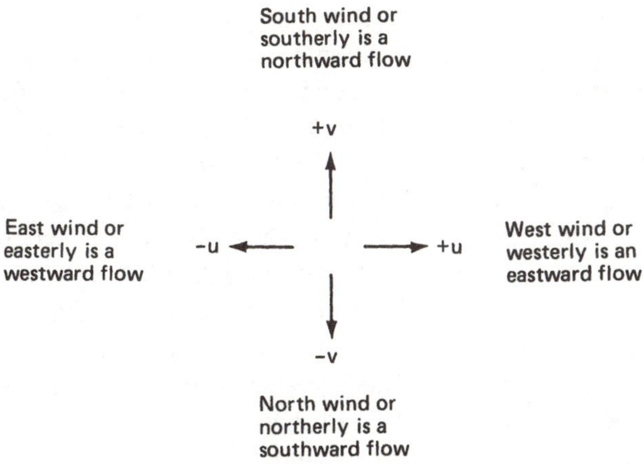

***Figure 20.7:** Nomenclature convention for describing wind flow directions.*

relative to inertial space. The air does, however, have an acceleration relative to the noninertial frame of the rotating earth. This is an example of a geostrophic flow as illustrated in Figure 7.2.

The other extreme of Coriolis phenomena is where the flow is constrained by some forcing mechanism to follow a specific path, for example, a flow of air through a long, deep, narrow canyon or a river constrained by gravity to remain in its channel. In a north-south flow of this type, the fluid is forced by the canyon walls or river channel to change its tangential (east-west) velocity. There is a real east-west force and a real acceleration relative to inertial space. This is an example of an ageostrophic flow as illustrated in Figure 7.1.

20.4 Planetary Flow Dynamics

The earth's atmospheric circulation patterns shown in Figure 20.5 are explained as a combination of geostrophic and ageostrophic flows driven by the solar-induced temperature differentials between the equatorial and polar regions. The flows are modified by flow instabilities and long-term mass, energy, and momentum conservation principles. Figure 20.5 shows north-south velocities to be of about the same size as east-west velocities. Actual east-west velocities associated with large-scale circulation patterns are much larger than north-south velocities, and a typical volume of air will go one or more times around the earth during a 30° movement north or south. Overall cloud patterns on the earth often are poor representations of general atmospheric circulation patterns as seen by comparing Figure 1.15 with the following description.

Beginning at the equator, air in contact with the surface (mostly ocean) acquires the earth's tangential velocity and thermal energy (as change of temperature and as latent heat of evaporation). Moisture condenses as the air rises, releasing its latent heat to the air and enabling it to reach altitudes ~12 km in the form of individual convective cells. These are evidenced by cumulus cloud structures. This upper atmospheric air flows to about 30° N and 30° S while being cooled by radiative losses to space. It then sinks back to the surface and flows equatorward to repeat the cycle. The closed cells so formed are the north and south Hadley cells. There is very little exchange of air between the Northern Hemisphere and the Southern Hemisphere, and only modest exchanges of air between the Hadley cells and higher latitudes.

This description explains the meridional flow (v) component of the Hadley cells but not the zonal (u) components. As the equatorial upper air flows away from the equator, it retains the equatorial tangential velocity so that it is moving eastward faster than the local surface. It becomes an upper atmospheric westerly (i.e., flowing west to east relative to the surface). The path of a parcel of air moves 1 to 1.5 times around the earth during the 30 days or so it takes to move from 0° to ±30°. Concentrated westerly jets of high-altitude air, called the *subtropical jets*, are formed at about 30° N and 30° S by this rapidly moving west wind.

As air sinks to and moves along the surface, it is slowed down by surface friction to approximately the tangential velocity of the earth at that latitude. Finally, as it moves back to the equator the process is reversed. The air is now transported to regions of earth surface that are moving faster. The air has an east to west motion relative to the earth and is called an easterly. These easterlies are the tropical trade winds and persist during the entire year, although the pattern shifts 10° or so north or south with the change of seasons.

The regions between 30° to 60° and between 60° to 90° are explained by similar arguments. In summary, the 30° to 60° Ferrel cells produce surface winds moving west to east that form the midlatitude westerlies. They also are trade winds because they persist throughout the year (with variations) and were known and used by commercial (trading) sailing ships. The polar front region separating the cold polar air and the midlatitude temperate air fluctuates widely. It typically appears as "fingers" of cold polar air moving away from the poles, alternating with adjacent parallel fingers of warmer temperate air moving toward the poles. The phenomena are exaggerated in the northern winter due to the large ratio of land area to ocean area (and larger seasonal temperature variations) in the northern midlatitudes. A second set of westerly upper atmospheric jets, called the *polar jets*, are formed near 60° N and 60° S. Jet aircraft flying west to east make use of these jets of high-velocity air while aircraft flying east to west must avoid them.

The Ferrel cell flow model used to describe atmospheric motions between 30° and 60° N or S is at best only a crude approximation. Poleward transport of momentum and heat in this region is better described using a model of convective and flow instabilities. Refer to a text on atmospheric physics or meteorology, e.g., Holton

(1979). Chapter 22 provides an introduction to more complex, subcontinental size, unstable flows, especially flows that are characterized by well-organized intense vortices.

20.5 Atmospheric Rossby Waves II

The study of Rossby waves in the atmosphere was initiated in Section 13.4 "Atmospheric Rossby Waves I." That section included a derivation of the equations used by Rossby et al. (1939) to explain stationary large-scale weather patterns over North America and Russia. This section summarizes Rossby's original data, reasoning, and conclusions to explain the wavelength and wavespeed of these atmospheric perturbations.

Rossby comments on changes in vorticity that must occur in vertical air columns that are displaced in latitude, and the fundamental role that such changes play in both atmospheric and oceanic circulation theories. He then reviews data on the existence of recurring nearly stationary atmospheric cells or eddies ~ 2000 miles (3200 km) in diameter. One is a high-pressure pattern over the United States (in August); others are the Icelandic and Aleutian Lows, and the Azores, Asiatic, and Pacific Highs. Some of these characteristically break up into two parts, with one part moving westward and the other eastward, after which the pattern may repeat from the original position.

Rossby uses a paper by Bjerknes (1937) to help explain pressure and velocity distributions and the transport of pressure cells, as well as the north-south transport of vorticity. See Application Project 20.9. As noted in this chapter and in Chapter 22, low-pressure cells in the Northern Hemisphere generate counterclockwise rotation of the surrounding air (cyclonic motion), while high-pressure cells generate clockwise rotation (anticyclonic), both due to Coriolis effects associated with air movements on the rotating earth. Coriolis forces here are so effective that motion is almost exactly perpendicular to the pressure gradient. As a consequence, the isobars (lines of constant pressure) are commonly used as the flow streamlines.

The derivation of (13.20) in Section 13.4 is structured after the derivation by Rossby. Rossby also extended the derivation to more complicated situations such as cell perturbations comprised of a sequence of wavelengths and the inclusion of forcing functions. His basic equation and application are as follows

$$c = U(1 - \lambda^2/\lambda_s^2) \tag{13.20}$$

where c is the eastward vorticity wave velocity relative to the earth (Rossby wave), U is the eastward velocity of a midlatitude trade wind, λ is the wavelength of an arbitrary vorticity wave, and λ_s is a critical wavelength for that latitude as given by (13.19) and (13.9). Since (13.20) shows the wave to be dispersive (speed a function of wavelength), the perturbations (waves) will move westward (upwind), be earth stationary, or move eastward (downwind) dependent upon their wavelength. A geographically stationary forcing function (e.g., a continent and its surrounding seas and oceans) can be expected to produce perturbations with a range of wavelengths. Those at or near ($\lambda = \lambda_S$) remain in the vicinity of the perturbation, those with ($\lambda > \lambda_S$) move west and dissipate; and those with ($\lambda < \lambda_S$) move east and dissipate. Rossby presents evidence that the Asiatic High (over Siberia), in one such event in early December 1938, formed over north central Siberia, moved slightly westward, and then split into two parts, one moving west over western Russia and the other east over northwest China.

A requirement of a valid theory, of course, is its consistency with experiments and/or observations. The test here is whether the sizes of the stationary high- or low-pressure cells are consistent with (13.20), where U is the zonal (east-west) flow velocity of the general circulation. Rossby shows the agreement to be very impressive, considering the rudimentary nature of the simplified theory and the extreme difficulty of measuring precise values of U at various elevations and latitudes in the 1930s. He estimates errors of from 10% to 25%. See Section 13.4 and his paper for additional details and numerical estimates.

20.6 Study and Research Projects

The projects suggested here are intended for individuals or groups. Some can be completed in a week or two, others are term or longer projects. Only a few will have exact answers. References to other texts and to professional journals are included.

Project 20.1: Review the earth's shape, orientation, and albedo (reflectivity) in a geophysics/meteorology text, e.g., Palmén and Newton (1969), and in an astronomy text, e.g., Abell (1982). Estimate solar heat input and reradiated heat as a function of latitude. If there were no oceans or atmosphere, what would equilibrium temperature be as a function of latitude? Obtain average temperatures of the earth as it is, compare these with the equilibrium temperatures, and estimate the total thermal energy transport of the oceans and atmosphere per unit time as a function of latitude. Estimate some rough north-south wind velocities. Average over seasons. How much thermal energy can be transported at latitude 45° N by the atmosphere for average atmospheric temperatures? Natural optimization and quasi-steady-state conditions of atmospheric energy/momentum transports have been analyzed, see, e.g., Paltridge (1979).

Project 20.2: The earth's daily spin rate, annual orbital motion, and inclination of its spin axis relative to its orbital plane introduce forcing functions to the global transport of its thermal energy and angular momentum. Review in the literature the relative importance of these forcing functions. What (approximately) would happen to latitudinal/longitudinal energy/momentum transport if the earth's spin rate (1) stopped, (2) decreased to the moon's orbital rate, (3) increased to three times its present speed, and (4) maintained its present rate but had no spin axis inclination? What would happen if the sidereal spin rate, inclination, and net heat input were the same as at present, but the earth had formed about a different star such that the earth's orbital period was 10 days? Review Stokes's solution for viscous flow near an oscillating plate and the similar problem of heat conduction through a rod when the temperature of one end is forced to vary periodically. How deep must one dig in the desert to reach a nearly constant temperature (± 2°C) when the surface varies daily by ± 20°C? How deep for the same ± 2°C if the surface varies annually by ± 20°C? Refer to Section 4.1 and a text on heat transfer.

Project 20.3: Both the large-scale circulation of the atmosphere and its smaller components are affected and are often dominated by the existence of water in its three states: ice, water, and vapor. Energy can be transported by motion of all three forms and can be absorbed and released through the heat of fusion for the water/ice transformation and the heat of evaporation for the water/vapor transformation. How much water would have to be evaporated each day to equal the total solar heat input each day? How much ice would have to melt to equal the same value? Assume a quantity of air with temperature 35°C and humidity 100% convects upward to a region where its temperature drops to 5°C; how much water is released and how much heat is transported per unit mass/volume? Suppose next a unit mass/volume of air is transported northward 1000 km, starting at a latitude of 10° N, and that the water (rain) drops out uniformly and continuously from saturated sea level air at 35°C to saturated air at 20°C. What is the average rain fall and how does this affect residual angular momentum of the air? Refer if necessary to a text on thermodynamics.

Project 20.4: Set up a laboratory experiment to observe horizontal transport of heat similar to that shown in Figure 20.4. Have the horizontal distance large enough to easily see the convective cells, say 24 inches. Vary water depth and temperature differences to get one cell, two cells, three cells, and then random motion. Observe the motion using neutrally buoyant particles or by injecting dye streaks. Investigate the availability of color/temperature-dependent liquid crystals that can be used to observe both temperature and motion. Reconstruct the experiment so the container is a pie-shaped segment when viewed from above. Heat the periphery and cool the vertex to simulate motion from the equator to a pole. Next, place the pie segment on a rotating table and repeat the experiment. Try to find parameters (ω, ΔT, depth-to-pie-radius ratios, etc.) to simulate three cells similar to the earth's general circulation pattern.

Project 20.5: Develop a computer program and display that shows the path of a parcel of air as it makes one complete circulation of a Hadley cell. Use a Lagrangian formulation and a simulation with time increments of one hour or less. Assume conditions at an equinox for your computation. Keep track of total internal energy, including evaporation and condensation. Consider also absorbed radiation (in clouds) and reradiated energy, and include some estimates of drag (viscous, form drag, and turbulence). Vary your parameters to achieve a model that compares favorably with the real flow. If using a color screen, experiment with display features that illustrate temperature by color and moisture content, and rain by symbols. How many times does a parcel circumnavigate the globe on its trip from 0° to 30° latitude, and how many times returning to the equator? For references see Garcia and Malone (1966), Palmén and Newton (1969), Gordon and Taylor (1975), Holton (1979), and Houghton (1986).

Project 20.6: Understanding weather phenomena in the middle latitudes and middle atmosphere is a major current task for meteorologists in data collection, analysis, computer modelling, and prediction. The complexity of the problem is enormous and is beyond the scope of any introductory text. For an introduction to the present state of knowledge, examine Scorer (1978), Holton (1979), Atkinson (1981a, 1981b), Manabe (1985), Pedlosky (1987), and Brown (1990). From these or equivalent texts, select a topic of interest, especially a topic related to vorticity generation and transport, and then write a brief report on current research as reported in professional journals and/or conferences. Relevant journals include *Journal of Meteorology*, *Quarterly Journal of the Royal Meteorological Society*, *Monthly Weather Review*, *Journal of Atmospheric Science*, *Journal of Applied Meteorology*, *Tellus*, and *Journal of Fluid Mechanics*.

Project 20.7: Investigate the errors anticipated by using a locally flat coordinate system over a 1000 km diameter region instead of using spherical coordinates. Examine the Navier-Stokes equations in spherical coordinates for a variable density and viscosity fluid, and apply it to the earth's atmosphere. See Hughes and Gaylord (1964) for the equations. Place the equations into a computer format and calculate the magnitude of each term for some typical winds. Perform the calculations for several latitudes, for winds up to 50 m/s, for several arbitrary directions, and over distances up to 1000 km. Do this also for some vertical winds with speeds up to, say, 30 m/s. For each calculation tabulate the contribution of each term, and then omit all terms that contribute less than several percent of the total. Are the remaining terms the same equivalent terms used in the locally flat (x, y, z) assumption? What are the units used in your calculation, and what is the consequence of omitting the smaller terms? Were all terms significant (say > 3%) in any calculation? Which terms are associated with Coriolis and centripetal accelerations?

Project 20.8: Find examples of geostrophic and ageostrophic flows in texts and articles discussing real atmospheric flows. Which flows fit the concepts and terminology used in Figures 7.1, 7.2, and 7.3? Locate, from the weather bureau, a pilot briefing room, or a text or article, a diagram of pressure distributions and wind velocities (at some elevation) and analyze these winds. Examine also winds in frontal systems and interacting frontal systems. Rapidly rotating winds (e.g., hurricanes and tornadoes) are covered in Chapter 22.

Project 20.9: In your own words and sketches, describe the topic of Rossby waves. You may wish to review a variety of texts and articles for data and concepts used by other researchers. Start with Rossby et al. (1939), and include, as available, Greenspan (1968), Palmén and Newton (1969), Scorer (1978), Holton (1979), Lugt (1983), Pedlosky (1987), Volland (1988), and Brown (1990) for descriptions of varying complexity and depth. Do Rossby waves compare in any way to surface waves, compression waves, Rayleigh stability, inertial waves, and/or Taylor-Proudman columns? How and in what way? A stationary Rossby wave, produced and filmed as a laboratory experiment (using water), can be seen in a video titled *Rotating Flows* by D. Fultz, distributed by Encyclopeadia Britannica Educational Corporation, Chicago, IL.

Project 20.10: As an individual or group project, schedule a visit to a weather bureau and to a pilot weather briefing room. Before going, read current literature on weather maps and symbols and their meaning. Check your library, book store, or a flight training center at a local airport for literature. Learn where the weather data originates, how it is collected and organized, and how it is used by weather forecasters and pilots. Note the emphasis for pilot weather reports on surface winds, winds aloft at various elevations, and cloud formations and their implications. How do pilots of small, single-engine aircraft, twin-engine aircraft with higher altitude and range capability, and large commercial jet aircraft use the information? Does the information on weather maps refer explicitly or implicitly to Hadley cells, Ferrel cells, polar fronts, midlatitude flows, meso-scale flows, Coriolis phenomena, latitudinal transport of energy and/or momentum, geostrophic flows, and/or Rossby waves?

Chapter 21

Oceanic Circulation

21.1 Flow Patterns and Energy Sources

This chapter assumes knowledge of the theory and observations of atmospheric flows presented in Chapter 20. Because winds supply most of the energy for large-scale oceanic flows, knowledge of atmospheric circulation patterns is critical for understanding oceanic circulation patterns. Major oceanic circulations are separated into surface currents and deep currents.

1. Surface currents with speeds of 3 to 6 km/day (and sometimes much larger) and depths of 100 to 200 m are driven by wind-water interaction at the sea surface. The surface currents, although much slower than the trade winds, for example, are yet much faster than the deep currents, with only a few exceptions.
2. Very deep currents are driven mostly by density gradients. These gradients arise as a result of temperature differences (direct solar heating) and salinity differences (evaporation, infusion of nonsaline water from rain and continental runoff, and freezing and melting of sea ice). These currents extend to the ocean bottoms and typically take 100 to 1000 years to complete a cycle.

Figure 21.1a shows a classic solution of air-water, laminar, viscous boundary layers forming as a result of air blowing over an otherwise stationary body of water. Stress and velocity are continuous at the surface due to the no-slip condition for viscous fluids, but the velocity gradient is not. In common applications, both the atmospheres and the oceans have moderate intensity turbulence, which greatly augments momentum transfer and thickens the boundary layer. Instabilities almost always develop that form water waves; these induce form drag, which further increases momentum transfer from the winds to the oceans, as in Figure 21.1b.

Velocities, directions, and flow patterns of large-scale oceanic surface flows result from an interaction of

1. Wind stress

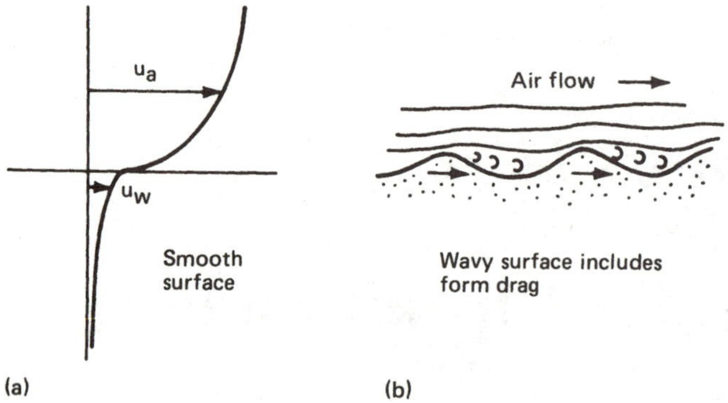

Figure 21.1: *(a) Interaction of air flow with surface water; u_a is air velocity, and u_w is water velocity. (b) Turbulence and waves greatly increase momentum transfer.*

2. Coriolis phenomena
3. Mass conservation within boundaries caused by continental margins (horizontal) and density stratification (vertical)
4. Long-term, steady-state demands of overall earth energy and momentum conservation

Figure 21.2 superimposes the mean-flow, surface trade wind patterns and the mean-flow, major oceanic circulation patterns. The dominant phenomenon in each of the five major oceans is a net rotation called a *gyre*. Gyres in the Northern Hemisphere (the North Pacific and North Atlantic gyres) rotate clockwise, and those in the Southern Hemisphere (the South Pacific, South Atlantic, and Indian Ocean gyres) rotate counterclockwise.

The low-latitude, easterly trade winds propel surface waters in a westward direction causing the North Equatorial Current (NEC) and the South Equatorial Current (SEC). These form the equatorial boundaries of the major oceanic gyres. The midlatitude westerlies propel surface waters in an eastward direction causing the North Pacific Drift (NPD), the North Atlantic Drift (NAD), and the Antarctic Circumpolar Current (West Wind Drift). These form the polar boundaries of the major oceanic gyres. The directions of rotation of the major oceanic gyres are consistent in each case with the major east-west flows, westward at low latitudes and eastward at high latitudes, leading to clockwise rotation in the Northern Hemisphere and counterclockwise in the Southern Hemisphere. This rotation direction is also consistent with Coriolis phenomena associated with geostrophic flows surrounding a high-pressure center, as detailed in Section 21.3.

A narrow Antarctic countercurrent, adjacent to the continent of Antarctica and flowing westward, is driven by the south polar easterlies. Refer to Figure 20.7 for nomenclature. Equivalent westward flows exist in the Arctic but are more fragmented

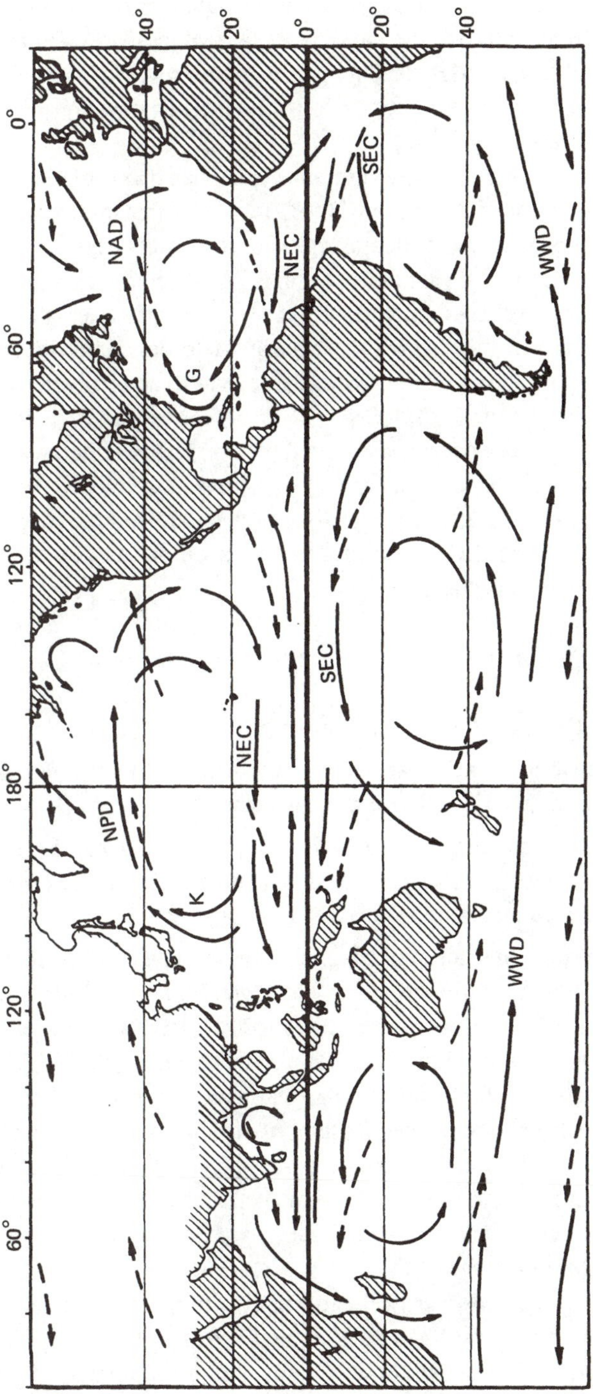

Figure 21.2: *Seasonally averaged ocean surface currents (solid) and trade winds (dashed). Anticylonic rotation exists in the North and South, Atlantic and Pacific, and Indian Oceans. NEC and SEC = North and South Equatorial Currents; NPD and NAD = North Pacific and Atlantic Drifts; K = Kuroshiro Current; G = Gulf Stream; and WWD = West Wind Drift.*

because of land obstructions. The narrow, shallow Equatorial Counter Currents oppose the trade wind directions and the adjacent equatorial currents. These counter currents form because the equatorial currents pile up masses of water at the eastern edge of continents, and the counter currents form a mass conservation return flow driven by the ocean height differential.

As westward-moving warm equatorial currents are deflected toward the poles they transport thermal energy from low to high latitudes. Total transport of heat from equatorial to polar regions is due to both atmospheric flows and oceanic flows. The relative contribution of each varies with latitude and is greatly complicated because of transfers of thermal energy between the oceans and the atmosphere as latent heat of evaporation and condensation. Oceanic currents appear to dominate the poleward transport of heat in equatorial regions ($< 20°$ latitude) while atmospheric flows are much more effective transporters of global heat at higher latitudes ($> 50°$). See Palmén and Newton (1969). The oceanic gyres, and the dominance of ocean areas over land mass areas, lend to both temporal and spatial variations in the relative importance of the oceans and atmosphere as transporters of global heat in the mid-latitudes. The heating of the Alaskan coast by the Kuroshiro Current and of the coast of Norway and Scotland by the Gulf Stream are examples.

21.2 Coriolis and Oceanic Gyres

Chapter 7 distinguished between two limiting cases:

1. Accelerations relative to inertial space where an object or fluid is constrained by real forces to follow a specific path in a rotating frame
2. Acceleration relative to a rotating, and therefore noninertial, frame. This case need not involve forces, because force-free motion in an inertial frame usually will appear to be accelerated motion relative to the rotating frame.

Case 2 is used to explain the oceanic gyres. Figures 21.3a and 21.3b show views of flows near the North Pole. The earth rotates at a rate Ω, and each latitude has a tangential velocity ($v_\phi = \Omega R \cos \phi$). An eastward flow ($+u$) has a real tangential velocity ($v_\phi + u$), and a westward flow ($-u$) has a real tangential velocity ($v_\phi - u$). The cross section distinguishes between the oblate earth and a sphere. The earth over its four-billion-year existence has reached its present equilibrium shape as a balance between gravitational and centrifugal forces. As a result, the difference between the sphere and the oblate earth represents the tilt of the earth's surface that just balances gravitational force (ρg) with the centrifugal force ($\rho v_\phi^2/R \cos \phi$) acting on a unit volume of material *stationary* on the earth.

If a parcel of fluid has an eastward velocity relative to the earth, its centrifugal force is no longer balanced by the tilted surface, and it begins to flow up the tilted surface toward the equator. A westward flow has a net tangential velocity less than v_ϕ

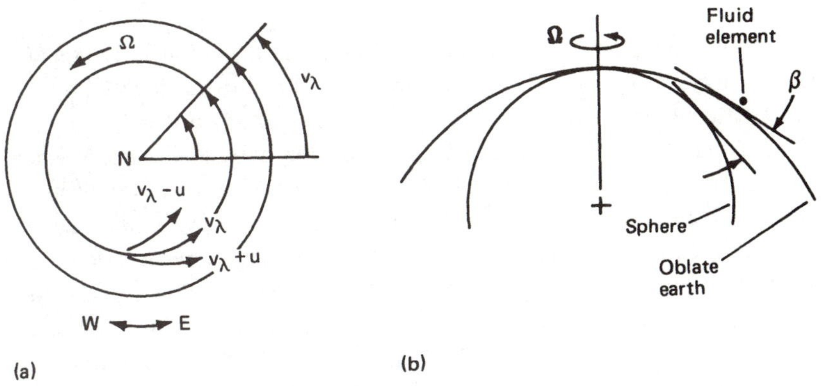

Figure 21.3: *Coriolis effect due to east or west flows on surface of oblate earth, (a) view from north, and (b) cross section.*

and cannot maintain its position on the tilted oblate surface. It begins to flow toward the pole. The effect is due to the horizontal component of $(\mathbf{\Omega} \times u\,\hat{\mathbf{i}})$ of Figure 20.6 and varies as $(\sin \phi)$.

Applied to oceanic gyres, this phenomenon causes the eastward-moving mid-latitude flows to bend toward the equator. The westward equatorial currents bend poleward to form the Kuroshiro Current, the Gulf Stream (Figure 1.16), and other poleward flows. To complete this self-consistent pattern, the poleward currents then convert low-latitude, earth tangential velocities toward the polar regions, and must therefore bend eastward. Currents flowing toward the equator do the reverse, blending in with the westward-moving North and South Equatorial Currents. Thus, the oceanic gyres are consistent with geostrophic Coriolis effects and also with seasonally averaged wind shear stresses. Note that major oceanic currents in the Northern Hemisphere always turn to the right and generate clockwise gyres, while those in the Southern Hemisphere always turn to the left and generate counterclockwise gyres.

Could wind shear stresses or adjacent flows induce gyres rotating in the opposite directions? The answer is, *Yes*. Water draining from a wash basin can be induced to rotate oppositely to earth Coriolis phenomena. The shape of the basin, or the initial angular momentum imparted to the water, can easily induce rotation in either direction. All that is required is a large enough forcing function. There are, in fact, several smaller oceanic gyres that are forced to rotate oppositely to the major gyres. See Figure 21.2. These oppositely rotating gyres have low-pressure centers rather than high-pressure centers.

21.3 Western Intensification and Elevated Centers

Major phenomena characteristic of the oceanic gyres include western displacement of the oceanic gyres with high-velocity flows along their western edges (western intensification) and geostrophic motion associated with the elevated masses of water (mounds) centered on the gyres. The western intensification is evidenced by the very narrow, swift, and deep currents along the western edge of the major oceanic gyres. The Gulf Stream and the Kuroshiro Current are the two most extreme examples, having speeds up to 120 km/day and depths to 1 km or more. The westward-flowing Equatorial Currents are at low latitudes where the Coriolis effect ($\sim \sin \phi$) is minimal. These currents impact against the eastern boundaries of continents, and their flows are deflected north or south as thin jets. By contrast, the eastward-flowing currents at the poleward boundaries of the gyres occur in regions where $\sin \phi$ is large. These flows are deflected toward the equator by Coriolis phenomena long before they reach the opposite continental boundary.

An additional factor aiding western intensification is the acquisition of the earth's angular velocity ($\Omega \sin \phi$) by each fluid element by remaining horizontal while flowing poleward. At the equator, a horizontal sheet of fluid, stationary on the earth, has no vertical component of vorticity. The same sheet of fluid, if simply rotated 90° by sliding it as a thin disk from the equator to a pole, still has no vertical component of vorticity (angular velocity) relative to inertial space. However, its zero vertical vorticity will now appear to an observer rotating with the earth to be a negative vorticity (rotation). This relative vorticity of fluid elements, which after integration around a closed curve and use of Stokes's theorem, provides additional circulation for low-latitude fluids transported poleward. The reverse, of course, is true for fluids transported toward the equator.

Figures 21.4a and 21.4b show a simplified Northern Hemisphere ocean basin and a central mound of elevated surface water. A question arises. Why an *elevated* central portion rather than a *depression*? All of the vortices and circular pathline flows of Chapters 10 and 12 had low-pressure central cores consistent with a depressed central region. The differences are these:

1. Centripetal acceleration was the dominant feature of vortex and circular pathline flows. Centrifugal force was in equilibrium with the surface height gradient, resulting in a central depression. In contrast, centripetal effects are negligible in slowly rotating oceanic gyres.
2. The vortex and circular pathline rotations occurred relative to (and on) an inertial surface. In a forceless motion, each fluid element would travel in a straight line. Oceanic gyres occur on the rotating earth where forceless motion is a curved path due to Coriolis phenomena. The forceless motion path typically curves more sharply than the curvature around the average periphery of an oceanic gyre. This forceless motion (natural) curvature is called the *circle of inertia*.

Consider an initial generation of such a gyre. Moving water does bend and flow sharply toward the center initially. Only after a mound is raised does the flow finally reach an equilibrium between its tendency to bend inward more sharply and the outflow

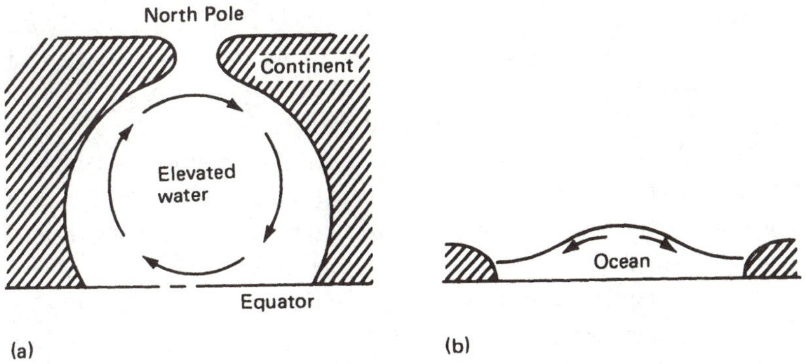

Figure 21.4: *Idealized northern oceanic gyre and geostrophic motion associated with high-pressure center, (a) view from above and (b) cross section.*

from the sloping surface of the mound. This mound of water (several meters high) provides an enormous energy reservoir for maintaining the gyral motion during periods of low or fluctuating wind stress. If wind stress went to zero, for example, the surface currents would slow down, and water from the mound would begin to flow radially outward. The outward moving flows would bend to the right for a northern gyre (left for a southern gyre) maintaining the gyral motion. In a perfect inviscid fluid, the mound and the gyral motion would continue indefinitely, each maintaining the other. Actually, of course, energy is continually dissipated and must be replenished by wind stress. Note that the raised central portion represents a high-pressure region. High-pressure regions always produce similar rotations whether in the atmosphere or the ocean. The smaller oceanic gyres that rotate oppositely to the major gyres have depressed (low-pressure) centers.

21.4 Localized Flow Phenomena

The oceanic gyres driven by averaged wind stresses, and deflected by Coriolis forces and continental boundaries, define the major oceanic flows. Superimposed on these gyres are numerous localized flow phenomena on a smaller scale and with shorter time constants. Some of these localized flow phenomena are affected by and show characteristics of rotating flows, while others do not. This section briefly describes several localized flow phenomena as illustrations of rotating flow theory with smaller dimensions and shorter characteristic times than the major gyres. Section 21.5 includes mathematical models for quantifying important phenomena.

Ekman analyzed the details by which oceanic surface currents are created by wind stress. Making use of Prandtl's then-recent boundary layer theory, Prandtl's mixing length turbulence theory, and the Coriolis phenomena, he computed that water very near the ocean surface should flow 45° to the right of the wind direction in the

Northern Hemisphere and 45° to the left in the Southern Hemisphere. By the same mechanism, stress between that layer and the next lower layer then results in another deflection, with each lower layer being deflected more and more. After a depth of about 100 m is reached, the induced flow is oppositely directed to the wind as shown in Figure 21.5. Also refer to Section 21.5. Ekman spiral flows can be created in the laboratory and have been observed in both atmospheric and oceanic flows. However, they are not observed routinely in nature because other factors often dominate. Figure 21.2 shows regions of atmospheric/oceanic flows consistent with the Ekman model, but it also shows regions where continuity of flow and other factors obviously dominate the flow directions.

Wave phenomena generally create or modify motions smaller in length and time scales than those discussed in this chapter. An exception are the tidal waves introduced in Chapter 19. Tidal waves have such long wavelengths that even in the deep oceans the ratio of wavelength to water depth is such that shallow water theory applies. The forward (east-to-west) wave crest motion of volume elements travels several kilometers, and then returns approximately to the starting position in the wave trough. As the crest flow moves westward, it is deflected toward a pole, and the return (trough) flow is deflected toward the equator. Viewed from high above, the surface water path (i.e., path of a floating object) is elliptical, an ellipse several km east-west and nearly as large north-south. Tidal currents in the open ocean exceed many surface currents, and they complicate the measurement of both.

As the Gulf Stream is deflected eastward by Coriolis force, it enters the main ocean as a meandering, swift, narrow jet as shown in Figure 1.15. Vortex rolls form along both edges. These and other oceanic distinct vortices in the mesoscale range (50 to 500 km in size and 2 to 20 days in time) are ubiquitous oceanic features that contribute significantly to local transport of heat, momentum, and vorticity. The Gulf Stream vortices (called *rings*) are 100 to 150 km in diameter. "Warm" rings form along the northern edge of the stream, and "cold" rings form along the southern edge; both translate westward, with the warm rings impacting on the continental shelf. Their diameter is much larger than the oceanic depth, indicating their possible extension to the ocean floor. Project 21.7 in Section 21.6 includes references and suggests further studies.

Other localized flow phenomena affect coastal regions and harbors. Some, such as the transverse tilt of swift tidal currents in long, narrow bays or river channels, are related to Coriolis acceleration (see Example 7.2). Others, such as the overturning of the south-moving littoral current along the California coast, may be related to the Ekman spiral mechanism. Many other local phenomena are too small in size or short-lived in time to show distinct effects of the earth's rotation; however, other rotating flow mechanisms, such as vorticity induced by local strain rates, are often apparent.

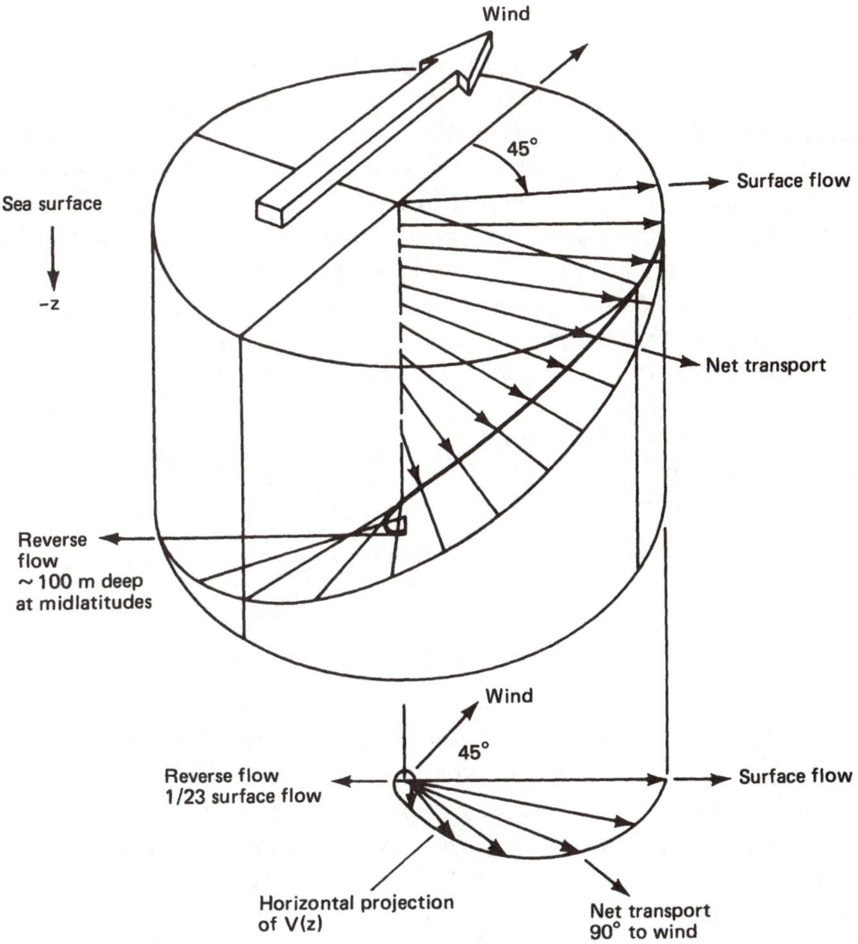

Figure 21.5: *Ekman spiral illustrates geostrophic oceanic flows induced by wind stress. Northern Hemisphere flow is shown.*

21.5 Some Quantitative Analyses

Quantitative analysis of atmospheric and oceanic flows are difficult for several reasons:

1. The complicated physical boundary conditions of continents and oceans
2. Interaction with other phenomena, especially the thermodynamics of air-water vapor and water-ice mixtures
3. Instability and turbulence
4. Subtle, but important, variations in external phenomena such as solar intensity
5. Measuring a complete set of three-dimensional (3-D) initial conditions over geographic space within a reasonably short time

In spite of these difficulties, much progress has been made. Weather satellites, aircraft, and ships provide useful (but not complete) sets of initial conditions, and computers are approaching the sophistication necessary for handling the computations. Predictions of specific weather patterns are always suspect, but some small-scale and some large-scale (space and time) predictions are successful.

All start with the general Navier-Stokes equations, the mass and energy conservation equations, and constitutive equations describing the fluid. The simplified analyses of this section will assume ($\rho = c$, $\mu = c$), so the energy equation is not needed. Use of the (u, v, w) approximation for velocities in a region centered at (λ, ϕ, r), and neglecting vertical transport phenomena ($w \sim 0$), yields

$$\frac{\partial u}{\partial t} + u\frac{\partial u}{\partial x} + v\frac{\partial u}{\partial y} - 2\Omega v \sin\phi = -\frac{1}{\rho}\frac{\partial p}{\partial x} + \frac{\mu}{\rho}\nabla^2 u \tag{21.1}$$

$$\frac{\partial v}{\partial t} + u\frac{\partial v}{\partial x} + v\frac{\partial v}{\partial y} + 2\Omega u \sin\phi = -\frac{1}{\rho}\frac{\partial p}{\partial y} + \frac{\mu}{\rho}\nabla^2 v \tag{21.2}$$

$$2\Omega u \cos\phi = \frac{1}{\rho}\frac{\partial p}{\partial z} \tag{21.3}$$

Refer to Figure 20.6 for derivation of the Coriolis terms. The term ($2\Omega u \cos\phi$) can be shown to be much less than typical ($\partial p/\partial z$) values and is typically neglected. In this case, the vertical equation reduces to $\partial p/\partial z = 0$, where p is the reduced pressure of (6.8) leading to a hydrostatic atmosphere or ocean. Here z is perpendicular to the oblate earth's surface, so g includes both gravitational and earth (angular velocity) centripetal effects.

$$\frac{\partial p}{\partial z} = -\rho g \tag{21.4}$$

Some older texts and articles on oceanography use a coordinate system with *+z down* and a *left-hand* coordinate system. All equations here use +z up and a right-hand coordinate system.

Flows in the atmosphere and ocean typically have only low-intensity turbulence, but, even so, the turbulent transfer of momentum (and other fluid properties) is much larger than that due to viscous diffusion. The viscous term ($\mu\nabla^2 V$) accordingly is replaced with symbols, say (K_x, K_y, K_z), representing combined viscous and turbulent interactions. Because ($2\Omega \sin\phi$) appears repeatedly, it has become convenient to refer to it as the Coriolis parameter

$$f = 2\Omega \sin\phi. \tag{21.5}$$

With these conventions, (21.1) and (21.2) reduce to

$$\dot{u} - fv = -\frac{1}{\rho}\frac{\partial p}{\partial x} + \frac{1}{\rho}K_x \tag{21.6}$$

$$\dot{v} + fu = -\frac{1}{\rho}\frac{\partial p}{\partial y} + \frac{1}{\rho}K_y \tag{21.7}$$

as the final equations for approximating important horizontal atmospheric and oceanic flows.

If a quantity of ocean water is given an arbitrary horizontal velocity, and if horizontal pressure gradients and viscous-turbulent interactions are negligible, (21.6) and (21.7) reduce to

$$\dot{u} = fv \tag{21.8}$$

$$\dot{v} = -fu \tag{21.9}$$

The solution is a horizontal circle of radius

$$r = V/f \tag{21.10}$$

where $V = (u^2 + v^2)^{1/2}$ is the horizontal speed. The circle is called the *circle of inertia* and represents the natural motion of a force-free flow on a rotating sphere. The time of rotation of one cycle (its period) is given as

$$\tau = 2\pi/f \tag{21.11}$$

The use of the symbol f for the Coriolis parameter is obvious here as it represents the frequency of the motion. Near the equator $f \sim \sin\phi$ is small, so r and τ are large. Near the poles $r \rightarrow V/2\Omega$ and $t \rightarrow \pi/\Omega$ with typical values of $r < 10$ km and $\tau \approx 12$ hours. This phenomenon was used in Section 21.3 to help explain the western intensification of the major oceanic gyres. The eastward polar edges of the gyres are attempting to curl sharply toward the equator, while the equatorial edges are more inclined to follow lines of constant latitude.

Geostrophic flows are analyzed by assuming a very slow-moving flow ($\dot{u} \sim \dot{v} \sim 0$) and negligible viscous-turbulent interaction so that (21.6) and (21.7) reduce to

$$\frac{\partial p}{\partial x} = \rho f v \tag{21.12}$$

$$\frac{\partial p}{\partial y} = -\rho f u \tag{21.13}$$

These equations indicate that a horizontal pressure gradient in one direction causes a flow in a perpendicular direction. In (21.12), a pressure increasing to the east (a force to the west) will cause a northward flow. The mound centered in each of the five major oceanic gyres is a region of high pressure, causing an outward radial force field. Equations (21.12) and (21.13) for the Northern Hemisphere (Ω and f positive) indicate a clockwise distribution of flows around the mound as observed. In the Southern Hemisphere the sign of f reverses, as viewed by an observer on the earth's surface, to produce a counterclockwise flow.

Rotating masses of fluid, such as gyres, are named *cyclonic* or *anticyclonic* based on whether they surround a low-pressure region (a depression) or a high-pressure region (an elevation). Flows about low-pressure regions, whether in the ocean or in the atmosphere, or in the Northern Hemisphere or Southern Hemisphere, are cyclonic; flows about high-pressure regions are anticyclonic. Although the direction of rotation of the northern and southern gyres are reversed relative to surface observers, in a larger sense they rotate in the same direction. An observer fixed in space will see that both the northern and southern major oceanic gyres have a component of rotation that opposes the earth's rotation. Cyclonic motion is always associated with a low-pressure region and always has a component of its rotation consistent with the earth's rotation. Anticyclonic motion is always associated with a high-pressure region and always has a component of its rotation opposing the earth's rotation (Figure 22.2).

The effect of wind stress in producing oceanic flows on the rotating earth is analyzed by reducing (21.6) and (21.7) to

$$fv = -\frac{1}{\rho} K_x \tag{21.14}$$

$$fu = \frac{1}{\rho} K_y \tag{21.15}$$

In this approximation everything is assumed negligible except Coriolis and friction (viscous-turbulent) effects. Equations (21.14) and (21.15) can be integrated, provided reasonable assumptions can be made regarding K_x and K_y. Ekman modelled wind shear using (μ_e) as an eddy "viscosity," which, times $\partial u/\partial z$, gives wind shear stress (τ_e) in the direction of u. Net force due to adjacent fluid elements on a unit volume then becomes

$$\frac{\partial}{\partial z}\left(\mu_e \frac{\partial u}{\partial z}\right) \quad \text{and} \quad \frac{\partial}{\partial z}\left(\mu_e \frac{\partial v}{\partial z}\right) \tag{21.16}$$

in the u and v directions, respectively. The equations to be solved become

$$\frac{\partial}{\partial z}\left(\mu_e \frac{\partial u}{\partial z}\right) + \rho f v = 0 \tag{21.17}$$

$$\frac{\partial}{\partial z}\left(\mu_e \frac{\partial v}{\partial z}\right) - \rho f u = 0 \tag{21.18}$$

On an assumption that μ_e is independent of depth, and that the ocean is not shallow, the solution is

$$u = V\, e^{-\pi z/D} \cos(\pi/4 - \pi z/D) \tag{21.19}$$

$$v = V\, e^{-\pi z/D} \sin(\pi/4 - \pi z/D) \tag{21.20}$$

where

$$V = 2\pi\tau_e/(D\rho f\sqrt{2}) \tag{21.21}$$

$$D = -\pi(2\mu_e/\rho f)^{1/2} \tag{21.22}$$

Figure 21.5 shows the solution pictorially. The surface water moves 45° to the right of the wind direction (Northern Hemisphere), and each successively deeper layer deflects progressively further to the right. Finally, at depth $z = D$, the solution shows the water flow direction to have completely reversed from the surface water flow direction, i.e., $\cos(\pi/4 - 0)$ compared to $\cos(\pi/4 - \pi)$ and to have reduced to $e^{-\pi}$ of its surface value (about 1/23). The quantity D is an estimate of the depth of penetration of wind shear.

These estimates are generally in agreement with the actual depth of typical ocean surface currents. Computed values, of course, are sensitive to the value or function chosen for μ_e. Ekman's solution assumes a constant eddy viscosity; in reality it varies, so that the angle between the surface wind stress and the surface current is 10° to 25° rather than 45°. The projection of horizontal velocity components at different heights onto a horizontal surface is called the *Ekman spiral.* Net water transport (integrated from $z = 0$ to $z = D$) is approximately 90° to the wind direction. This result is independent of the choice for eddy viscosity. In water much shallower than D, the solution is modified, and surface water flow and net transport are more aligned with the wind direction.

21.6 Study and Research Projects

The projects suggested here are intended for individuals or groups. Some can be completed in a week or two, others are term or longer projects. Only a few will have exact answers. References to other texts and to professional journals are included.

Project 21.1: Quantify air-sea interactions over a range of possible atmospheric pressures, wind velocities, and wave heights. Air-sea flow interaction between a horizontal wind in laminar flow and a perfectly smooth sea can be estimated from a solution by Lock (1951). Laminar flow perpendicular to a smooth sinusoidal surface can be approximated by the same theory as long as turbulence and flow separation do not occur. When separation occurs, as is common, form drag is introduced and can be estimated from data and empirical equations for flows past shapes similar to wave shapes. Quantities of surface water nearly reach wind speeds for high winds and breaking waves. Wave heights can reach extreme dimensions. In a North Atlantic hurricane in December 1945, winds greatly exceeded 100 mph and waves approached 100 ft trough-to-peak. Several ships were sunk and/or broken in half, a riveted Liberty ship (the George Westinghouse) was at times completely under water (like a submarine) as the helmsman kept it headed into the waves, and an aircraft carrier's flight deck was crushed and partly twisted from its hull *(New York Times*, stories on p. 1 of December 24 and December 26, 1945, and photo on p. 14 of January 6, 1946). Next, estimate possible sea-level height distortions due to atmospheric pressure variations--i.e., between an atmospheric high and low. Chapter 22 discusses details of hurricane dynamics.

Project 21.2: Estimate the north-south transport of thermal energy (per unit time) by oceanic flows in the North Atlantic at latitude 30°N, for average oceanic temperatures. How does this compare with estimates for atmospheric transport of energy discussed in Project 20.1? Estimate the amount of angular momentum transported by the same flows. What is the total angular momentum (relative to the earth) stored in the South Atlantic gyre and in the Antarctic circumpolar drift? How large are these compared to the earth's total angular momentum? Many introductory, intermediate, and advanced texts are available for reference, as well as professional journals. See, e.g., Sverdrup (1942), Neumann and Pierson (1966), Palmén and Newton (1969), Gross (1971), Stowe (1983), Maul (1985), Manabe (1985), Pond and Pickard (1986), Nihoul and Jamart (1987), and Pedlosky (1987).

Project 21.3: As a laboratory experiment/demonstration fabricate an "inside out" ocean as follows. Place a large circular pan on a large turntable and fill about half full with water. A 3 ft diameter pan is adequate, but larger is better. Spin the turntable at a rapid enough speed that the water surface is a clearly discernable paraboloidal surface. After a steady state is reached, you have a motionless ocean capable of illustrating Coriolis phenomena, oceanic gyres, and other oceanic phenomena. Introduce at some arbitrary point an east-west flow (circumferential, relative to the rotating pan) and then a north-south flow (radial). Observe the Coriolis effects with a dye stream or suspended particles. Next, introduce a low-pressure site (extract water at some point) and a high-pressure site, and observe the flows. Then constrain a portion of the water within a sheet metal ring with one-fourth or less the diameter of the pan. Create an arbitrary flow perturbation inside the metal ring. Is an "oceanic" gyre created? Is the flow distorted to one side or the other of the ring? For a similar experiment see von Arx (1962).

Project 21.4: Model the experiment of Project 21.3 as a computer exercise. Model the tests described in Project 21.3 as prescribed perturbations to the rotating flow. If the experiment is also available, attempt to predict the experimental results. For a different and simpler laboratory model see Pedlosky and Greenspan (1967).

Project 21.5: Coriolis phenomena on the rotating earth dictates that major oceanic flows (gyres) will preferentially rotate clockwise in the Northern Hemisphere and counterclockwise in the Southern Hemisphere. Some smaller gyres oppose this rule, e.g., a flow in the North Pacific that carries water northward along western Canada, west and south below Alaska and the Aleutians, and returning eastward at about a latitude of 45°N. Assume, for conceptual and quantitative simplicity, a circular gyre of about that size, strength, and location. Will flows along the eastern edge or the western edge be intensified? Estimate by how much. Will the center be elevated or depressed? By how much?

Project 21.6: Perform a literature search for theory and examples of oceanic Ekman spirals and flows that correspond to circles of inertia. See Sverdrup (1942). Examine charts showing seasonal wind velocities and oceanic flow velocities, and attempt to predict where Ekman spiral phenomena should occur. Then study the references listed in the previous projects and finally search the professional journals, including *Journal of Marine Research*, *Journal of Physical Oceanography*, *Journal of Fluid Mechanics*, *Dynamics of Atmospheres and Oceans*, and *Oceanographic Monthly Summary*. Nature rarely accomodates by displaying perfect examples of these phenomena but rather by hints that the mechanism is perturbing a larger flow field in a way consistent with the mechanism. If the experiment described in Project 21.3 is available, float a small object on the water surface and attempt to get it to move in a circle of inertia. Attempt to create an Ekman spiral by mounting a small fan on the rotating pan and having it blow horizontally over the surface in a small region.

Project 21.7: The Gulf Stream and the Kuroshiro Current are sometimes shown as smooth flowing streams in introductory texts, but in actuality they leave Florida and Japan as meandering streams that enclose and transport eddies (vortices) along either edge. Some vortices are hot and others are cold relative to the average ocean and are seen best on satellite photographs in the infrared. These vortices are known to dominate oceanic motions in the range 50 to 500 km in space and 2 to 20 days in time. The journal *EOS (Transactions American Geophysical Union)* includes research contributions on this topic and also notices for meetings. See, for example, the article by Robinson et al. (1989) and the meeting notice in *EOS*, "Modelling Distinct Vortices of the Ocean" (Cushman-Roison 1987) that describes one of a series of such meetings held at Florida State University. On the same page of *EOS*, a report by Munk (1987) illustrates the Gulf Stream, and the associated eddies and rings. See Figure 1.14 in the Introduction of this text. Review the references in the Robinson et al. article and search the literature for more recent research results. Include the journals listed in Project 21.6 and *EOS*. Given the size and strength of these vortices, the depth of the ocean, and the theorem that vortices cannot simply end in an unbounded fluid, what is the probability that they extend to the ocean floor? From data given in the literature, estimate their vortex strength. Why are some hot and some cold? Do they cause vertical mixing?

Project 21.8: Examine details of the generation of tides, their effect on oceanic circulation, and near-shore phenomena associated with tides. To what extent do these phenomena react to Coriolis or other rotating forcing functions, and to what extent does vorticity generated by the tidal motions influence them? See Pugh (1987) and Marchuk and Kagan (1989) for broader and more detailed coverage of both tides and waves, and Brosche and Sündermann (1982) for the cumulative effect of tidal motion on the earth's rotation rate.

Project 21.9: Perform typical calculations using the theory and equations of Section 21.6. Find realistic examples for the application of (21.8) and (21.9), (21.12) and (21.13), and (21.14) and (21.15). Review the several assumptions and simplifying restrictions used in their derivation, and describe situations where their use would be inappropriate. Remember that even the general Navier-Stokes equations in spherical coordinates are also an approximation; an oblate spheroidal coordinate system would be more correct. See Hughes and Gaylord (1964).

Project 21.10: Survey the professional literature and write a report on the use of computers to model the oceans from global circulation to turbulence. The entire problem, of course, should include solar and subsurface heat input and interactions with the atmosphere. Holland and McWilliams (1987) review the use of supercomputers to achieve realistic models in the journal *Physics Today*. Read their article and update it in the more recent literature. Include research projects to model global meteorology. Note especially the frequent need to include isolated vortices that can appear and disappear on small and medium scales. Some of these transient local phenomena can have major impacts on subglobal and even global phenomena. Hurricanes, covered in Chapter 22, are important examples.

Chapter 22

Intense Atmospheric Vortices

22.1 Whirlwinds and Dust Devils

Rotating masses of air, in sizes of interest to this chapter, cover a range of

1. Small whirlwinds, such as might form behind the sharp edge of a building
2. Dust devils commonly seen on a hot, flat desert
3. Highly destructive tornadoes (water spouts over the ocean)
4. Large-scale vortex storms (squall lines, dust storms, and thunderstorms)
5. Hurricanes, also known as tropical cyclones and/or typhoons

These all rotate relative to the earth and, with few exceptions, with their axes nominally vertical. They have in common that they are intense, concentrated, atmospheric vortices. They differ in their size and duration, why and how they form, how they are maintained, and how they finally end.

Small whirlwinds form behind buildings, hills, and other obstructions. The viscous boundary layer that forms along the obstruction has insufficient energy to overcome the adverse pressure gradient at the rear of the obstruction. The free stream separates from the surface, and the vorticity produced in the boundary layer forms a vortex in the wake of the obstacle. At appropriate Reynolds numbers, the vortices are shed from the region. If the vortex core has one end on a solid surface, e.g., the ground, a radial secondary flow forms along the surface. The radial flow moves inward and then up into the low-pressure core, providing a vertical component of motion. Whirlwinds form in nearly every wind and behind nearly every obstacle, but only if dust or leaves are entrained in the upward-moving air do they have a visible core.

The energy comes from the energy of the free stream, often as a gust of wind. No other energy is available, and the whirlwind dissipates within seconds. Viscous dissipation at the ground, the introduction of low-vorticity air into its core from the bottom surface, and diffusion and dissipation into the surrounding upper air dissipate its momentum and energy. Helmholtz's theorem, that a vortex cannot end in a fluid, holds strictly for an inviscid, constant-density fluid. In a viscous fluid, an open-ended vortex can be initiated, but like the whirlwind, cannot persist long unless it is maintained by external sources of energy and angular momentum. The direction of rotation, clockwise or counterclockwise viewed from above, depends upon the side of the obstacle that caused the whirlwind. Pairs of whirlwinds, rotating in opposite directions, can form simultaneously. The concepts *cyclonic* and *anticyclonic* do not apply because whirlwinds are too small and occur too quickly for earth Coriolis effects to influence their dynamics.

The other four atmospheric vortices have various other causes. The causes often involve direct thermal energy sources during the initial formation and/or maintenance of the phenomena. The dust devil is an atmospheric vortex with a central core 1 to 10 meters in diameter and from 50 to several hundred meters tall. Dust devils are concentrations of vorticity originating from wind shear, either shear between the wind and the earth or shear between opposing winds. The forming vortex, if initially horizontal, will tend to assume a more stable vertical orientation (see Figures 10.10 and 10.11). Favorable winds and/or thermal convection can intensify the vortex by longitudinal stretching.

Their persistence, from many minutes to more than an hour, as a usually gentle column of rotating dust and sand moving slowly across the desert indicates a continuing energy input. The low-pressure core may feed on a thin layer of heated air not yet risen from the desert floor; other researchers have suggested that sunlight heats the dust-filled core directly. In either case, the central convecting core is able to maintain itself as a low-pressure region. As with a whirlwind, the dust devil always has a low-pressure core, but it is small enough that it need not by cyclonic. However, cyclonic motion (counterclockwise in the Northern Hemisphere) is consistent with earth Coriolis phenomena for a low-pressure core and provides a needed source of angular momentum for the larger and more persistent dust devils.

22.2 Large-Scale Vortex Storms

Tornadoes develop from vortex storms and hurricanes are the most intense of all atmospheric vortices, and are discussed last. Large-scale vortex-generating storms, more often identified as *squall lines* or *dust storms*, vary from tens to thousands of kilometers in size and from a few hours to a few days in duration. They are often associated with frontal systems (divisions between large cold and warm masses of air). If moisture is present, temperature-pressure variations are usually adequate to produce visible condensation; in dry air the velocities are often large enough to transport visible sand, loose soil, or vegetation. They are called dust storms in desert areas, and provide the mechanism for erosion and transport of sand. They can be much more effective transporters of material than large rivers. Their appearance often includes very smooth, rounded surfaces, indicating the presence of intense rotation. The leading

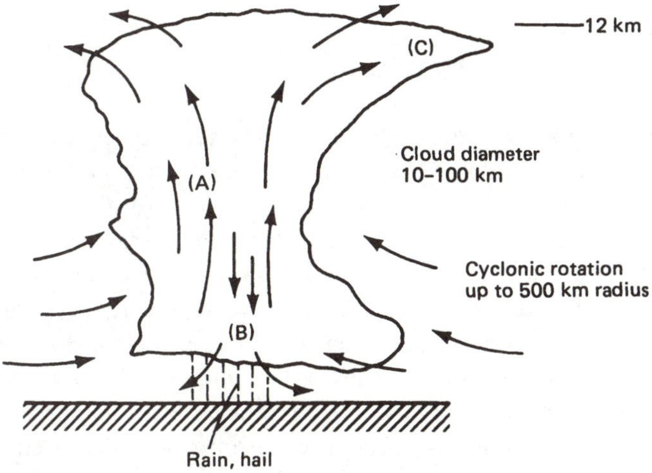

Figure 22.1: *Mature thunderstorm. (a) Updrafts of 30 m/s. (b) Downdrafts of 10 m/s with rain, hail. (c) Anvil shape produced by light upper-atmospheric winds.*

edge of a dust storm sometimes shows a series of parallel, nearly vertical vortices. Large vortex storms produce individual tornadoes, and/or a line of thunderstorms, which in turn may produce tornadoes.

Thunderstorms (cumulonimbus cloud formations) can be produced in a squall line or as an isolated convection cell. The classic and largest example produces an upper anvil cloud as shown in Figure 22.1. Wind shear may augment their formation, but they are basically large thermal convective cells. Most of their energy derives from the condensation of moisture in rising, unstable, warm, moist air currents. Strong vertical velocities in the convecting central region extend to the stratosphere. The rising air creates a low-pressure system drawing in air from distances of several hundred kilometers. The final structure resembles an inviscid vortex with sink. A well-defined circular core rarely develops, and the surface wind velocities are not usually fast enough to be destructive.

Size and time scales are consistent with earth Coriolis phonomena, and the converging air always produces a cyclonic rotation with tangential velocities approximating an inviscid vortex ($v_\theta \sim r^{-1}$). The vertical winds produce toroidal vortices concentrated at the forward-moving edge of the anvil cloud. These horizontal vortices sometimes interact with the major vertical vortex of the thunderstorm to produce what is called a *rotor* or *squall cloud*. This cloud appears to be associated with the tornadoes that sometimes accompany thunderstorm systems. Vortex storms, including squall lines and thunderstorms, are mostly mid-latitude phenomena. Alternating meridional "fingers" of cold and warm air (Figure 20.5) augment and/or create both high- and low-pressure regions. As noted earlier, low-pressure systems are cyclonic, while high-pressure regions are anticyclonic. These vortex storms are sometimes called *extratropical cyclones* meaning "away from the tropics."

22.3 Hurricanes (Typhoons)

Tropical cyclones originate between 10° and 20° of the equator, are low-pressure systems, and, when fully developed, are called *hurricanes* in the Atlantic and *typhoons* in the Pacific.

Hurricanes and tornadoes are the most violent and destructive of atmospheric phenomena. Hurricanes tend to be ~1000 km in diameter, to persist for several weeks or longer, and to have surface winds ~200 km/hr. Tornadoes are much smaller but are more intense. A tornado diameter of 0.1 km is large, and a time scale of one hour is long, but the velocity of near-surface winds has been estimated for some tornadoes to be near supersonic. In some cases where direct measurements might have been possible, the measuring instruments and/or the meteorological stations were destroyed. Hurricanes are a direct effect of earth Coriolis phenomena. Tornadoes are not dependent on earth Coriolis phenomena, but rather represent intense concentrations of vorticity that is concentrated, stored, and intensified by other mechanisms--squall lines, thunderstorms, hurricanes, and even erupting volcanoes and forest fires (fire storms).

Figure 22.2 illustrates how air converging on a thermally induced "sink" concentrates earth angular momentum. Figure 22.2 applies to large thunderstorm convective cells and to hurricanes. The phenomenon can be explained on several levels of sophistication. The simplest is that used earlier to explain major atmospheric and oceanic circulations. Assume a convective cell (a sink) at 30° N latitude. Air from the south approaches the cell with a larger earth transverse velocity than the transverse velocity of the earth at 30°. As it approaches the cell it is moving eastward relative to the earth. Air approaching from the north has a smaller transverse velocity and, to an earth-fixed observer, is an easterly (i.e., moves westward).

Air from the west has a larger velocity than the earth at that latitude. It is no longer stable relative to the rotating earth's inclined oblate surface at that latitude and slides "uphill" toward the equator. See Figure 21.3 and remember that the oblate earth long ago adjusted its shape to balance centripetal and gravitational accelerations of *stationary* fluids. A stationary mass of air has no net north-south (or east-west) force acting on it, but a mass of air with larger tangential velocity than the earth will deflect toward the equator, and a slower-moving mass will deflect toward a pole. Therefore, westerlies move south, and easterlies move north in this example. The net effect is a cyclonic vortex with sink.

All other explanations say the same thing but in a different fashion. A briefer explanation is that Coriolis phenomena on the earth causes all flows in the Northern Hemisphere to veer to the right of the forcing function, which here is the radially directed pressure gradient from the surrounding high-pressure area toward the central low-pressure sink. In the Northern Hemisphere, a low-pressure area therefore produces

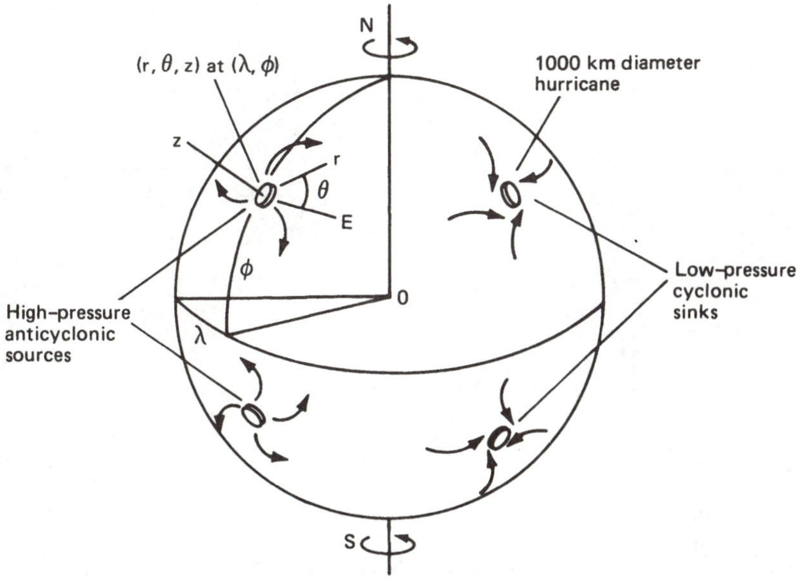

Figure 22.2: *Cyclonic rotation, north and south, is consistent with earth rotation, which appears counterclockwise viewed from the north and clockwise viewed from the south. Hurricane approximately to scale.*

a counterclockwise rotation of the surrounding air. In the Southern Hemisphere, the signs are reversed and all flows veer to the left. A low-pressure central core there produces clockwise rotation (viewed from above) of the converging winds. Remember that the earth's rotation viewed from above the South Pole also is clockwise, so low-pressure cores, north or south, always produce flows consistent with the earth's rotation (cyclonic).

A more elegant explanation, no more accurate but more adapted to a mathematical model, makes use of the coordinate system of Figure 22.2. A cylindrical coordinate system with its z axis vertical is placed at the center of the convective cell located at some longitude and latitude (λ, ϕ). The earth is assumed locally flat over the approximately 1000 km diameter of the hurricane. A hurricane this size subtends one-tenth of the distance from the equator to a pole, or 9° of latitude. The most active central portion of the hurricane is only a few hundred kilometers in diameter, so the errors in assuming this portion of the earth to be flat are very small. The explanation continues as follows.

A developing hurricane has two initial needs, energy and angular momentum. Massive amounts of each are required. Availability of these determine when and if a hurricane will develop. Solar energy stored over several months in warm ocean water provides the energy, and the earth's rotation provides the angular momentum. The North and South Equatorial Currents are capable of providing a continuous supply of energy over the week or two required for the formation of a mature hurricane. These

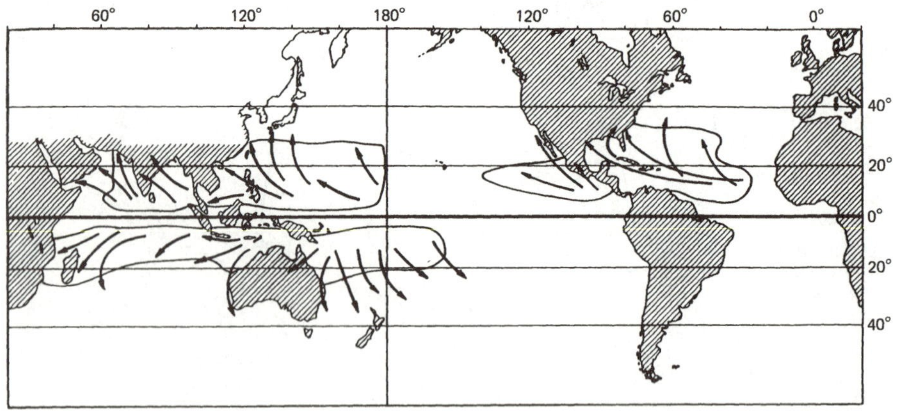

Figure 22.3: *Location of warm-water regions that spawn tropical cyclones and typical tracks (September north of the equator, and March south of the equator). They are called hurricanes in the Atlantic and typhoons in the Pacific. Note complete absence of major tropical cyclones east or west of South America. Data from Bergeron (1954), Gray (1979), and Harding and Kotsch (1965).*

currents are also far enough from the equator at 10° to 20° latitude for the Coriolis parameter ($f = 2\Omega \sin \phi$) to have a finite value. The Coriolis parameter, of course, is a measure of the local vertical component of the earth's rotation (vorticity). It is this vorticity that is concentrated by air flowing toward the central convective cell. At higher latitudes f is greater, but, except for a few isolated regions, the ocean water is too cool. No hurricane has been observed to form over water whose temperature is less than 26.5°C (80°F), and even warmer temperatures are more typical.

Figure 22.3 shows the regions of initial genesis points and typical tracks of most hurricanes. All form while travelling several thousand kilometers westward over an equatorial current, blown by the prevailing easterly trade winds. Many then veer poleward following especially the Gulf Stream (hurricanes) or the Kuroshiro (Japanese) Current (typhoons). These warm currents continue to provide the necessary energy, and as the tropical cyclone moves poleward, the Coriolis parameter and the hurricane rotation rate increase. Other typhoons occur in the South Pacific and in the northern Indian Ocean. Hurricanes are not known in South America, presumably because the South Atlantic Equatorial Current is too cool and/or short for a mature hurricane to develop.

Theories have been proposed to explain why or how hurricanes (typhoons) maintain their position over narrow, poleward-moving currents. There may be stability mechanisms at work that help to guide the hurricane's path. Pudov (1979), for example, using observations by Pavlov (1978) that typhoons frequently move along "tongues" of warm water, suggests a stability mechanism based on asymmetries in the pressure and velocity fields caused by departures of storms from the warmest paths. Passage of a tropical cyclone cools the ocean surface typically by several degrees, and Pudov's stability mechanism is corroborated by Pavlov's observation that the tracks of subsequent typhoons rarely coincide with those of immediately previous ones.

Another factor is simply that storm paths that coincide with warm water have a chance of developing into hurricanes--or if already hurricanes, of continuing. Those that leave warm water soon dissipate. It is well known, for example, that a hurricane moving over a large area of land quickly dissipates. Increased surface friction is only a small contributor to the dissipation; a much larger contributor is loss of the energy source--warm water. Also, as a hurricane (typhoon) moves poleward, it moves to regions of increased earth vorticity, causing its angular momentum collection ability to increase. If it happens to move toward the equator, it may have adequate convective energy, but it loses its source of angular momentum.

In summary, the 80 or so tropical cyclones that reach maturity each year seem to progress through a typical sequence of events.

1. There is a region of thermally unstable air over the central or eastern portion of an ocean between latitudes 10° and 20°.
2. Water temperature is equal to and probably greater than 26.5°C.
3. The lower and middle troposphere is warm, moist, and near saturation. These three conditions typically occur in late summer when the thermal equator is displaced north (or south) of the equator.
4. Convective cells begin. Precipitation, with release of latent energy, is heavy throughout the troposphere, leading to the growth of thunderstorms. Note that dry, middle tropospheric air would dry out the ascending air, preventing rain and release of latent heat.
5. Radial inflow begins, coalescing a group of thunderstorms into one giant "mesoscale" thunderstorm. The inflowing surface air is warmed by the ocean, and more importantly, acquires latent heat by evaporation. High winds and waves increase evaporation rate. The mesoscale storm begins to create its own weather system.
6. The normal vorticity at that latitude ($f = 2\Omega \sin \phi$) is concentrated by the converging air. Rotation accelerates. Rotation can be augmented by favorable large-scale wind shear conditions.
7. After about a week of continuous operation, the tangential velocities near the middle of the well-defined rotational pattern reach hurricane strength $\geq$115 km/hr.

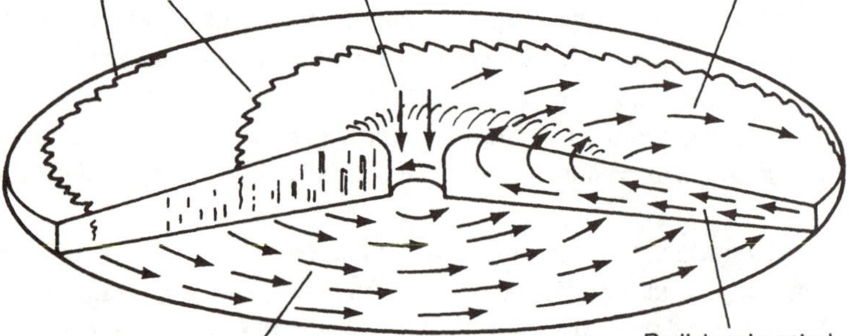

Figure 22.4: *Cross section of northern hemisphere hurricane. Diameter ~ 1000 km, eye diameter ~ 40 km, eye wall height ~ 10 km. Surface pressure in eye ~ 900 mb, eye wall rainfall ≥ 50 cm/day.*

8. The hurricane "eye" forms as shown in Figure 22.4. See also Figure 1.17. Until the eye forms, the center is a region of nearly rigid rotation, blending with the inward and upward motion of the rapidly rotating larger vortex. The central angular velocity eventually becomes so large that the radial pressure gradient is insufficient to maintain the centripetal acceleration. The spinning inner wall enlarges its diameter until an equilibrium is reached between pressure differential and centripetal acceleration.

9. In a completely formed eye, the low-pressure core reaches the stratosphere and pulls dry, nonrotating stratospheric air downward. As the air descends, it is heated adiabatically (no outside gain or loss of heat energy) by compression. Its relative humidity decreases. The eye can be 10 to 100 km in diameter, and its air is often clear, dry, warm, and nearly motionless. This air slowly diffuses outward into the wall of the eye.

10. The maximum tangential and upward velocities occur just outside the eye. Rainfall is extreme, often far in excess of 50 cm/day. Condensation, of course, is the release mechanism for the latent energy of evaporation from the ocean surface. This energy maintains the upward convection, which maintains the radial inflow, which in turn maintains the large tangential velocities by concentrating earth vorticity through the Coriolis mechanism.

11. A fully developed hurricane can persist as long as its energy source persists, and can survive for a day or so from its own stored kinetic energy and angular momentum. In 1966 a hurricane persisted for a month while passing over two Caribbean islands, the tip of Florida, and parts of Mexico by renewing its strength over the warm Caribbean waters of its route.

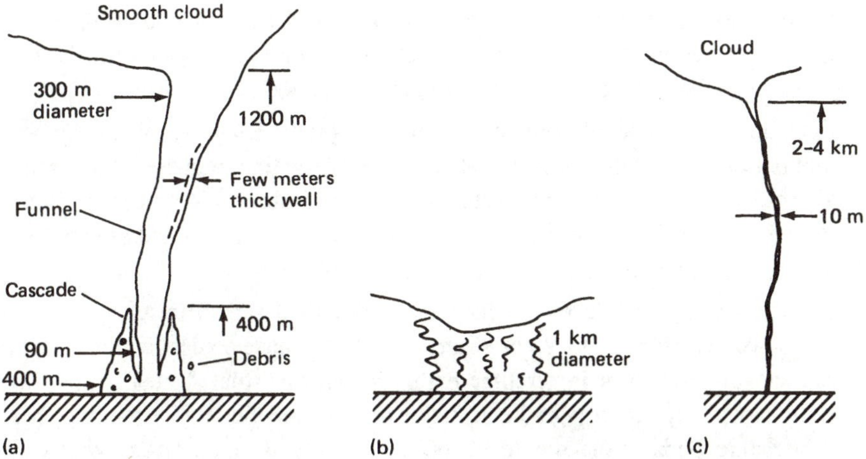

Figure 22.5: *(a) Classical tornado with typical smooth funnel and cascade, (b) low, wide, diffuse tornadoes often are the most destructive, (c) high, narrow whip tornado.*

22.4 Tornadoes

Figures 22.5a, 22.5b, and 22.5c show three general forms in which tornadoes appear. Forms (a) and (b) are the most destructive. Form (c) is sometimes observed as the tornado dissipates. Details of tornado phenomena have been much more difficult to measure and model than hurricanes because of the rapidity with which tornadoes appear and disappear (usually minutes), their small size (~10 to 100 m diameter), and their extreme destructiveness (speeds possibly approaching the speed of sound). Important features of a tornado's life cycle, held together with a modest amount of speculation to yield a crude descriptive model, is as follows.

1. An enormous quantity of angular momentum and rotational kinetic energy has been concentrated by another mechanism (hurricane, thunderstorm, squall line, volcano, forest fire storm, etc.) into a low-altitude ($\lesssim$ 2 km high) cloud. The cloud is sometimes called a *rotor cloud* and often shows smooth contours consistent with a rapidly rotating mass of air. Rotor clouds appear to be able to persist for some hours without major inputs of energy or momentum.

2. Rotating columns of air called *funnels* reach down from the cloud toward the earth. See Figures 1.18 and 1.19. This phenomenon may be related to Kelvin's theorem concerning the inability of a vortex to end in a fluid. An instability exists between the rotor cloud and the earth that promotes these extensions. If the funnels have enough energy and momentum, they reach the earth. Their rotation is intensified by vertical elongation.

3. Energy and momentum are fed to the tornado from the rotor cloud. If the tornado funnel reaches the earth, the central pressure will be low enough for the radial pressure gradient to balance the centrifugal force of the rotating sheath of air.
4. Sometimes air inside the core does not rotate rapidly. Likewise, air can be nearly motionless at small distances outside the rapidly rotating, thin cylindrical shell of air (funnel). There are many reports of persons observing quiescent cores in tornadoes from the safety of a shelter and also reports of unharmed fragile structures and objects just outside a passing tornado.
5. Note that it is possible for air in a thin cylindrical sheath to have supersonic tangential velocities that are not observed and that cause no destruction. Because the sheath air moves in circular paths, it can be isolated at all times from stationary air by velocity gradients inside and outside the rapidly rotating sheath. Supersonic phenomena would be observed only if some nonmoving object penetrated the sheath. Small fragile objects (e.g., birds) have been found showing portions with extreme destruction separated by a distance of a centimeter or so from portions with no damage. These have been used to support estimates of speeds and velocity gradients.
6. The tornado column is visible only if it condenses moisture and/or picks up dust, debris, water, etc. Often the lower portion of the column carries large masses of foreign material. A base cloud sometimes forms from surface air, dust, debris, water, etc., set in motion by the column as in Figure 22.5a.
7. Destruction can be caused by the low-pressure core, especially when the core passes over a hollow frame building. The building explodes due to the pressure gradient between the building's normal internal pressure and the reduced internal pressure of the tornado core. People are told to leave their windows and doors open if a tornado approaches. More generally, destruction is caused by encounters with the high-velocity wind in the funnel sheath.
8. Occasionally the parent rotor cloud contacts the earth as in Figure 22.5b. In one extreme case, a steel truss bridge, 75 m long and weighing 108 tons, was lifted from its foundation, deformed into a compact twisted mass of metal, and dropped in its river at Irving, KS (Finley 1881).
9. Rotation direction is determined by random conditions and can be either cyclonic or anticyclonic. Pairs of tornadoes rotating in opposite directions have been recorded.

A classic inviscid vortex with a central sink provides an approximate model for a thunderstorm and a tropical cyclone. There are significant differences, however. In the mathematical model of an inviscid vortex $v_\theta \sim r^{-1}$, while in a hurricane or thunderstorm, outside the central core, $v_\theta \sim r^{-0.5}$ to $r^{-0.7}$. The model has a line sink at the origin, while real phenomena diffuse the sink over a large central area of the vortex. Because of turbulence, eddy viscosity seriously compromises the inviscid approximation for air. Also, vertical flows transport air to the central core from the stratosphere, and/or surface, to cause deviations in the rotational pattern.

Caution must be used when approximating the rotational patterns of whirlwinds, dust devils, and tornadoes with an inviscid vortex model. The inviscid vortex models conservation of angular momentum as fluid elements are transported radially inward. The whirlwind and dust devil are wind shear vortices, and might be approximated better as rigidly rotating columns of air with reduced core pressure. The air external to these vortices need not have tangential velocities except as momentum is diffused radially outward or as may be transferred by secondary currents at a surface.

The rotational pattern of a tornado is determined by vorticity concentrated within a parent cloud. The responsible mechanism and structural details are still uncertain. In some cases the funnel appears to be a very thin, rapidly rotating sheath modelled by v_θ = max at $r = r_o$ and $v_\theta \sim 0$ for $r \lessgtr r_o$. See Example 12.2. Other cases indicate near-rigid rotation within the funnel and a rapid drop-off of v_θ external to r_o. Secondary currents and diffusion often create additional rotating clouds of air and debris near the base.

22.5 Study and Research Projects

The projects suggested here are intended for individuals or groups. Some can be completed in a week or two, others are term or longer projects. Only a few will have exact answers. References to other texts and to professional journals are included.

Project 22.1: Whirlwinds that form behind obstacles, such as buildings, are easy to produce in the laboratory. Quickly insert and then remove a thin, flat surface (e.g., sheet metal) perpendicular to a uniform flow of water in a channel so that 1/4 to 1/3 of the channel on one side is briefly obstructed. A vortex "whirlwind" will form at the free edge of the sheet and propagate downstream. See Figure 10.16 and discussion. A similar whirlwind can be created in a wind tunnel. Remember that this procedure would produce nothing in a perfectly inviscid flow. An inviscid fluid would pass around the edge while the obstruction is in place, and would immediately return to uniform flow when the obstruction is removed (a potential flow solution). No vortex would form. Examine carefully where and how the vortex forms in water. Try to estimate $v_\theta(r)$. Observe the persistence of the water vortex and the reduced pressure at its center. Place particles on the channel bottom. Does the vortex "whirlwind" pick them up as it traverses downstream? Do you get the same effect by briefly translating the sheet with constant velocity along the side of a channel of nonmoving water? Would that merely represent a transformation from one inertial coordinate system to another?

Project 22.2: Read the article by Mullen and Maxworthy (1977), and repeat their experiment with dust devils. You should be able to get observable results with a very simple setup. Repeat their measurements where practical, and use techniques to aid in observing the flow field. Where do the angular momentum and energy required by their "dust devil" originate in their experiment? Are there equivalent sources of each for naturally occurring dust devils? Would the center of the dust devil have an upward velocity regardless of a heated subsurface? Dust devils are occasionally discussed by authors writing books about hurricanes and/or tornadoes. Locate different opinions on the topic. If you live near a high, open desert, observe some and attempt to measure their parameters. Even large persistent dust devils do not appear to be dangerous, but measuring them is not easy because they are transient and moving. They are abundant along U.S. Route 95 on the western edge of Nevada in late summer and early autumn. When a dust devil intercepts an obstacle about its diameter, e.g., a convertible automobile with its top down, it dissipates and drops its load of sand. What happens?

Project 22.3: High-velocity, surface-level, frontal systems occur with regularity in North Africa and the high plains of central Asia. They are often preceded by a region of vertically oriented intense vortices. Extremely large amounts of dirt, dust, and debris are transported. Read the Russian book titled *Hurricanes, Storms, and Tornadoes* by Nalivkin (1983) as translated to English. His Part II details dust storms and provides data that dust and sand (meters thick) can be transported hundreds and thousands of kilometers. He asserts that infrequent major dust storms in central Asia are much more effective than water and large rivers in the creation of sedimentary plains. Read his book, with photographs, for extensive detailed semiquantitative discussions and analyses of atmospheric vortices.

Project 22.4: Middle-latitude squall lines associated with frontal systems, and especially interacting frontal systems, typically produce thunderstorms and tornadoes. Examine Project 20.6 and Project 20.8 on midlatitude atmospheric phenomena and continue those studies by analyzing details of instabilities. Manabe (1985) and AMS (1988), in addition to the references listed in Project 20.6, are useful sources of information. In a report, illustrate the details of frontogenesis, and the flow and thermal instabilities that result in squall lines and vortex-like phenomena.

Project 22.5: Computer model a large isolated thunderstorm of the type that occurs in northern Texas and vicinity. Heights above 10 km are feasible, and high-velocity winds extending for a radius of 500 km are possible in late summer. Their structure, including flow velocities, temperature and moisture distributions, and electrical discharge phenomena, are detailed by Magono (1980). Examine the many roles that vorticity and vortices play in the generation and maintenance of a thunderstorm. What is the origin of its energy and angular momentum? How does it differ from a whirlwind and a dust devil. As a first iteration, model only the velocities, then include energy sources and dissipation, then angular momentum, and finally phenomena such as electrical discharges. Much remains to be learned about details so make cautious assumptions as needed. Suppose a large isolated thunderstorm is just north of Amarillo, TX, and you are in a light plane (cruise speed 180 mph) over Oklahoma City, OK, headed for Albuquerque, NM. Safe maximum altitude without oxygen is 12,000 ft. What can/should you do?

Project 22.6: Write a report on tropical cyclones as a weather phenomena, including detailed descriptions of, say, six different ones. Select two that occurred in the Caribbean/North Atlantic, two from China/Japan, one from Australia, and one from India. Include their effects on land and sea, and their contributions to weather on a scale larger than itself. To what extent do people on the ground and on the sea perceive the cyclone as a rotational phenomena? What could a careful observer detect about its overall vertical and horizontal structure from a fixed position on the ground? The book by Nalivkin (1983) includes many descriptive narratives on past tropical cyclones as does the book titled *Heavy Weather Guide* by Harding and Kotsch (1965) who, as ship captains, describe tropical cyclones at sea. See also Project 21.1.

Project 22.7: Develop a rough computer model of a hurricane as a simulation, beginning with a convective pattern west of North Africa. Use a Lagrangian formulation and a cylindrical coordinate system. Let the origin drift westward at the trade wind velocity for 20° N. Use time increments of several hours. If reasonable parameters are chosen, a hurricane should develop before it reaches the Caribbean. Discontinue heat/moisture input and see how rapidly the hurricane dissipates. See Anthes (1982), Bengtsson and Lighthill (1982), Bortkovskii (1983), and AMS (1989) for useful information.

Project 22.8: Search the professional literature and write a report on tornadoes of the American Midwest. Can you find references to tornadoes in other countries? Useful references include Ludlum (1976), Bengtsson and Lighthill (1982), Nalivkin (1983), and AMS (1988). Use the references in AMS (1988) to locate relevant journals and then systematically review four or five of the apparent best for the last decade. Examine each volume, read the table of contents, then read the abstract if the title looks good, and finally read the text if the abstract is relevant. Tabulate characteristic parameters and structures. Include presence or absence of precipitation and electrical activity. Also see Thorarinsson and Vonnegut (1964), Vaughan and Vonnegut (1976), and Watkins et al. (1978). How quickly does $v_\theta(r)$ decrease? What would be a minimum safe distance for a person, a car, and a strong shelter? Where and how is vorticity concentrated, or is its structure more like a vortex ($v_\theta = \kappa/r$) whose vorticity is zero? What are the dynamics that cause funnels to drop from a cloud to make contact with the earth? How does entrained mass affect a tornado's wind velocity, angular momentum, and energy, e.g., several houses in a tornado or water in a waterspout?

Project 22.9: A tornado-like vortex can be created in the laboratory as follows. Fill a cylindrical transparent cylinder, say 6 cm diameter by 40 cm tall, with clear carbonated water (a carbonated soft drink) and then rotate the cylinder about its vertical axis, preferably on a turntable. A vortex, made visible with bubbles of the carbonated water, will appear and continue as long as bubbles form. Explain the phenomena in detail. How closely does this vortex resemble a tornado? How about a whirlwind, dust devil, or hurricane? Smith (1980) describes a similar experiment that uses air bubbles injected at the bottom center of the rotating cylinder. Either would be relatively easy to model as a computer exercise.

Project 22.10: Intense atmospheric vortices such as tornadoes and hurricanes often exhibit internal wave structures, or at least structures that may be related to wave phenomena. Not all are well understood at present. Hurricanes, for example, can develop spiral arm structures where local wind activity and precipitation are more intense than at other places at the same distance from the eye. Tall, thin funnel tornadoes sometimes develop "kinks" along the column similar to travelling waves and kinks observed along the axis of a cavitation vortex in water. See Van Dyke (1982, photograph 101) for an illustration of inertial waves on a cavitation vortex. From the references listed in earlier projects, select a typhoon or tornado structure of interest and analyze it for rotating flow wave phenomena. Greenspan (1968) and Pedlosky (1987) provide classic reviews of wave phenomena in rotating geophysical flows for initiating a study.

Appendices

Appendix A Mathematical Relationships

A.1 Vectors

For S any scalar and **A**, **B**, **C** any vectors

$$S\mathbf{A} = \mathbf{A}S$$

$$\mathbf{A} \cdot \mathbf{B} = \mathbf{B} \cdot \mathbf{A}$$

$$\mathbf{A} \times \mathbf{B} = -\mathbf{B} \times \mathbf{A}$$

$\mathbf{A} \cdot \mathbf{B} \times \mathbf{C}$, or any cyclic permutation, equals the volume of a parallelepiped formed by edges **A**, **B**, **C**.

$$(\mathbf{A} \times \mathbf{B}) \times \mathbf{C} = \mathbf{B}(\mathbf{A} \cdot \mathbf{C}) - \mathbf{C}(\mathbf{A} \cdot \mathbf{B})$$

A.2 Matrices

With (A) any 1×3 row matrix, {B} any 3×1 column matrix, and [L], [M], [N] any 3×3 matrices

$$\mathbf{A} \cdot \mathbf{B} = (A)\{B\}$$

$$\mathbf{A} \times \mathbf{B} = [\tilde{A}]\{B\}$$

where $[\tilde{A}]$ is a skew-symmetric matrix composed of the elements of **A** (in Cartesian coordinates only) as

$$\begin{bmatrix} 0 & -A_3 & A_2 \\ A_3 & 0 & -A_1 \\ -A_2 & A_1 & 0 \end{bmatrix}$$

Use of $[\tilde{A}]$ rather than $\mathbf{A}\times$ is an aid in many programming languages.

Except for special cases, e.g., the identity matrix $[1] = \begin{bmatrix} 1 & 0 & 0 \\ 0 & 1 & 0 \\ 0 & 0 & 1 \end{bmatrix}$

$$(A)[M] \neq [M]\{A\}$$
$$[M][N] \neq [N][M]$$

The associative law of continued products states

$$[M]([N]\{A\}) = ([M][N])\{A\}$$
$$[L]([M][N]) = ([L][M])[N]$$

$[M]^T$, the transpose of [M], interchanges rows and columns of [M].
A row matrix can be written as the transpose of a column matrix, $(A) = \{A\}^T$.

Division by vectors and matrices is undefined.
$[M]^{-1}$ is the inverse of [M] in the sense of $[M][M]^{-1} = [1]$, but it is *not* 1/[M].

A.3 Direction Cosine Matrices

An arbitrary orientation of one reference frame (coordinate system) relative to another cannot be represented as a scalar or a vector. A matrix or its equivalent is required. It is mathematically convenient to subdivide an arbitrary orientation into a plurality of single axis rotations. A systematic procedure simplifies the process. Consider in Figure 18.2 the arbitrary orientation of the b frame relative to the n frame.

First establish the orientation of the unit vectors $(\hat{b}_1, \hat{b}_2, \hat{b}_3)$ as a function of the $(\hat{e}_1, \hat{e}_2, \hat{e}_3)$ in matrix form.

$$\begin{bmatrix} \hat{b}_1 \\ \hat{b}_2 \\ \hat{b}_3 \end{bmatrix} = \begin{bmatrix} C\psi & S\psi & 0 \\ -S\psi & C\psi & 0 \\ 0 & 0 & 1 \end{bmatrix} \begin{bmatrix} \hat{e}_1 \\ \hat{e}_2 \\ \hat{e}_3 \end{bmatrix}$$

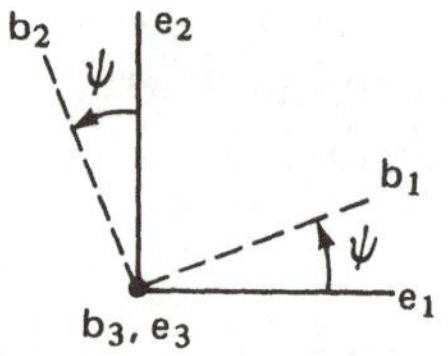

This can be abbreviated as $\{\hat{b}_i\} = [C^{be}]\,\{\hat{e}_i\}$. The matrix $[C^{be}]$ was quickly obtained by the following routine procedure.

1. $\hat{b}_3 = \hat{e}_3$, therefore a "1" was inserted in the matrix for element (3, 3).

2. The row and the column containing this "1" could only have zeros for the other elements.

3. Place cosine angle ($C\psi$) on the other two main diagonal elements and sine angle ($S\psi$) in the remaining two off-diagonal elements.

4. Axis b_1 is between e_1 and e_2, and consists of only positive components of e_1 and e_2, so $S\psi$ opposite $\hat{b}_1$ is positive, and the other $S\psi$ is negative. In summary, $[C^{be}]$ was constructed in the following sequence.

$$\begin{bmatrix} & & \\ & & \\ & & 1 \end{bmatrix} \rightarrow \begin{bmatrix} & & 0 \\ & & 0 \\ 0 & 0 & 1 \end{bmatrix} \rightarrow \begin{bmatrix} C\psi & & 0 \\ & C\psi & 0 \\ 0 & 0 & 1 \end{bmatrix} \rightarrow \begin{bmatrix} C\psi & S\psi & 0 \\ S\psi & C\psi & 0 \\ 0 & 0 & 1 \end{bmatrix} \rightarrow \begin{bmatrix} C\psi & S\psi & 0 \\ -S\psi & C\psi & 0 \\ 0 & 0 & 1 \end{bmatrix}$$

Orientation of e relative to g is an angle θ about the g_1, e_1 axes and is obtained as $\{\hat{e}_i\} = [C^{eg}]\,\{\hat{g}_i\}$.

$$\begin{bmatrix} \hat{e}_1 \\ \hat{e}_2 \\ \hat{e}_3 \end{bmatrix} = \begin{bmatrix} 1 & 0 & 0 \\ 0 & C\theta & S\theta \\ 0 & -S\theta & C\theta \end{bmatrix} \begin{bmatrix} \hat{g}_1 \\ \hat{g}_2 \\ \hat{g}_3 \end{bmatrix}$$

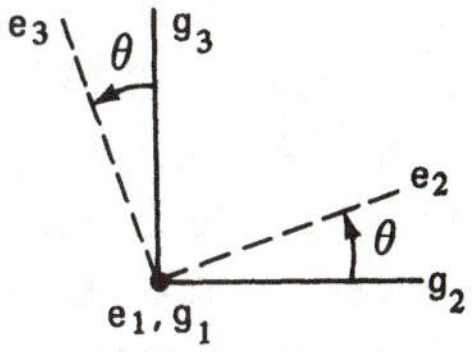

Substitution using matrix algebra yields, e.g.,

$$\{\hat{b}_i\} = [C^{be}]\,\{\hat{e}_i\} = [C^{be}]\,[C^{eg}]\,\{\hat{g}_i\} = [C^{bg}]\,\{\hat{g}_i\}$$

and if $[C^{gn}]$ has been obtained

$\{\hat{b}_i\} = [C^{bg}]\,[C^{gn}]\,\{\hat{n}_i\} = [C^{bn}]\,\{\hat{n}_i\}$, where $[C^{bn}]$ provides the orientation of the b frame relative to the n frame.

Unique rules apply to direction cosine matrices not applicable to matrices in general. Their determinate always equals +1, and their inverse always equals their transpose; therefore, e.g., $[C^{nb}] = [C^{bn}]^{-1} = [C^{bn}]^T$.

Time rates of change of angular orientation are skew-symmetric tensors or matrices $[\tilde{\omega}]$, which, having only three independent components, are used as vectors ω, with $[\tilde{\omega}]$ being equivalent to $(\omega\times)$.

A.4 Dyadics and Tensors

Dyadics, and indices notation for tensors, are avoided in this text. Indices notation is similar to matrix notation as used in FORTRAN, for example,

$$A_i = A(I)$$
$$M_{ij} = M(I,J)$$

Indices notation requires special symbols for scalar and vector products,

$$\mathbf{A} \cdot \mathbf{B} = A_i \, B_j \, \delta_{ij} \qquad \delta = 1 \text{ if } i = j$$
$$= 0 \text{ if } i \neq j$$

$$\mathbf{A} \times \mathbf{B} = A_i \, B_j \in_{ijk} \qquad \in = +1 \text{ for cyclic subscripts}$$
$$= -1 \text{ for anticyclic subscripts}$$
$$= 0 \text{ for repeated subscripts}$$

Dyadic notation, developed by Gibbs in 1890 as a part of vector mathematics, is seldom used today. A dyadic is sometimes written as $\mathbb{D}$ or in two parts as $\mathbb{D} = \mathbf{AB}$, where **AB** corresponds to the *outer* product $\{A\}(B)$ which produces a 3×3 matrix. Note that the inner product $(A)\{B\}$ produces a scalar. A dyadic represents the same concept as a matrix or a tensor.

$$\mathbb{D} = [D] = D_{ij}$$

$\mathbb{D}$ combines with vectors in the same manner as 3×3 matrices combine with row or column matrices, e.g., with $\mathbb{I}$ as a dyadic and $\boldsymbol{\omega}$ as a vector

$$\boldsymbol{\omega} \cdot \mathbb{I} \cdot \boldsymbol{\omega} = (\omega)[I]\{\omega\} = \text{a scalar}$$
$$\boldsymbol{\omega} \times \mathbb{I} \cdot \boldsymbol{\omega} = [\tilde{\omega}][I]\{\omega\} = \text{a vector}$$

A.5 The Del Operator ∇

With ∇ the spatial derivative, sometimes written as $\partial/\partial\mathbf{R}$

$$\nabla S = \text{gradient } S$$
$$\nabla \cdot \mathbf{A} = \text{divergence } \mathbf{A}$$
$$\nabla \times \mathbf{A} = \text{curl } \mathbf{A}$$

$\nabla \cdot \nabla = \nabla^2$, the Laplacian

$\nabla \times \nabla S = 0$ and $\nabla \cdot \nabla \times \mathbf{A} = 0$, always

$\nabla \cdot (S\mathbf{A}) = \nabla S \cdot \mathbf{A} + S(\nabla \cdot \mathbf{A})$

$\nabla \times (S\mathbf{A}) = (\nabla S) \times \mathbf{A} + S(\nabla \times \mathbf{A})$

$\nabla(\mathbf{A} \cdot \mathbf{B}) = (\mathbf{B} \cdot \nabla)\, \mathbf{A} + (\mathbf{A} \cdot \nabla)\, \mathbf{B} + \mathbf{B} \times (\nabla \times \mathbf{A}) + \mathbf{A} \times (\nabla \times \mathbf{B})$

$\nabla \cdot (\mathbf{A} \times \mathbf{B}) = \mathbf{B} \cdot \nabla \times \mathbf{A} - \mathbf{A} \cdot \nabla \times \mathbf{B}$

$\nabla \times (\mathbf{A} \times \mathbf{B}) = (\mathbf{B} \cdot \nabla)\, \mathbf{A} - (\mathbf{A} \cdot \nabla)\, \mathbf{B} + \mathbf{A}(\nabla \cdot \mathbf{B}) - \mathbf{B}(\nabla \cdot \mathbf{A})$

$\nabla \times (\nabla \times \mathbf{A}) = \nabla(\nabla \cdot \mathbf{A}) - \nabla^2 \mathbf{A}$

$(\mathbf{A} \cdot \nabla)$ = component of ∇ in direction of **A**, times A, i.e., a scalar product of **A** and ∇. It is not the same as $\nabla \cdot \mathbf{A}$, the divergence of **A**.

$(\mathbf{A} \cdot \nabla)\, \mathbf{A}$ is a pseudovector in that it computes correctly only in Cartesian coordinates. It can be converted to true vector form using

$$(\mathbf{A} \cdot \nabla)\mathbf{A} = \nabla\left(\frac{A^2}{2}\right) - \mathbf{A} \times (\nabla \times \mathbf{A})$$

A.6 Operations Using ∇

∇S, $\nabla^2 S$, $\nabla \cdot \mathbf{A}$, $\nabla \times \mathbf{A}$, and $D/Dt = (\partial/\partial t + \mathbf{V} \cdot \nabla)$ in Cartesian (x, y, z), cylindrical (r, θ, z), and spherical (r, θ, ϕ) coordinate systems are

$$\nabla S = \frac{\partial S}{\partial x}\hat{\mathbf{x}} + \frac{\partial S}{\partial y}\hat{\mathbf{y}} + \frac{\partial S}{\partial z}\hat{\mathbf{z}} \qquad \text{Cartesian}$$

$$\nabla S = \frac{\partial S}{\partial r}\hat{\mathbf{r}} + \frac{1}{r}\frac{\partial S}{\partial \theta}\hat{\boldsymbol{\theta}} + \frac{\partial S}{\partial z}\hat{\mathbf{z}} \qquad \text{Cylindrical}$$

$$\nabla S = \frac{\partial S}{\partial r}\,\hat{\mathbf{r}} + \frac{1}{r}\,\frac{\partial S}{\partial \theta}\,\hat{\boldsymbol{\theta}} + \frac{1}{r\,\sin\,\theta}\,\frac{\partial S}{\partial \phi}\,\hat{\boldsymbol{\phi}} \qquad \text{Spherical}$$

$$\nabla^2 S = \frac{\partial^2 S}{\partial x^2} + \frac{\partial^2 S}{\partial y^2} + \frac{\partial^2 S}{\partial z^2} \qquad \text{Cartesian}$$

$$\nabla^2 S = \frac{1}{r}\,\frac{\partial}{\partial r}\left(r\frac{\partial S}{\partial r}\right) + \frac{1}{r^2}\,\frac{\partial^2 S}{\partial \theta^2} + \frac{\partial^2 S}{\partial z^2} \qquad \text{Cylindrical}$$

$$\nabla^2 S = \frac{1}{r^2}\,\frac{\partial}{\partial r}\left(r^2\,\frac{\partial S}{\partial r}\right) + \frac{1}{r^2 \sin\theta}\,\frac{\partial}{\partial \theta}\left(\sin\theta\,\frac{\partial S}{\partial \theta}\right) + \frac{1}{r^2 \sin^2\theta}\,\frac{\partial^2 S}{\partial \phi^2} \qquad \text{Spherical}$$

$$\nabla \cdot \mathbf{A} = \frac{\partial A_x}{\partial x} + \frac{\partial A_y}{\partial y} + \frac{\partial A_z}{\partial z}$$ Cartesian

$$\nabla \cdot \mathbf{A} = \frac{1}{r}\frac{\partial}{\partial r}(rA_r) + \frac{1}{r}\frac{\partial A_\theta}{\partial \theta} + \frac{\partial A_z}{\partial z}$$ Cylindrical

$$\nabla \cdot \mathbf{A} = \frac{1}{r^2}\frac{\partial}{\partial r}(r^2 A_r) + \frac{1}{r\sin\theta}\frac{\partial}{\partial \theta}(A_\theta \sin\theta) + \frac{1}{r\sin\theta}\frac{\partial A_\phi}{\partial \phi}$$ Spherical

$$\nabla \times \mathbf{A} = \left(\frac{\partial A_z}{\partial y} - \frac{\partial A_y}{\partial z}\right)\hat{\mathbf{x}} + \left(\frac{\partial A_x}{\partial z} - \frac{\partial A_z}{\partial x}\right)\hat{\mathbf{y}} + \left(\frac{\partial A_y}{\partial x} - \frac{\partial A_x}{\partial y}\right)\hat{\mathbf{z}}$$ Cartesian

$$\nabla \times \mathbf{A} = \left(\frac{1}{r}\frac{\partial A_z}{\partial \theta} - \frac{A_\theta}{\partial z}\right)\hat{\mathbf{r}} + \left(\frac{\partial A_r}{\partial z} - \frac{\partial A_z}{\partial r}\right)\hat{\boldsymbol{\theta}} + \frac{1}{r}\left[\frac{\partial}{\partial r}(rA_\theta) - \frac{\partial A_r}{\partial \theta}\right]\hat{\mathbf{z}}$$ Cylindrical

$$\nabla \times \mathbf{A} = \frac{1}{r\sin\theta}\left[\frac{\partial}{\partial \theta}(A_\phi \sin\theta) - \frac{\partial A_\theta}{\partial \phi}\right]\hat{\mathbf{r}} + \left[\frac{1}{r\sin\theta}\frac{\partial A_r}{\partial \phi} - \frac{1}{r}\frac{\partial}{\partial r}(rA_\phi)\right]\hat{\boldsymbol{\theta}}$$

$$+ \frac{1}{r}\left[\frac{\partial}{\partial r}(rA_\theta) - \frac{\partial A_r}{\partial \theta}\right]\hat{\boldsymbol{\phi}}$$ Spherical

$$D/Dt = \frac{\partial}{\partial t} + v_x\frac{\partial}{\partial x} + v_y\frac{\partial}{\partial y} + v_z\frac{\partial}{\partial z}$$ Cartesian

$$D/Dt = \frac{\partial}{\partial t} + v_r\frac{\partial}{\partial r} + \frac{v_\theta}{r}\frac{\partial}{\partial \theta} + v_z\frac{\partial}{\partial z}$$ Cylindrical

$$D/Dt = \frac{\partial}{\partial t} + v_r\frac{\partial}{\partial r} + \frac{v_\theta}{r}\frac{\partial}{\partial \theta} + \frac{v_\phi}{r\sin\theta}\frac{\partial}{\partial \phi}$$ Spherical

Appendix B Stream Functions and Velocity Potentials

B.1 Summary

Chapter 3 outlines the theory and some uses of stream functions (ψ) and velocity potentials (ϕ). These functions have the useful properties that (1) any sum of (ψ), or of (ϕ), yields new two-dimensional (2-D) flow solutions; and (2) ϕ and ψ can, by definition, be interchanged to create even more flow solutions, i.e., let ψ be ϕ, and let ϕ be ψ. This appendix lists a useful set of such functions. For the use of equivalent functions in complex variables, see Milne-Thomson (1960). Uniform flow is given in Cartesian coordinates (x, y), with the remainder in polar coordinates (r, θ). They are related by the expressions $x = r\cos\theta$, $y = r\sin\theta$. Velocity (u, v) or (v_r, v_θ) at some point P(x, y) or P(r, θ) is obtained using

Cartesian	**Polar**
$u = \dfrac{\partial\psi}{\partial y} = -\dfrac{\partial\phi}{\partial x}$	$v_r = \dfrac{1}{r}\dfrac{\partial\psi}{\partial\theta} = -\dfrac{\partial\phi}{\partial r}$
$v = -\dfrac{\partial\psi}{\partial x} = -\dfrac{\partial\phi}{\partial y}$	$v_\theta = -\dfrac{\partial\psi}{\partial r} = -\dfrac{1}{r}\dfrac{\partial\phi}{\partial\theta}$

B.2 Basic Flows

Uniform Flow

$$\psi = Uy$$

$$\phi = -Ux$$

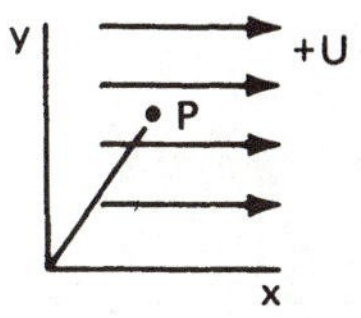

Line Source

$$\psi = \frac{Q}{2\pi}\theta$$

Q = volume flow rate per unit depth

$$\phi = -\frac{Q}{2\pi}\ln r$$

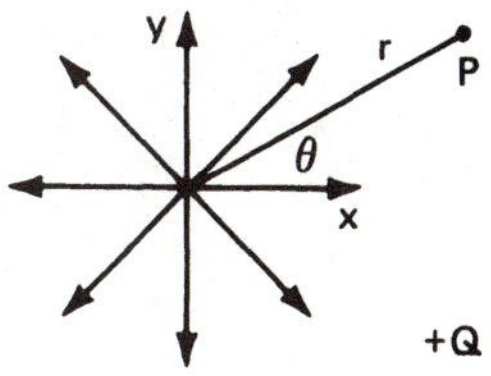

Line Sink

$$\psi = -\frac{Q}{2\pi}\theta$$

$$\phi = \frac{Q}{2\pi}\ell n\, r$$

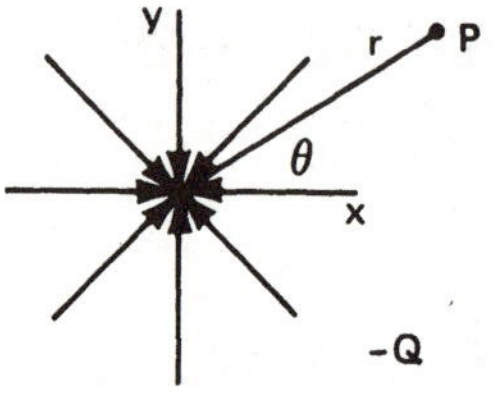

Line Vortex

$$\psi = -\kappa\, \ell n\, r$$

Vortex strength
$\kappa = v_\theta / r$

$$\phi = -\kappa\theta$$

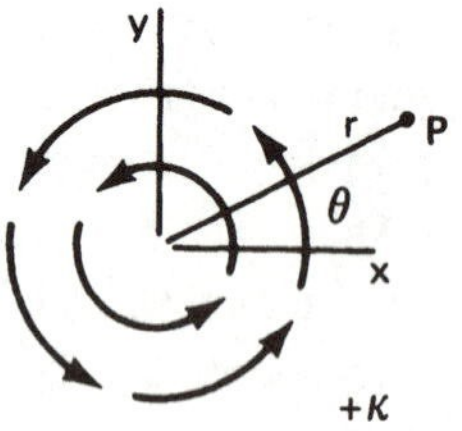

B.3 Sums of Two Flows

Source Plus Vortex

$$\psi = \frac{Q}{2\pi}\theta - \kappa\, \ell n\, r$$

$$\phi = -\frac{Q}{2\pi}\ell n\, r - \kappa\theta$$

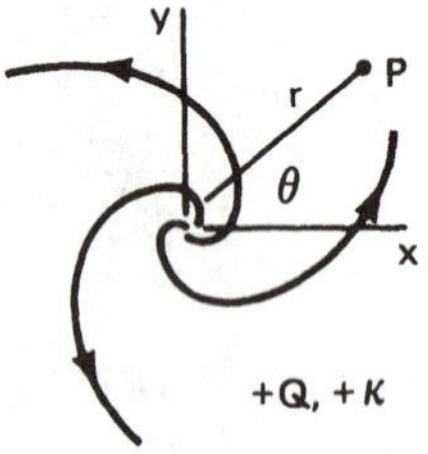

Sink Plus Vortex

$$\psi = -\frac{Q}{2\pi}\theta - \kappa\, \ell n\, r$$

$$\phi = \frac{Q}{2\pi}\ell n\, r - \kappa\theta$$

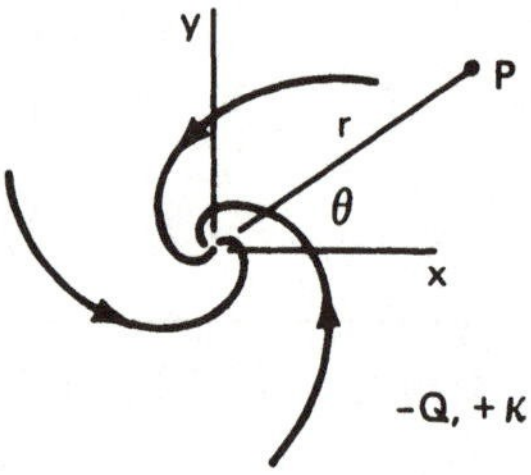

Source and Uniform Flow

$$\psi = \frac{Q}{2\pi}\,\theta + Ur\sin\theta$$

$$\phi = -\frac{Q}{2\pi}\,\ell n\, r - Ur\cos\theta$$

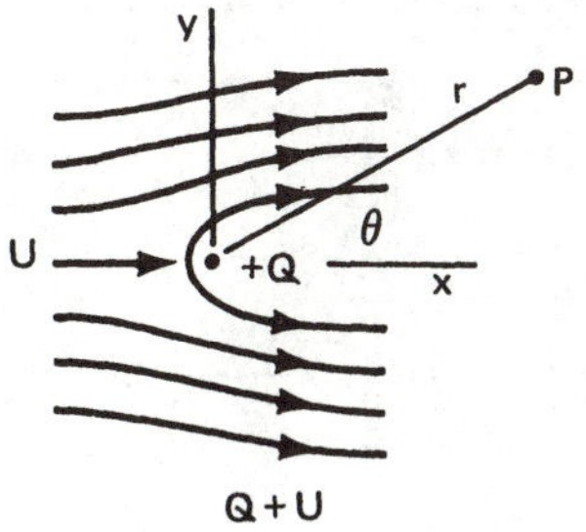

Source and Sink

$$\psi = \frac{Q}{2\pi}(\theta_1 - \theta_2)$$

$$\phi = \frac{Q}{2\pi}\,\ell n\,\frac{r_2}{r_1}$$

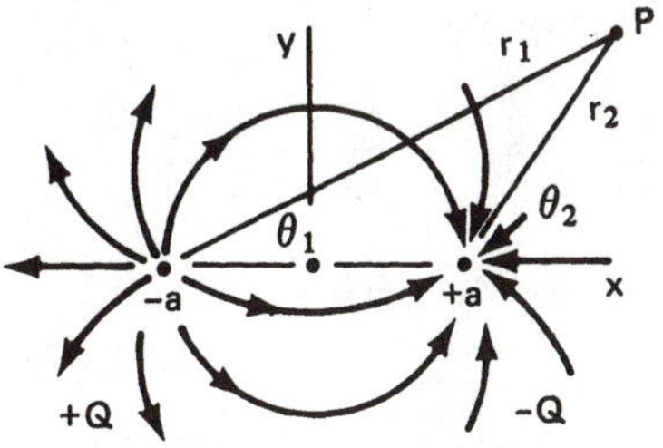

B.4 Cylinders and Vortex Pairs

Doublet

$$\psi = -\frac{k}{r}\sin\theta$$

$$\phi = -\frac{k}{r}\cos\theta$$

Doublet strength
$k = Qa/\pi$

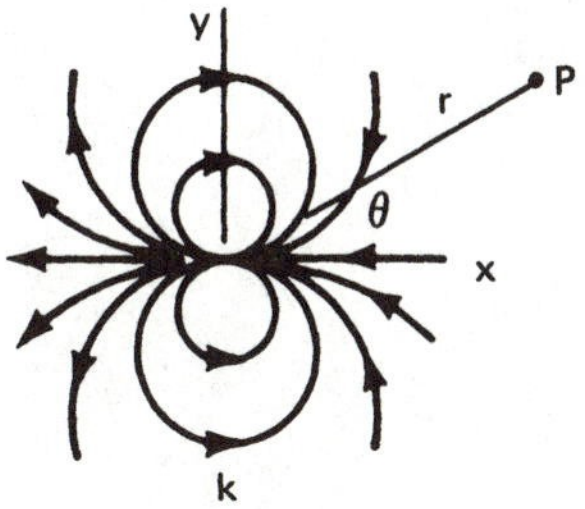

Doublet and Uniform Flow

$$\psi = Ur\left(1 - \frac{R^2}{r^2}\right)\sin\theta$$

$$r = \sqrt{k/U}$$

$$\phi = -Ur\left(1 + \frac{R^2}{r^2}\right)\cos\theta$$

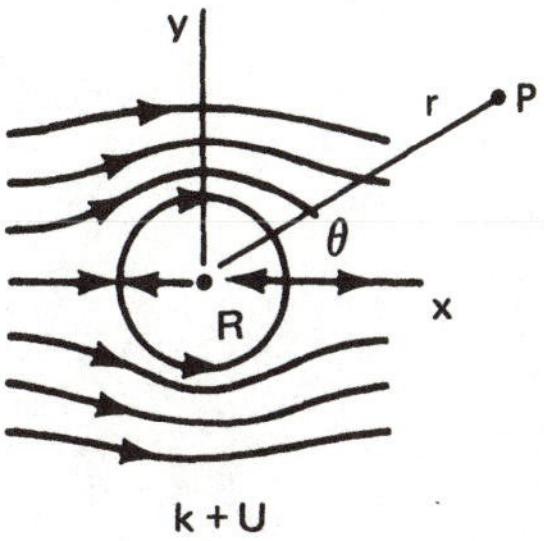

Doublet, Uniform Flow, and Vortex (cw)

$$\psi = Ur\left(1 - \frac{R^2}{r^2}\right)\sin\theta + \kappa\,\ell n\, r$$

$$\phi = -Ur\left(1 + \frac{R^2}{r^2}\right)\cos\theta + \kappa\theta$$

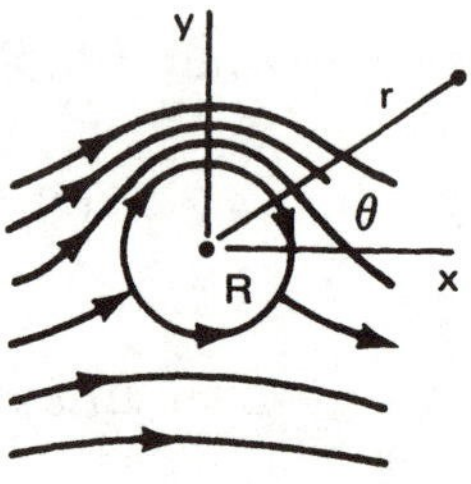

Vortex Pair ($\kappa_2 = -\kappa_1$)

$$\psi = \kappa\,\ell n\,\frac{r_2}{r_1}$$

$$\phi = \kappa(\theta_2 - \theta_1)$$

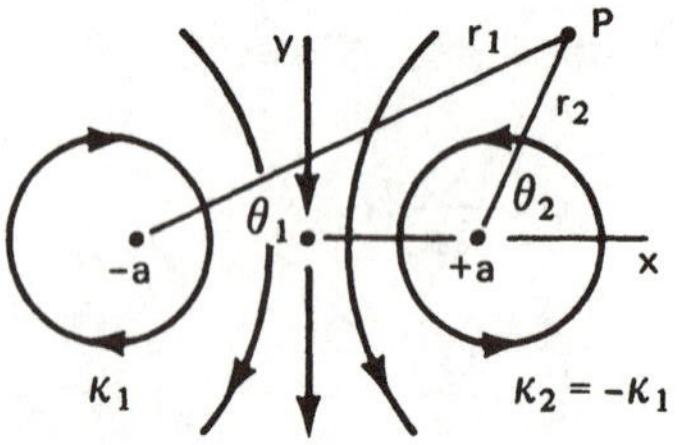

As shown, the fluid at infinity has $\mathbf{V} = (\kappa/2a)\,\hat{\mathbf{j}}$. If the fluid at infinity is stationary, the pair has $\mathbf{V} = -(\kappa/2a)\,\hat{\mathbf{j}}$. Exchange ϕ for ψ and ψ for ϕ, and compare with "Source and Sink."

Appendix C Equations of Motion

The continuity equation and the Navier-Stokes equations in an inertial frame for a Newtonian, constant-viscosity, constant-density fluid are given in Cartesian, cylindrical, and spherical coordinate systems. Newtonian indicates viscous shear is linear with rate of strain, $\tau = \mu\, du/dy$. Constant viscosity means $\mu \neq f(x, y, z, t)$, and constant density means $\rho \neq f(x, y, z, t)$.

C.1 Cartesian

$$\frac{\partial v_x}{\partial x} + \frac{\partial v_y}{\partial y} + \frac{\partial v_z}{\partial z} = 0$$

$$\rho\left(\frac{\partial v_x}{\partial t} + v_x\frac{\partial v_x}{\partial x} + v_y\frac{\partial v_x}{\partial y} + v_z\frac{\partial v_x}{\partial z}\right) = F_x - \frac{\partial P}{\partial x} + \mu\left(\frac{\partial^2 v_x}{\partial x^2} + \frac{\partial^2 v_x}{\partial y^2} + \frac{\partial^2 v_x}{\partial z^2}\right)$$

$$\rho\left(\frac{\partial v_y}{\partial t} + v_x\frac{\partial v_y}{\partial x} + v_y\frac{\partial v_y}{\partial y} + v_z\frac{\partial v_y}{\partial z}\right) = F_y - \frac{\partial P}{\partial y} + \mu\left(\frac{\partial^2 v_y}{\partial x^2} + \frac{\partial^2 v_y}{\partial y^2} + \frac{\partial^2 v_y}{\partial z^2}\right)$$

$$\rho\left(\frac{\partial v_z}{\partial t} + v_x\frac{\partial v_z}{\partial x} + v_y\frac{\partial v_z}{\partial y} + v_z\frac{\partial v_z}{\partial z}\right) = F_z - \frac{\partial P}{\partial z} + \mu\left(\frac{\partial^2 v_z}{\partial x^2} + \frac{\partial^2 v_z}{\partial y^2} + \frac{\partial^2 v_z}{\partial z^2}\right)$$

C.2 Cylindrical

$$\frac{1}{r}\frac{\partial}{\partial r}(r v_r) + \frac{1}{r}\frac{\partial v_\theta}{\partial \theta} + \frac{\partial v_z}{\partial z} = 0$$

$$\rho\left(\frac{\partial v_r}{\partial t} + v_r\frac{\partial v_r}{\partial r} + \frac{v_\theta}{r}\frac{\partial v_r}{\partial \theta} + v_z\frac{\partial v_r}{\partial z} - \frac{v_\theta^2}{r}\right)$$

$$= F_r - \frac{\partial P}{\partial r} + \mu\left(\frac{\partial^2 v_r}{\partial r^2} + \frac{1}{r}\frac{\partial v_r}{\partial r} + \frac{1}{r^2}\frac{\partial^2 v_r}{\partial \theta^2} + \frac{\partial^2 v_r}{\partial z^2} - \frac{v_r}{r^2} - \frac{2}{r^2}\frac{\partial v_\theta}{\partial \theta}\right)$$

$$\rho\left(\frac{\partial v_\theta}{\partial t}+v_r\frac{\partial v_\theta}{\partial r}+\frac{v_\theta}{r}\frac{\partial v_\theta}{\partial \theta}+v_z\frac{\partial v_\theta}{\partial z}+\frac{v_r v_\theta}{r}\right)$$

$$=F_\theta-\frac{1}{r}\frac{\partial P}{\partial \theta}+\mu\left(\frac{\partial^2 v_\theta}{\partial r^2}+\frac{1}{r}\frac{\partial v_\theta}{\partial r}+\frac{1}{r^2}\frac{\partial^2 v_\theta}{\partial \theta^2}+\frac{\partial^2 v_\theta}{\partial z^2}+\frac{2}{r^2}\frac{\partial v_r}{\partial \theta}-\frac{v_\theta}{r^2}\right)$$

$$\rho\left(\frac{\partial v_z}{\partial t}+v_r\frac{\partial v_z}{\partial r}+\frac{v_\theta}{r}\frac{\partial v_z}{\partial \theta}+v_z\frac{\partial v_z}{\partial z}\right)$$

$$=F_z-\frac{\partial P}{\partial z}+\mu\left(\frac{\partial^2 v_z}{\partial r^2}+\frac{1}{r}\frac{\partial v_z}{\partial r}+\frac{1}{r^2}\frac{\partial^2 v_z}{\partial \theta^2}+\frac{\partial^2 v_z}{\partial z^2}\right)$$

C.3 Spherical

$$\frac{1}{r^2}\frac{\partial}{\partial r}(r^2 v_r)+\frac{1}{r\sin\theta}\frac{\partial}{\partial \theta}(v_\theta \sin\theta)+\frac{1}{r\sin\theta}\frac{\partial v_\phi}{\partial \phi}=0$$

$$\rho\left(\frac{\partial v_r}{\partial t}+v_r\frac{\partial v_r}{\partial r}+\frac{v_\theta}{r}\frac{\partial v_r}{\partial \theta}+\frac{v_\phi}{r\sin\theta}\frac{\partial v_r}{\partial \phi}-\frac{v_\theta^2+v_\phi^2}{r}\right)$$

$$=F_r-\frac{\partial P}{\partial r}+\mu\left[\frac{1}{r^2}\frac{\partial}{\partial r}\left(r^2\frac{\partial v_r}{\partial r}\right)+\frac{1}{r^2\sin\theta}\frac{\partial}{\partial \theta}\left(\sin\theta\frac{\partial v_r}{\partial \theta}\right)\right.$$

$$\left.+\frac{1}{r^2\sin^2\theta}\frac{\partial^2 v_r}{\partial \phi^2}-\frac{2v_r}{r^2}-\frac{2}{r^2}\frac{\partial v_\theta}{\partial \theta}-\frac{2v_\theta\cot\theta}{r^2}-\frac{2}{r^2\sin\theta}\frac{\partial v_\phi}{\partial \phi}\right]$$

$$\rho\left(\frac{\partial v_\theta}{\partial t}+v_r\frac{\partial v_\theta}{\partial r}+\frac{v_\theta}{r}\frac{\partial v_\theta}{\partial \theta}+\frac{v_\phi}{r\sin\theta}\frac{\partial v_\theta}{\partial \phi}+\frac{v_r v_\theta}{r}-\frac{v_\phi^2\cot\theta}{r}\right)$$

$$=F_\theta-\frac{1}{r}\frac{\partial P}{\partial \theta}+\mu\left[\frac{1}{r^2}\frac{\partial}{\partial r}\left(r^2\frac{\partial v_\theta}{\partial r}\right)+\frac{1}{r^2\sin\theta}\frac{\partial}{\partial \theta}\left(\sin\theta\frac{\partial v_\theta}{\partial \theta}\right)\right.$$

$$\left.+\frac{1}{r^2\sin^2\theta}\frac{\partial^2 v_\theta}{\partial \phi^2}+\frac{2}{r^2}\frac{\partial v_r}{\partial \theta}-\frac{v_\theta}{r^2\sin^2\theta}-\frac{2\cos\theta}{r^2\sin^2\theta}\frac{\partial v_\phi}{\partial \phi}\right]$$

$$\rho\left(\frac{\partial v_\phi}{\partial t}+v_r\frac{\partial v_\phi}{\partial r}+\frac{v_\theta}{r}\frac{\partial v_\phi}{\partial \theta}+\frac{v_\phi}{r\sin\theta}\frac{\partial v_\phi}{\partial \phi}+\frac{v_\phi v_r}{r}+\frac{v_\theta v_\phi \cot\theta}{r}\right)$$

$$=F_\phi-\frac{1}{r\sin\theta}\frac{\partial P}{\partial \phi}+\mu\left[\frac{1}{r^2}\frac{\partial}{\partial r}\left(r^2\frac{\partial v_\phi}{\partial r}\right)+\frac{1}{r^2\sin\theta}\frac{\partial}{\partial \theta}\left(\sin\theta\frac{\partial v_\phi}{\partial \theta}\right)\right.$$

$$\left.+\frac{1}{r^2\sin^2\theta}\frac{\partial^2 v_\phi}{\partial \phi^2}-\frac{v_\phi}{r^2\sin^2\theta}+\frac{2}{r^2\sin^2\theta}\frac{\partial v_r}{\partial \phi}+\frac{2\cos\theta}{r^2\sin^2\theta}\frac{\partial v_\theta}{\partial \phi}\right]$$

For the Navier-Stokes and continuity equations with variable density and/or viscosity, see Hughes and Gaylord (1964)

Appendix D Fluid Properties

D.1 Units

$(^\circ F) = 9/5(^\circ C) + 32$ $kg/m^3 = (g/cm^3) \times 10^3$

$N\text{-}s/m^2 = \text{poise} \times 10^{-1}$ $\text{poise} = \text{dyne-s}/cm^2 = g/cm\text{-}s$

$m^2/s = \text{stokes} \times 10^{-4}$ $\text{stokes} = cm^2/s$

μ and ν are often specified in centipoise (cp) and centistokes (cs), respectively.

D.2 Properties of Air (sea-level pressure)

T (°C)	ρ (g/cm^3)	μ (poise)	ν (stokes)
−20	1.40×10^{-3}	1.61×10^{-4}	1.16×10^{-1}
−10	1.34×10^{-3}	1.67×10^{-4}	1.24×10^{-1}
0	1.29×10^{-3}	1.72×10^{-4}	1.33×10^{-1}
10	1.25×10^{-3}	1.76×10^{-4}	1.41×10^{-1}
20	1.20×10^{-3}	1.81×10^{-4}	1.51×10^{-1}
30	1.17×10^{-3}	1.86×10^{-4}	1.60×10^{-1}
40	1.13×10^{-3}	1.91×10^{-4}	1.69×10^{-1}
50	1.09×10^{-3}	1.95×10^{-4}	1.79×10^{-1}
60	1.06×10^{-3}	2.00×10^{-4}	1.89×10^{-1}
70	1.03×10^{-3}	2.04×10^{-4}	1.99×10^{-1}
80	1.00×10^{-3}	2.09×10^{-4}	2.09×10^{-1}

D.3 Properties of Water (sea-level pressure)

T (°C)	ρ (g/cm³)	μ (poise)	ν (stokes)
0	1.00	1.79×10^{-2}	1.79×10^{-2}
10	1.00	1.31×10^{-2}	1.31×10^{-2}
20	0.998	1.00×10^{-2}	1.00×10^{-2}
30	0.996	0.80×10^{-2}	0.80×10^{-2}
40	0.992	0.65×10^{-2}	0.66×10^{-2}
50	0.988	0.55×10^{-2}	0.55×10^{-2}
60	0.983	0.47×10^{-2}	0.47×10^{-2}
70	0.978	0.40×10^{-2}	0.41×10^{-2}
80	0.972	0.35×10^{-2}	0.36×10^{-2}
90	0.965	0.32×10^{-2}	0.33×10^{-2}
100	0.958	0.28×10^{-2}	0.29×10^{-2}

D.4 Common Gases and Liquids (sea-level pressure)

	T (°C)	ρ (g/cm³)	μ (poise)	ν (stokes)
Hydrogen	15	8.51×10^{-5}	8.60×10^{-5}	1.01
Oxygen	15	1.35×10^{-3}	2.03×10^{-4}	0.15
Methane	15	6.78×10^{-4}	1.08×10^{-4}	0.16
Ethyl alcohol	20	0.799	1.2×10^{-2}	1.5×10^{-2}
Sea water[1]	10	1.026	1.4×10^{-2}	1.4×10^{-2}
SAE 30 oil	40	0.88	1.0	1.1
Mercury	20	13.55	1.5×10^{-2}	1.2×10^{-3}

[1]Sea water with 3.3% salinity.

All viscous properties vary significantly with temperature.

D.5 Compressible Fluids

The flow properties of an incompressible fluid are completely determined by its viscosity (μ) and density (ρ). When analyzing an incompressible fluid, properties such as specific heat and conductivity are used only to analyze heat transport by the fluid from or to external sources.

When flow energy is adequate to generate significant changes in fluid density, the fluid must be treated as compressible. An energy equation is needed, as are equations that define (p, ρ, T) for that fluid and for the thermodynamic process involved (e.g., adiabatic, isentropic). Such analyses typically involve the quantities listed next.

Various approximations are made when very small, thermally driven density variations provide the driving force for a nearly incompressible convecting fluid.

R_u = 8.314 J/K mole = universal gas constant. The mass of 1 mol of any gas at 1 atm and 273 K has a volume of 22.4 liters.

$R = C_p - C_v$ = gas constant

R = 287.1 J/K kg for dry air

C_p = 1004 J/K kg for dry air at constant pressure

C_v = 717 J/K kg for dry air at constant volume

$\gamma = C_p/C_v$ = ratio of specific heats

Monatomic gases $\gamma = 5/3 = 1.67$

He and Ar are monatomic.

Diatomic gases $\gamma = 7/5 = 1.40$

Air and H_2, O_2, N_2, etc., are diatomic.

Polyatomic gases $\gamma \approx 1.30$

Gases like CO_2, SO_2, and NH_3 are polyatomic.

$c = \sqrt{\gamma RT}$ = speed of sound in fluid with absolute temperature T.

Appendix E Geophysical Data

E.1 The Earth

Pole to equator = 10,000 km, (definition of km)
Equatorial radius (a) = 6378 km
Polar radius (b) = 6357 km
    Ellipticity = $(a - b)/a \approx 1/298$
Radius of liquid core = 3485 km
    Ellipticity of liquid core ≈ 1/400
Radius of central solid core = 1215 km
Area of crust ≈ 29.2% of total earth surface
Mass (total) = 5.977×10^{24} kg
    Continental crust ≈ 1.95×10^{22} kg
    Mantle ≈ 4.02×10^{24} kg
    Liquid and solid core ≈ 1.94×10^{24} kg
Total moment of inertia ≈ 8.02×10^{37} kg m^2 (assuming a spherical, completely rigid earth)
Obliquity of the ecliptic (1900) = 23° 27' 8" (tilt of earth's spin axis relative to the plane of its orbit)
Astronomical unit (AU) = 1.49598×10^8 km (average distance sun to earth)
Solar day = 24 hr (by definition of mean solar time)
Sidereal day = $23^h\, 56^m\, 4.1^s$ (in solar hours)
Sidereal year = $365^d\, 6^h\, 9^m\, 10^s$ (in solar days)

E.2 The Sun

Diameter = 1.392×10^6 km
Mass = 1.989×10^{30} kg

E.3 The Moon

Diameter = 3476 km
Mass = 7.35×10^{22} kg
Distance to earth (average) = 3.84×10^5 km
Sidereal period = 27.32 days (solar days)
Synodic month ≈ 29.5 days (full moon to full moon)

E.4 The Planets

Planet	Mass (in earth masses)	Diameter (in earth diameters)	Spin Period (in earth days)	Orbital Radius (in AU[1])	Orbital Period (in earth years)
Mercury	0.055	0.382	58.8	0.387	0.241
Venus	0.82	0.949	−244	0.723	0.615
Earth	1.000	1.000	1.000	1.000	1.000
Mars	0.107	0.533	1.029	1.524	1.881
Jupiter	317.8	11.19	0.41	5.203	11.86
Saturn	94.3	9.41	0.43	9.538	29.46
Uranus	14.6	4.11	−0.65	19.19	84.07
Neptune	17.2	3.96	0.72	30.06	164.8
Pluto	0.003	0.20	6.40	39.53	248.6

[1] Astronomical unit (AU) is the average distance, over an orbit, of the earth from the sun.

E.5 The Oceans

About 98% of water outside the solid earth is in the oceans
Ocean area = 70.8% of total earth surface of 5.10×10^8 km^2
= 81% of Southern Hemisphere
= 61% of Northern Hemisphere
Pacific Ocean ≈ 50% of total ocean area of 3.61×10^8 km^2
Atlantic Ocean ≈ 30% of total ocean area
Indian Ocean ≈ 20% of total ocean area
Average depth of all oceans = 3.73 km
Maximum depth (Marianas Trench) = 11 km
Typical salinity is 3.30% to 3.50%
Typical freezing point = −1.9°C
Typical boiling point = 100.6°C
Heat capacity ≈ 0.96 of distilled water
About 93% of all ocean water is below 10°C.
Average ocean temperature below about 100 m is 3.5°C, near the point of maximum water density. Water expands (and rises) below this temperature.

Typical ocean densities increase from about 1.024 g/cm^3 at the surface, to 1.0265 g/cm^3 at 200 m depth, and to 1.0283 g/cm^3 at 1200 m and deeper.

Density changes in the oceans are due to salinity, temperature, and pressure (1% change/2 km depth).

E.6 The Standard Atmosphere

Height (km)	T (°C)	p/p_o	ρ/ρ_o
−0.5	18.4	1.061	1.049
0	15.2	1.000[1]	1.000[2]
0.5	11.9	0.942	0.953
1	8.7	0.887	0.908
1.5	5.4	0.835	0.864
2	2.2	0.785	0.822
3	−4.3	0.692	0.742
4	−10.8	0.609	0.669
5	−17.3	0.533	0.601
6	−23.8	0.466	0.539
8	−36.8	0.352	0.429
10	−49.7	0.262	0.338
12	−56.3	0.192	0.255
20	−56.3	0.0546	0.0726
40	−22.6	0.0028	0.0033
60	−17.2	2×10^{-4}	2×10^{-4}
90	−92.3	2×10^{-6}	2×10^{-6}

[1] $p_o = 1.0133 \times 10^5$ N/m^2, [2] $\rho_o = 1.225\times10^{-3}$ g/cm^3.

Temperature drops to a minimum of about −56°C between 11 and 20 km, then climbs to −17°C at 60 km before dropping again. Density drops nearly exponentially with height. Approximately half the total air mass is below 6 km. Auxiliary oxygen is recommended for small aircraft pilots above about 4 km, but climbers have reached the top of Mount Everest at about 9 km without using auxiliary oxygen. Commercial jet aircraft typically fly in the vicinity of 11 km, and special high-altitude jets can reach 30 km. Satellites in low-earth orbit are at or above about 200 km. Any air molecules at this elevation are too far apart to be considered a fluid. See Chapters 20 and 22 for heights of typical atmospheric phenomena.

References

References are cited in the text on the pages [ ] indicated.

Abell GO, 1982, *Exploration of the Universe* (4th ed), Saunders College Publ., Philadelphia, 721 pp. [344]

Abramson HN, 1966, *The Dynamic Behavior of Liquids in Moving Containers*, NASA SP-106, NASA, Washington, DC, 467 pp. [230, 306, 317]

Aller LH, 1963, *The Atmospheres of the Sun and Stars* (2nd ed), Ronald Press, New York, 650 pp. [330, 331]

AMS, 1988, *15th Conference on Severe Local Storms*, Am. Meteorological Soc., Boston, 549 pp. [376, 377]

AMS, 1989, *18th Conference on Hurricanes and Tropical Meteorology* (Extended Abstracts), Am. Meteorological Soc., Boston, 288 pp. [377]

Anderson JD Jr, 1985, *Introduction to Flight* (2nd ed), McGraw-Hill Book Co., New York, 560 pp. [275, 276, 300]

Andrews JA, 1985, True Polar Wander: An Analysis of Cenozoic and Mesozoic Paleomagnetic Poles, *J. Geophys. Res.* **90** 7737–7750. [330]

Anthes RA, 1982, *Tropical Cyclones: Their Evolution, Structure and Effects*, Am. Meteorological Soc., Boston, 208 pp. [377]

ASCE, 1983, *Rivers '83 Abstracts*, (Conf. Proc., New Orleans, Oct. 1983), Am. Soc. Civil Eng., New York. [248]

ASHRAE, 1977, *ASHRAE Handbook*, Am. Soc. Heat. Refrig. and Air-Cond. Eng., New York, 532 pp. [242]

Aston D, 1978, The Evolution of Liquid and Gas Centrifuges, *Endeavour New Series* **2** 142–148. [257, 263]

Atkinson BW, 1981a, *Meso-Scale Atmospheric Circulations*, Academic Press, New York, 495 pp. [333, 346]

Atkinson BW (Ed), 1981b, *Dynamical Meteorology*, Methuen, New York, 228 pp. [346]

Avery DG and Davies E, 1973, *Uranium Enrichment by Gas Centrifuge*, Mills and Boon Limited, London, 96 pp. [263]

Babcock HW, 1961, The Topology of the Sun's Magnetic Field and the 22-Year Cycle, *Astrophys. J.* **133** 572–587. [330]

Babcock KL, Ahlers G, and Cannell DS, 1991, Noise-Sustained Structure in Taylor-Couette Flow with Through Flow, *Phys. Rev. Lett.* **67** 3388–3391. [262]

Balje OE, 1981, *Turbomachines: A Guide to Design, Selection, and Theory*, John Wiley & Sons, New York, 513 pp. [301]

Bario F, Leboeuf F, and Papailiou KD, 1982, Study of Secondary Flows in Blade Cascades of Turbomachines, *J. Eng. Pwr.* **104** 497–509. [302]

Beij KH, 1938, Pressure Losses for Fluid Flow in 90° Pipe Bends, *J. Res. Nat. Bur. Std.* **21** (Also see Streeter 1961). [242]

Bellanca CP (Ed), 1990, *Advances in Steam Turbine Technology for Power Generation*, (Conf., Boston, Oct. 1990), Am. Soc. Mechanical Eng., New York 269 pp. [303]

Bengtsson L and Lighthill J (Eds), 1982, *Intense Atmospheric Vortices*, Springer-Verlag, New York, 326 pp. [377]

Bergeron T, 1954, The Problem of Tropical Hurricanes, *Quart. J. Roy. Meteor. Soc.* **80** 131–164. [370]

Berker H, 1963, *Handbuch der Physik*, Springer-Verlag, KG, Berlin, **VII** 1–384. [248]

Bertin JJ and Smith ML, 1989, *Aerodynamics for Engineers* (2nd ed), Prentice-Hall, Englewood Cliffs, NJ, 576 pp. [275, 276, 278]

Bertoglio JP, 1982, Homogeneous Turbulent Field within a Rotating Frame, *AIAA J.* **20** 1175–1181. [302, 318]

Bjerknes J, 1937, Die Theorie der Aussertropischen Zyklonenbildung, *Meteorol. Zeit.* **54** 460–466. [343]

Bondi H and Lyttleton RA, 1953, The Effect of Precession on the Motion of the Liquid Core, *Proc. Camb. Phil. Soc.* **49** 498–515. [331]

Bortkovskii RS, 1983, *Air-Sea Exchange of Heat and Moisture During Storms*, D. Reidel (Kluwer Acad. Publ., Norwell, MA), 194 pp. [377]

Boyce MP, 1982, *Gas Turbine Engineering Handbook*, Gulf Publ., Houston, 603 pp. [302]

Brosche P and Sündermann J (Eds), 1982, *Tidal Friction and the Earth's Rotation II*, (Conf., Oct. 1981, Bielefeld), Springer-Verlag, New York, 345 pp. [364]

Brown RA, 1990, *Fluid Mechanics of the Atmosphere*, Academic Press, New York, 489 pp. [333, 346, 347]

Busse FH, 1968, Steady Fluid Flow in a Precessing Spheroidal Shell, *J. Fluid Mech.* **33** 739–751. [331]

Busse FH, 1970a, Differential Rotation in Stellar Convective Zones, *Astrophys. J.* **159** 629–639. [330]

Busse FH, 1970b, Thermal Instabilities in Rapidly Rotating Systems, *J. Fluid Mech.* **44** 441–460. [325, 331]

Busse FH, 1978, Magnetohydrodynamics of the Earth's Dynamo, *Annu. Rev. Fluid Mech.* **10** 435–462. [331]

Cardin P and Olson P, 1992, An Experimental Approach to Thermochemical Convection in the Earth's Core, *Geophys. Res. Lett.* **19** 1995–1998. [331]

Carrigan CR and Busse FH, 1983, An Experimental and Theoretical Investigation of the Onset of Convection in Rotating Spherical Shells, *J. Fluid Mech.* **126** 287–305. [331]

Chen LD and Dixon SL, 1986, Growth of Secondary Flow Losses Downstream of a Turbine Blade Cascade, *J. Eng. Gas Turb. Pwr.* **108** 270–276. [302]

Chow VT, 1959, *Open Channel Hydraulics*, McGraw-Hill Book Co., New York, 680 pp. [239]

Church EF, 1950, *Steam Turbines* (3rd ed), McGraw-Hill Book Co., New York, 531 pp. [303]

Clark BJ and Scott JR, 1986, *Coupled Aerodynamic and Acoustical Predictions for Turboprops*, NASA TM-87094, NASA, Washington, DC. [300]

Clarke AC, 1973, *Rendezvous with Rama*, Ballantine Books, New York, 276 pp. [318]

Clauser LM and Clauser F, 1937, *The Effect on Curvature on the Transition from Laminar to Turbulent Boundary Layer*, NACA TN 613, NASA, Washington, DC. [249]

Cochran WG, 1934, The Flow Due to a Rotating Disk, *Proc. Camb. Phil. Soc.* **30** 365–375. [262]

Cohen H, Rogers GFC, and Saravananuttoo HIH, 1973, *Gas Turbine Theory*, John Wiley & Sons, New York, 337 pp. [302]

Conlisk AT, Foster MR, and Walker JDA, 1982, Fluid Dynamics and Mass Transfer in a Gas Centrifuge, *J. Fluid Mech.* **125** 283–317. [263]

Creager WP and Justin JD, 1963, *Hydroelectric Handbook* (2nd ed), John Wiley & Sons, New York, 1151 pp. [301]

Crow SC, 1970, Stability Theory for a Pair of Trailing Vorticies, *AIAA J.* **8** 2172–2179. [9]

Csanady GT, 1964, *Theory of Turbomachines*, McGraw-Hill Book Co., New York, 378 pp. [299, 300, 302]

Cushman-Roison B, 1987, Modeling Distinct Vortices of the Ocean, *EOS Trans. AGU* (Aug. 4) 677. [364]

D'Amico WP, 1984, Instabilities of a Gyroscope Produced by Rapidly Rotating, Highly Viscous Liquid, *J. Guid. Cntl. Dyn.* **7** 443–449. [318]

Dixon SL, 1984, *Fluid Mechanics, Thermodynamics of Turbomachinery* (3rd ed), Pergamon Press, Elmsford, NY, 263 pp. [302]

Donnelly R, 1991, Taylor-Couette Flow: The Early Days, *Phys. Today* (Nov. 1991) 32–39. [225, 262]

Dryden HL, Murnaghan FD, and Bateman H, 1956, *Hydrodynamics*, Dover Publ., Mineola, NY, 197 pp. [248]

Durney BR and Sofia S (Eds), 1987, *The Internal Solar Angular Velocity*, (8th Natl. Solar Observ. Symp., August 1986), Kluwer Acad. Publ., Norwell, MA, 374 pp. [330]

Finley JP, 1881, Report on the Tornadoes of May 29 and 30, 1879, in Kansas, Nebraska, *Prof. Paper of the Signal Service 4 1881*, 116 pp. (summarized in Nalivkin 1983). [374]

Flores RM, Ethridge FG, Miall AD, Galloway WE, and Fouch TD, 1985, *Recognition of Fluvial Depositional Systems and Their Resource Potential*, Soc. Econ. Paleontologists and Mineralogists, Tulsa, OK, 290 pp. [248]

Fox RW and McDonald AT, 1992, *Introduction to Fluid Mechanics* (4th ed), John Wiley & Sons, New York, 741 pp. [68, 275, 277, 300]

Fujita H, 1975, *Foundations of Ultracentrifugal Analysis*, John Wiley & Sons, New York, 459 pp. [263]

Fultz D, 1959, A Note on Overstability and the Elastoid-Inertia Oscillations of Kelvin, Solberg, and Bjerkness, *J. Meteorol.* **16** 199–208. [230, 232, 347]

Garcia RV and Malone TF (Eds), 1966, *Problems of Atmospheric Circulation*, (Conf., Argentina, May 1965), Spartan Books (MacMillan), Washington, DC, 186 pp. [346]

Garg SC, Furumoto N, and Vanyo JP, 1986, Spacecraft Nutational Instability Prediction by Energy-Dissipation Measurements, *J. Guid. Cntl. Dyn.* **9** 357–362. [318]

Gold T, 1955, Instability of the Earth's Axis of Rotation, *Nature* **175** 526–529. [330]

Goldstein H, 1965, *Classical Mechanics*, Addison-Wesley Pub. Co., Reading, MA, 399 pp. [316]

Gordon AH and Taylor RC, 1975, *Computations of Surface Layer Air Parcel Trajectories, and Weather, in the Oceanic Tropics*, Univ. Press of Hawaii, Honolulu, 112 pp. [346]

Görtler H, 1941, Über eine Dreidimensionale Instabilität Laminarer Grenzschichten an Konkaven Wänden, *ZAMM* **21** 250–252. [249]

Granger RA, 1988, *Experiments in Fluid Mechanics*, Holt, Rinehart & Winston, Orlando, FL, 486 pp. [248, 261, 276, 277]

Gray WM, 1979, Hurricanes: Their Formation, Structure, and Likely Role in the Tropical Circulation, in *Meteorology Over the Tropical Oceans*, Shaw DB (Ed), Roy. Meteor. Soc., 155–218. [370]

Greenspan HP, 1968, *The Theory of Rotating Fluids*, Cambridge Univ. Press, Cambridge, 327 pp. [156, 191, 193, 198, 199, 221, 226, 230, 231, 253, 261, 317, 347, 378]

Gregory-Smith DG, 1982, Secondary Flows and Losses in Axial Flow Turbines, *J. Eng. Pwr.* **104** 819–822. [302]

Gross MG, 1971, *Oceanography* (2nd ed), Merrill Publ., Westerville, OH, 150 pp. [362]

Guibert JP, 1987, A Free-Fall Facility to Investigate Liquid Slosh Effects on Spinning Satellite, *Space Comm. Broadcast* **5** 281–291. [318]

Gupta AK, Lilley DG, and Syred N, 1984, *Swirl Flows*, Abacus Press, Kent, England, 475 pp. [263]

Hamed A and Baskharone E, 1980, Analysis of a Three-Dimensional Flow in a Turbine Scroll, *J. Fluids Eng.* **102** 297–301. [302]

Harding ET and Kotsch WJ, 1965, *Heavy Weather Guide*, U.S. Naval Inst., Annapolis, MD, 209 pp. [370, 377]

Harrison JA, Garg SC, and Furumoto N, 1983, *A Free-Fall Technique to Measure Nutation Divergence, and Applications*, (AAS/AIAA Astrodynamics Specialist Conf., Lake Placid, NY, Aug. 22–25), AAS Publ., San Diego. [318]

Hart JE, Glatzmaier GA, and Toomre J, 1986, Space-Laboratory and Numerical Simulations of Thermal Convection in a Rotating Hemispherical Shell with Radial Gravity, *J. Fluid Mech.* **173** 519–544. [331]

Herring HJ, Soler A, and Steltz WG (Eds), 1979, *Flow in Primary Non-Rotating Passages in Turbomachines*, Am. Soc. Mechanical Eng., New York, 188 pp. [248, 302]

Holland WR and McWilliams JC, 1987, Computer Modeling in Physical Oceanography from the Global Circulations to Turbulence, *Phys. Today* (Oct. 1987) 51–57. [364]

Holton JR, 1979, *An Introduction to Dynamic Meteorology* (2nd ed), Academic Press, New York, 391 pp. [226, 333, 342, 346, 347]

Hough SS, 1895, The Oscillations of a Rotating Ellipsoidal Shell Containing Fluid, *Phil. Trans. R. Soc.* **186** 469–506. [331]

Houghton JT, 1986, *The Physics of the Atmosphere* (2nd ed), Cambridge Univ. Press, Cambridge, 271 pp. [333, 346]

Howard JHG, Patankar SV, and Bordynulk RM, 1980, Flow Prediction in Rotating Ducts Using Coriolis-Modified Turbulence Models, *J. Fluids Eng.* **102** 456–461. [302]

Hughes WF and Gaylord EW, 1964, *Basic Equations of Engineering Science*, Schaum's Outline Series, McGraw-Hill Book Co., New York, 163 pp. [247, 346, 364, 391]

Idel'chik IE, 1966, *Handbook of Hydraulic Resistance-Coefficients of Local Resistance and of Friction*, Israel Prog. for Scientific Transl. (A. Barouch, Ed.). [242]

Ikeda S and Parker G (Eds), 1989, *River Meandering*, Am. Geophys. Union, Washington, DC, 485 pp. [248]

IMechE, 1979, *Steam Turbines for the 1980s*, (Conf., London, Oct. 1979), Inst. of Mechanical Eng., Suffolk, England, 419 pp. [303]

IMechE, 1986, *Turbocharging and Turbochargers*, (Conf., London, May 1986), Inst. of Mechanical Eng., Suffolk, England. [302]

IMechE, 1990, *Steam Plant for the 1990s*, (Conf., London, April 1990), Inst. of Mechanical Eng., Suffolk, England, 350 pp. [303]

Inversion AR, 1986, *Micro-Hydropower Sourcebook*, NRECA Intl. Foundation, Washington, DC, 285 pp. [301]

Jacobs JA, 1987, *The Earth's Core* (2nd ed), Academic Press, New York, 416 pp. [330, 331]

Jenson VG, 1959, Viscous Flow Around a Sphere at Low Reynolds Numbers (<40), *Proc. R. Soc. London* **A249**, 346–366. [148, 149]

Johnston JP, Halleen RM, and Lezius DK, 1972, Effects of Spanwise Rotation on the Structure of Two-Dimensional Fully Developed Turbulent Channel Flow, *J. Fluid Mech.* **56** 533–557. [317]

Jordan TH and Anderson DL, 1974, Earth Structure from Free Oscillations and Travel Times, *Geophys. J. R. Astron. Soc.* **36** 411–459. [329]

Kitchens Jr. CW, Gerber N, and Sedney R, 1978, Spin Decay of Liquid-filled Projectiles, *J. Spacecraft* **15** 348–354. [318]

Koyama H, Masuda S, Ariga I, and Watanabe I, 1979, Stabilizing and Destabilizing Effects of Coriolis Force on Two-Dimensional Laminar and Turbulent Boundary Layers, *J. Eng. Power* **101** 23–31. [318]

Kückemann D and Weber J, 1953, *Aerodynamics of Propulsion*, McGraw-Hill Book Co., New York, 340 pp. [300]

Lamb H, 1932, *Hydrodynamics* (6th ed), Republished 1945 by Dover Publ., Mineola, NY, 738 pp. [230]

Lambeck K, 1980, *The Earth's Variable Rotation*, Cambridge Univ. Press, Cambridge, 449 pp. [329, 330]

Langston LS, 1980, Crossflows in a Turbine Cascade Passage, *J. Eng. Pwr.* **102** 866–874. [302]

Langston LS, Nice ML, and Hooper RM, 1977, Three-Dimensional Flow within a Turbine Cascade Passage, *J. Eng. Pwr.* **99** 21–28. [302]

Lazarkiewicz S and Troskolanski AT, 1965, *Impeller Pumps*, Pergamon Press, Elmsford, NY, 648 pp. with additional charts and graphs. [298, 301]

Liepmann HW, 1943, *Investigations on Laminar Boundary Layer Stability and Transition on Curved Boundaries*, ARC RM 7802, U.S. Govt. Printing Ofc, Washington, D.C. [249]

Liepmann HW, 1945, *Investigation of Boundary Layer Transition on Concave Walls*, NACA Wartime Rep. W-87, NASA, Washington, DC. [249]

Likins PW, 1973, *Elements of Engineering Mechanics*, McGraw-Hill Book Co., New York, 538 pp. [124, 316]

Litt JS, DeLaat JC, and Merrill WC, 1989, *A Real-Time Simulation of a Turbofan Engine*, NASA TM 100869, NASA, Washington, DC. [303]

Lobanoff VS and Ross RR, 1985, *Centrifugal Pumps: Design and Application*, Gulf Publ., Houston, 374 pp. [301]

Lock RC, 1951, The Velocity Distribution in the Laminar Boundary Layer between Parallel Streams, *Quart. J. Mech. Appl. Math.* **4** 42–63. [362]

Logan E Jr., 1981, *Turbomachinery: Basic Theory and Applications*, Marcel Dekker, New York, 119 pp. [301]

Loper DE, 1970, On Viscous Flow within a Rotating Spheroidal Container, *Quart. J. Mech. Appl. Math.* **23** 119–125. [331]

Loper DE, 1975, Torque Balance and Energy Budget for the Precessionally Driven Dynamo, *Phys. Earth Planetary Interiors* **11** 43–60. [331]

Loper DE, 1989, Dynamo Energetics and the Structure of the Outer Core, *Geophys. Astrophys. Fluid Dyn.* **49** 213–219. [331]

Ludlum DM, 1976, *Early American Tornadoes*, Am. Meteorological Soc., Boston, 219 pp. [377]

Lugt HJ, 1983, *Vortex Flow in Nature and Technology*, John Wiley & Sons, New York, 297 pp. [249, 347]

Magarvey RH and MacLatchy CS, 1964, The Formation and Structure of Vortex Rings, *Can. J. Phys.* **42** 678–683. [6]

Magono C, 1980, *Thunderstorms*, Elsevier, North-Holland Publ. Co., Amsterdam, 261 pp. [376]

Malkus WVR, 1968, Precession of the Earth as the Cause of Geomagnetism, *Science* **160** 259–264. [318, 331]

Manabe S (Ed), 1985, *Issues in Atmospheric and Oceanic Modeling* (Adv. in Geophys. vol. 28), Academic Press, New York, Part A, 591 pp; Part B, 432 pp. [346, 362, 376]

Marchuk GI and Kagan BA, 1989, *Dynamics of Ocean Tides*, Kluwer Acad. Publ., Norwell, MA, 327 pp. (Transl. from the 1983 Russian ed). [364]

Maul GA, 1985, *Introduction to Satellite Oceanography*, Kluwer Acad. Publ., Norwell, MA, 606 pp. [362]

McCall JS and Potter BJ, 1973, *Ultracentrifugation*, Bailliere Tindall (Williams and Wilkins, Baltimore, MD), 126 pp. [263]

Melchior PJ, 1986, *The Physics of the Earth's Core*, Pergamon Press, Elmsford, NY, 256 pp. [330]

Melchior PJ and Ducarme B, 1986, Detection of Inertial Gravity Oscillations in the Earth's Core with a Superconducting Gravimeter at Brussels, *Phys. Earth Planetary Interiors* **42** 129–134. [331]

Milne-Thomson LM, 1960, *Theoretical Hydrodynamics* (4th ed), Macmillan Publ. Co., New York, 660 pp. [48, 160, 226, 227, 269, 275, 385]

Moore J and Adhye RY, 1985, Secondary Flows and Losses Downstream of a Turbine Cascade, *J. Eng. Gas Turb. Pwr.* **107** 961–968. [302]

Morris WD, 1981, *Heat Transfer and Fluid Flow in Rotating Coolant Channels*, Research Studies Press (John Wiley & Sons, New York), 228 pp. [302]

Mullen JB and Maxworthy T, 1977, A Laboratory Model of Dust Devil Vortices, *Dyn. Atmos. Oceans* **1** 181–214. [376]

Munk WH, 1966, Variation of the Earth's Rotation in Historical Time, in *The Earth-Moon System*, Marsden BG and Cameron AGW (Eds), Plenum Press, New York, pp. 52–69. [328]

Munk WH, 1987, ONR Honors Gordon Hamilton, *EOS Trans. AGU*, (Aug. 4) 677. [18, 364]

Munk WH and MacDonald GJF, 1975, *The Rotation of the Earth*, Cambridge Univ. Press, Cambridge, 323 pp. [329]

Murakami M, Kikuyama K, and Asakura E, 1980, Velocity and Pressure Distributions in the Impeller Passages of Centrifugal Pumps, *J. Fluids Eng.* **102** 420–426. [302]

Murray BT, McFadden GB, and Coriell SR, 1990, Stabilization of Taylor-Couette Flow Due to Time-Periodic Outer Cylinder Oscillations, *Phys. Fluids A* **2** 2147–2156. [262]

Nalivkin DV, 1983, *Hurricanes, Storms, and Tornadoes*, Balkema, Rotterdam (Transl. from the 1969 Russian ed), 579 pp. [376, 377]

Nelson JM and Smith JD, 1989a, Flow in Meandering Channels with Natural Topography, in *River Meandering*, (Ikeda S and Parker G, Eds), pp. 69–102. [248]

Nelson JM and Smith JD, 1989b, Evolution and Stability of Erodible Channel Beds, in *River Meandering*, (Ikeda S and Parker G, Eds), pp. 321–377. [248]

Neumann G and Pierson WJ Jr, 1966, *Principles of Physical Oceanography*, Prentice-Hall, Englewood Cliffs, NJ, 545 pp. [362]

Nihoul JCJ and Jamart BM (Eds), 1987, *Three-Dimensional Models of Marine and Estuarine Dynamics*, Elsevier/North-Holland Publ. Co., Amsterdam, 629 pp. [362]

Nikuradse J, 1926, Untersuchungen über die Geschwindigkeitsverteilung in turbulenten Strömungen. Dissertation Göttingen. (See Schlichting 1979, pp. 613–614.) [237, 247, 248]

Nikuradse J, 1930, Tubulente Strömung in nichtkreisförmigen Rohren. *Ing.-Arch.* **1** 306-332. (See Schlichting 1979, pp. 613–614.) [237, 247, 248]

Ning L, Tveitereid M, Ahlers G, and Cannell DS, 1991, Taylor-Couette Flow Subjected to External Rotation, *Phys. Rev. A* **44** 2505–2513. [262]

O'Dell CR, 1987, The Physics of Aerobatic Flight, *Phys. Today* (Nov. 1987) 24–30. [277]

Olfe DB, 1987, *Fluid Mechanics Programs for the IBM PC*, McGraw-Hill Book Co., New York, 174 pp. and software. [276]

Olson RM, 1980, *Essentials of Engineering Fluid Mechanics* (4th ed), Harper & Row Publ., New York, 583 pp. [248, 301]

Palmén E and Newton CW, 1969, *Atmospheric Circulation Systems*, Academic Press, New York, 603 pp. [333, 344, 346, 347, 352, 362]

Paltridge GW, 1979, Climate and Thermodynamic Systems of Maximum Dissipation, *Nature* **279** 630–631. [344]

Pavlov NI, 1978, The Relation of the Intensity and Nature of Typhoon Travel to the Thermal Regime of the Surface Layer of the Ocean, *Taifun-75* **2** 14–22. (See Bortkovskii 1983, pp. 160 and 178.) [371]

Pedlosky J, 1987, *Geophysical Fluid Dynamics* (2nd ed), Springer-Verlag, New York, 710 pp. [226, 346, 347, 362, 378]

Pedlosky J and Greenspan HP, 1967, A Simple Laboratory Model for the Oceanic Circulation, *J. Fluid Mech.* **27** 291–304. [222, 363]

Peltier WR, 1986, Slow Changes in the Earth's Shape and Gravitational Field: Constraints on the Glaciation History and Internal Viscoelastic Stratification, in *Space Geodesy and Geodynamics* (Anderson AJ & Cazenave A, Eds) Academic Press, New York, pp.75–109. [330]

Peterson RL, 1976, *Air-Bearing Spin Facility for Measuring Energy Dissipation*, NASA Tech Note (TN D-8346), NASA, Washington, DC, 21 pp. [318]

Physics Today, 1993, Special Issue: High-Performance Computing and Physics, March, 1993. [170]

Plateau J, 1863, The Figures of Equilibrium of a Liquid Mass, *Smithsonian Rep.* 207–285. [331]

Poincaré MH, 1910, Sur la Précession des Corps Déformables, *Bull. Astronomique* **27** 321–367. [331]

Pond S and Pickard GL, 1986, *Introductory Dynamical Oceanography* (2nd ed), Pergamon Press, Elmsford, NY, 329 pp. [362]

Priest ER, 1982, *Solar Magnetohydrodynamics*, D. Reidel, Kluwer Acad. Publ., Norwell, MA, 469 pp. [331]

Priest ER and Hood AW (Eds), 1991, *Advances in Solar System Magnetohydrodynamics*, Cambridge Univ. Press, Cambridge, 435 pp. [331]

Pudov VD, 1979, The Effect of the Heat Reserve of the Upper Ocean Layer on Typhoon Trajectories, *Okeanologiya* **19** 1002–1007. (See Bortkovskii 1983, pp. 160 and 178.) [370]

Pugh DT, 1987, *Tides, Surges, and Mean Sea-Level*, John Wiley & Sons, New York, 472 pp. [364]

Reiter GS and Lee DA, 1966, Zero Gravity Stability Testing of a Liquid-Filled Space Vehicle, *Chem. Eng. Progress Symp. Series* **62** 178–183. [318]

Roache PJ, 1982, *Computational Fluid Dynamics*, Hermosa Publ., Albuquerque, NM, 446 pp. [276]

Roberson JA and Crowe CT, 1990, *Engineering Fluid Mechanics* (4th ed) Houghton Mifflin, Boston, 785 pp. [248, 277]

Roberts PH, 1968, On the Thermal Instability of a Self-Gravitating Fluid Sphere Containing Heat Sources, *Phil. Trans. R. Soc. London* **A263** 93–117. [331]

Roberts PH, 1988, Future of Geodynamo Theory, *Geophys. Astrophys. Fluid Dyn.* **44** 3–32. [331]

Roberts PH and Stewartson K, 1965, On the Motion of Liquid in a Spheroidal Cavity of a Precessing Rigid Body II, *Proc. Camb. Phil. Soc.* **61** 279–288. [331]

Robinson AR, Glenn SM, Spall MA, Walstad LJ, Gardner GM, and Leslie WG, 1989, Forecasting Gulf Stream Meanders and Rings, *EOS Trans. AGU* **70** 1464–1465. [364]

Rochester MG, Jacobs JA, Smylie DE, and Chong KF, 1975, Can Precession Power the Geomagnetic Dynamo?, *Geophys. J. R. Astron. Soc.* **43** 661–678. [331]

Roluti MJ (Ed), 1985, *Waterpower '85*, (Conf., Las Vegas, Sept. 1985) Am Soc. Civil Eng., New York, 2280 pp. in 3 vol. [301]

Rossby CG et al., 1939, Relation Between Variations in the Intensity of the Zonal Circulation of the Atmosphere and the Displacements of the Semi-permanent Centers of Action, *J. Marine Res.* **2** 38–55. [215, 343, 347]

Rubin DM, 1987, *Cross-Bedding, Bedforms, and Paleocurrents*, Soc. Econ. Paleontologists and Mineralogists, Tulsa, OK, 187 pp. [248]

Schlichting H, 1979, *Boundary-Layer Theory* (7th ed), McGraw-Hill Book Co., New York, 817 pp. [33, 55, 68, 75, 248, 249, 253, 261, 262]

Schultz-Grunow F and Behbahani D, 1975, Boundary Layer Stability at Longitudinally Curved Walls, *ZAMP* **26** 493–495. [249]

Scorer RS, 1978, *Environmental Aerodynamics*, Halsted Press (John Wiley & Sons, New York), 488 pp. [275, 346, 347]

Shevell RS, 1983, *Fundamentals of Flight*, Prentice-Hall, Englewood Cliffs, NJ, 405 pp. [276, 300]

Shukry A, 1950, Flow Around Bends in an Open Flume, *Trans. ASCE* **115** 751–779. [242, 243, 248]

Sieverding CH, 1985, Recent Progress in the Understanding of Basic Aspects of Secondary Flows in Turbine Blade Passages, *J. Eng. Gas Turb. Pwr.* **107** 248–257. [302]

Sieverding CH and Van Den Bosche P, 1983, The Use of Coloured Smoke to Visualize Secondary Flows in a Turbine-Blade Cascade, *J. Fluid Mech.* **134** 85–89. [302]

Slabinski VJ, 1981, A Dynamical Objection to the Inversion of the Earth on its Spin Axis, *J. Phys. A: Math. Gen.* **14** 2503–2507. [330]

Smith JD and McLean SR, 1984, A Model for Flow in Meandering Streams, *Water Resour. Res.* **20** 1301–1315. [248]

Smith R, 1980, Untwisting the Mysteries of Tornadoes, *New Scientist* (Feb. 28), 650–652. [378]

Sparrow EM and Gregg JL, 1960, Mass Transfer, Flow and Heat Transfer about a Rotating Disk, *J. Heat Trans.* **82** 294–302. [186, 262]

Stenning AH, 1980, Rotating Stall and Surge, *J. Fluids Eng.* **102** 14–20. [302]

Stowe K, 1983, *Ocean Science*, John Wiley & Sons, New York, 673 pp. [362]

Streeter VL (Ed), 1961, *Handbook of Fluid Dynamics*, McGraw-Hill Book Co., New York. [242]

Sverdrup HU, 1942, *Oceanography for Meteorologists*, Prentice-Hall, Englewood Cliffs, NJ, 235 pp. with charts for surface temperature, salinity, and velocity. [362, 363]

Szeri S, Schneider SJ, Labbe F, and Kaufman HN, 1983, Flow Between Rotating Disks, *J. Fluid Mech.* **134** (Part 1. Basic Flow) 103–131, (Part 2. Stability) 133–154. [262]

Tabakoff W, Crowe CT, and Cale DB, 1982, *Particulate Laden Flows in Turbomachinery*, (Conf., St. Louis, June 1982), Am. Soc. Mech. Eng., New York, 150 pp. [302]

Tabakoff W, Sheoran Y, and Kroll K, 1980, Flow Measurements in a Turbine Scroll, *J. Fluids Eng.* **102** 290–296. [302]

Takeda Y, Kobaski K, and Fischer WE, 1990, Observation of the Transient Behavior of Taylor Vortex Flow Between Rotating Concentric Cylinders After Sudden Start, *Exp. Fluids* **9** 317–319. [262]

Thomson WT, 1963, *Introduction to Space Dynamics*, John Wiley & Sons, New York, 317 pp. [316]

Thorarinsson S and Vonnegut B, 1964, Whirlwinds Produced by the Eruption of Surtsey Volcano, *Bull. Am. Meteor Soc.* **45** 440–444. [377]

Tritton DJ, 1978, Turbulence in Rotating Fluids, in *Rotating Fluids in Geophysics* (Roberts PH and Soward AM, Eds), Academic Press, New York, pp. 105–138. [302, 318]

Tritton DJ, 1985, Turbulence in Rotating Fluids, in *Turbulence and Predictability in Geophysical Fluid Dynamics and Climate Dynamics*, Soc. Italiana di Fisica, Bologna, Italy, pp. 172–192. [225, 318]

Van Dyke M, 1982, *An Album of Fluid Motion*, The Parabolic Press, Stanford, CA, 176 pp. (Continued in succeeding years as an annual "Gallery of Fluid Motion" in *Phys. Fluids A*.) [170, 248, 249, 261, 276, 378]

Vanyo JP, 1973, Measurement of System Inertia Tensors in Precessing Nonrigid Systems, *AIAA J.* **11** 577–578. [316]

Vanyo JP, 1984, Earth Core Motions: Experiments with Spheroids, *Geophys. J. R. Astron. Soc.* **77** 173–183. [331]

Vanyo JP, 1991, A Geodynamo Powered by Luni-Solar Precession, *Geophys. Astroph. Fluid Dyn.* **59** 209–234. [318, 326, 331]

Vanyo JP and Awramik SM, 1985, Stromatolites and Earth-Sun-Moon Dynamics, *Precamb. Res.* **29** 121–142. [328]

Vanyo JP and Likins PW, 1971, Measurement of Energy Dissipation in a Liquid-Filled, Precessing, Spherical Cavity, *J. Appl. Mech.* **38** 674–682. [318]

Vanyo JP and Likins PW, 1972, Rigid-Body Approximations to Turbulent Motion in a Liquid-Filled, Precessing, Spherical Cavity, *J. Appl. Mech.* **39** 18–24. [307, 312, 316, 317]

Vanyo JP, Wilde P, Cardin P, and Olson P, 1992, Flow Structures Induced in the Liquid Core by Precession, *EOS Trans. AGU* (Oct. 27 suppl.) 63. [331]

Vaughan OH and Vonnegut B, 1976, Luminous Electrical Phenomena Associated with Nocturnal Tornadoes in Huntsville, Ala., 3 April 1974, *Bull. Am. Meteor Soc.* **57** 1220–1224. [377]

Volland H, 1988, *Atmospheric Tidal and Planetary Waves*, Kluwer Acad. Publ., Norwell, MA, 348 pp. [347]

von Arx WS, 1962, *An Introduction to Physical Oceanography*, Addison-Wesley Publ. Co., Reading, MA, 422 pp. [363]

von Kármán T, 1921, Über Laminare und Turbulent Reibung, *ZAMM* **1** 233–252, also NACA TM 1092 dated 1946. [262]

von Kármán T, 1963, *From Low-Speed Aerodynamics to Astronautics*, Macmillan Publ. Co., New York, 82 pp. [272, 278]

Warlow P, 1978, Geomagnetic Reversals?, *J. Phys. A: Math. Gen.* **11** 2107–2130 (Note: p. 2112 is located at p. 2012). [330]

Watkins DC, Cobine JD, and Vonnegut B, 1978, Electrical Discharges Inside Tornadoes, *Science* **199** 171–174. [377]

White FM, 1974, *Viscous Fluid Flow*, McGraw-Hill Book Co., New York, 725 pp. [248]

White FM, 1986, *Fluid Mechanics* (2nd ed), McGraw-Hill Book Co., New York, 732 pp. [248, 277]

Zedd MF and Dodge FT, 1985, *Energy Dissipation of Liquids in Nutating Spherical Tanks Measured by Forced Motion-Spin Table*, Report 8932, Naval Res. Lab., Washington, DC, 17 pp. [318]

Zhang K, 1992, Spiraling Columnar Convection in Rapidly Rotating Spherical Fluid Shells, *J. Fluid Mech.* **236**, 535–556. [331]

Index

PHOTOGRAPHIC SKETCHBOOK OF THE CIVIL WAR, Alexander Gardner. 100 photos taken on field during the Civil War. Famous shots of Manassas Harper's Ferry, Lincoln, Richmond, slave pens, etc. 244pp. 10⅝ x 8¼. 22731-6 Pa. $10.95

FIVE ACRES AND INDEPENDENCE, Maurice G. Kains. Great back-to-the-land classic explains basics of self-sufficient farming. The one book to get. 95 illustrations. 397pp. 5⅜ x 8½. 20974-1 Pa. $7.95

SONGS OF EASTERN BIRDS, Dr. Donald J. Borror. Songs and calls of 60 species most common to eastern U.S.: warblers, woodpeckers, flycatchers, thrushes, larks, many more in high-quality recording. Cassette and manual 99912-2 $9.95

A MODERN HERBAL, Margaret Grieve. Much the fullest, most exact, most useful compilation of herbal material. Gigantic alphabetical encyclopedia, from aconite to zedoary, gives botanical information, medical properties, folklore, economic uses, much else. Indispensable to serious reader. 161 illustrations. 888pp. 6½ x 9¼. 2-vol. set. (Available in U.S. only.) Vol. I: 22798-7 Pa. $10.95
Vol. II: 22799-5 Pa. $10.95

HIDDEN TREASURE MAZE BOOK, Dave Phillips. Solve 34 challenging mazes accompanied by heroic tales of adventure. Evil dragons, people-eating plants, blood-thirsty giants, many more dangerous adversaries lurk at every twist and turn. 34 mazes, stories, solutions. 48pp. 8¼ x 11. 24566-7 Pa. $2.95

LETTERS OF W. A. MOZART, Wolfgang A. Mozart. Remarkable letters show bawdy wit, humor, imagination, musical insights, contemporary musical world; includes some letters from Leopold Mozart. 276pp. 5⅜ x 8½. 22859-2 Pa. $9.95

BASIC PRINCIPLES OF CLASSICAL BALLET, Agrippina Vaganova. Great Russian theoretician, teacher explains methods for teaching classical ballet. 118 illustrations. 175pp. 5⅜ x 8½. 22036-2 Pa. $6.95

THE JUMPING FROG, Mark Twain. Revenge edition. The original story of The Celebrated Jumping Frog of Calaveras County, a hapless French translation, and Twain's hilarious "retranslation" from the French. 12 illustrations. 66pp. 5⅜ x 8½. 22686-7 Pa. $4.95

BEST REMEMBERED POEMS, Martin Gardner (ed.). The 126 poems in this superb collection of 19th- and 20th-century British and American verse range from Shelley's "To a Skylark" to the impassioned "Renascence" of Edna St. Vincent Millay and to Edward Lear's whimsical "The Owl and the Pussycat." 224pp. 5⅜ x 8½. 27165-X Pa. $5.95

COMPLETE SONNETS, William Shakespeare. Over 150 exquisite poems deal with love, friendship, the tyranny of time, beauty's evanescence, death and other themes in language of remarkable power, precision and beauty. Glossary of archaic terms. 80pp. 5$\frac{3}{16}$ x 8¼. 26686-9 Pa. $1.00

THE BATTLES THAT CHANGED HISTORY, Fletcher Pratt. Eminent historian profiles 16 crucial conflicts, ancient to modern, that changed the course of civilization. 352pp. 5⅜ x 8½. 41129-X Pa. $9.95

THE INFLUENCE OF SEA POWER UPON HISTORY, 1660–1783, A. T. Mahan. Influential classic of naval history and tactics still used as text in war colleges. First paperback edition. 4 maps. 24 battle plans. 640pp. 5⅜ x 8½. 25509-3 Pa. $14.95

THE STORY OF THE TITANIC AS TOLD BY ITS SURVIVORS, Jack Winocour (ed.). What it was really like. Panic, despair, shocking inefficiency, and a little heroism. More thrilling than any fictional account. 26 illustrations. 320pp. 5⅜ x 8½. 20610-6 Pa. $8.95

FAIRY AND FOLK TALES OF THE IRISH PEASANTRY, William Butler Yeats (ed.). Treasury of 64 tales from the twilight world of Celtic myth and legend: "The Soul Cages," "The Kildare Pooka," "King O'Toole and his Goose," many more. Introduction and Notes by W. B. Yeats. 352pp. 5⅜ x 8½. 26941-8 Pa. $8.95

BUDDHIST MAHAYANA TEXTS, E. B. Cowell and others (eds.). Superb, accurate translations of basic documents in Mahayana Buddhism, highly important in history of religions. The Buddha-karita of Asvaghosha, Larger Sukhavativyuha, more. 448pp. 5⅜ x 8½. 25552-2 Pa. $12.95

ONE TWO THREE . . . INFINITY: Facts and Speculations of Science, George Gamow. Great physicist's fascinating, readable overview of contemporary science: number theory, relativity, fourth dimension, entropy, genes, atomic structure, much more. 128 illustrations. Index. 352pp. 5⅜ x 8½. 25664-2 Pa. $9.95

EXPERIMENTATION AND MEASUREMENT, W. J. Youden. Introductory manual explains laws of measurement in simple terms and offers tips for achieving accuracy and minimizing errors. Mathematics of measurement, use of instruments, experimenting with machines. 1994 edition. Foreword. Preface. Introduction. Epilogue. Selected Readings. Glossary. Index. Tables and figures. 128pp. 5⅜ x 8½. 40451-X Pa. $6.95

DALÍ ON MODERN ART: The Cuckolds of Antiquated Modern Art, Salvador Dalí. Influential painter skewers modern art and its practitioners. Outrageous evaluations of Picasso, Cézanne, Turner, more. 15 renderings of paintings discussed. 44 calligraphic decorations by Dalí. 96pp. 5⅜ x 8½. (Available in U.S. only.) 29220-7 Pa. $5.95

ANTIQUE PLAYING CARDS: A Pictorial History, Henry René D'Allemagne. Over 900 elaborate, decorative images from rare playing cards (14th–20th centuries): Bacchus, death, dancing dogs, hunting scenes, royal coats of arms, players cheating, much more. 96pp. 9¼ x 12¼. 29265-7 Pa. $12.95

MAKING FURNITURE MASTERPIECES: 30 Projects with Measured Drawings, Franklin H. Gottshall. Step-by-step instructions, illustrations for constructing handsome, useful pieces, among them a Sheraton desk, Chippendale chair, Spanish desk, Queen Anne table and a William and Mary dressing mirror. 224pp. 8⅛ x 11¼. 29338-6 Pa. $16.95

THE FOSSIL BOOK: A Record of Prehistoric Life, Patricia V. Rich et al. Profusely illustrated definitive guide covers everything from single-celled organisms and dinosaurs to birds and mammals and the interplay between climate and man. Over 1,500 illustrations. 760pp. 7½ x 10⅛. 29371-8 Pa. $29.95

Prices subject to change without notice.

Available at your book dealer or write for free catalog to Dept. GI, Dover Publications, Inc., 31 East 2nd St., Mineola, N.Y. 11501. Dover publishes more than 500 books each year on science, elementary and advanced mathematics, biology, music, art, literary history, social sciences and other areas.